Molecular Biology of Plant Stress Responses

Molecular Biology of Plant Stress Responses

Edited by

Muhammad Waseem

and

Pingwu Liu

School of Tropical Agriculture and Forestry, Hainan University, China
School of Breeding and Multiplication (Sanya Institute of Breeding and Multiplication), Hainan University, China

CABI

CABI is a trading name of CAB International

CABI
Nosworthy Way
Wallingford
Oxfordshire OX10 8DE
UK

Tel: +44 (0)1491 832111
E-mail: info@cabi.org
Website: www.cabi.org

CABI
200 Portland Street
Boston
MA 02114
USA

Tel: +1 (617)682-9015
E-mail: cabi-nao@cabi.org

A catalogue record for this book is available from the British Library, London, UK.

Library of Congress Cataloging-in-Publication Data

Names: Waseem, Muhammad, 1988- editor | Liu, Pingwu, 1980- editor
Title: Molecular biology of plant stress responses / editors, Muhammad Waseem, Pingwu Liu.
Description: First edition. | Boston, MA : CAB International, 2025. | Includes bibliographical references and index. | Summary: "This book reviews the latest research and advancements in the field of plant stress biology, presenting them in a coherent and accessible manner for researchers, academics, and students. Topics include signal transduction, gene expression regulation, metabolic adjustments, and the role of epigenetics in stress responses"-- Provided by publisher.
Identifiers: LCCN 2025037916 (print) | LCCN 2025037917 (ebook) | ISBN 9781800628809 hardback | ISBN 9781800628816 ebook | ISBN 9781800628823 epub
Subjects: LCSH: Plants--Effect of stress on | Plant molecular biology
Classification: LCC QK754 .M64 2026 (print) | LCC QK754 (ebook)
LC record available at https://lccn.loc.gov/2025037916
LC ebook record available at https://lccn.loc.gov/2025037917

ISBN-13: 9781800628809 (hardback)
9781800628816 (ePDF)
9781800628823 (ePub)

DOI: 10.1079/9781800628823.0000

Commissioning Editor: Ward Cooper
Editorial Assistant: Theresa Regueira
Production Editor: Shankari Wilford

Typeset by Straive, Pondicherry, India
Printed and bound in the UK by CPI Group (UK) Ltd, Croydon, CR0 4YY

Contents

Contributors

Javed Abbas, Department of Plant Sciences, Faculty of Biological Sciences, Quaid-i-Azam University, Islamabad 45320, Pakistan. Email: abbasjaved869@gmail.com

Banzeer Ahsan Abbasi, Department of Botany, Rawalpindi Women University, 6th Road, Satellite Town, Rawalpindi 46300, Pakistan. Email: benazirahsanabbasi786@gmail.com

Muhammad Abubakr, Department of Botany, Division of Science and Technology, University of Education, Lahore, Pakistan. Email: raoabubakr789@gmail.com

Riaz Ahmad, Department of Horticulture, The University of Agriculture, Dera Ismail Khan 29111, Pakistan. Email: riazahmadbzu@gmail.com

Asma Ajmal, Department of Plant Sciences, Faculty of Biological Sciences, Quaid-i-Azam University, Islamabad 45320, Pakistan. Email: asmaajmal681@gmail.com

Kamran Akbar, Department of Botany, Bacha Khan University, Charsadda 24420, Khyber Pakhtunkhwa, Pakistan. Email: durrani_kamran0911@yahoo.com

Muhammad Ahsan Altaf, Key Laboratory for Quality Regulation of Tropical Horticultural Crops of Hainan Province, School of Breeding and Multiplication (Sanya Institute of Breeding and Multiplication), Hainan University, Sanya 572025, China; Key Laboratory for Quality Regulation of Tropical Horticultural Crops of Hainan Province, School of Tropical Agriculture and Forestry, Hainan University, Haikou 570228, China. Email: ahsanaltaf8812@gmail.com

Muhammad Mohsin Altaf, School of Tropical Agriculture and Forestry (School of Agricultural and Rural Affairs, School of Rural Revitalization), Hainan University, Haikou 570228, China. Email: mohsinaltaf641@gmail.com

Muhammad Anas, Department of Soil Science and Plant Nutrition, University of Selcuk Campus, Konya 42079, Türkiye; Department of Plant Sciences, Faculty of Biological Sciences, Quaid-i-Azam University, Islamabad 45320, Pakistan. Email: muhammed.anas@selcuk.edu.tr

Muhammad Anwar, School of Tropical Agriculture and Forestry (School of Agricultural and Rural Affairs, School of Rural Revitalization), Hainan University, Haikou 570228, China. Email: anwar_uaar@yahoo.com

Sana Basharat, College of Tropical Agriculture and Forestry, Hainan University, Danzhou, Hainan, China. Email: sanabasharat143@gmail.com

Rifat Bashir, Department of Plant Sciences, Faculty of Biological Sciences, Quaid-i-Azam University, Islamabad 45320, Pakistan. Email: rifatbashir@bs.qau.edu.pk

Ifra Batool, Department of Plant Sciences, Faculty of Biological Sciences, Quaid-i-Azam University, Islamabad 45320, Pakistan. Email: ifrabatool9510@gmail.com

Hajira Bibi, Department of Biochemistry, Faculty of Biological Sciences, Quaid-i-Azam University, Islamabad 45320, Pakistan. Email: hajira.bibi@bs.qau.edu.pk

Afshan Farid, Department of Botany, PMAS Arid Agricultural University, Rawalpindi 46300, Punjab, Pakistan. Email: afshanfarid2017@gmail.com

Waleed Hussain, Department of Botany, University of Narowal, Punjab, Pakistan. Email: mwaleed@gmail.com

Muhammad Awais Ibrahim, Department of Plant Sciences, Faculty of Biological Sciences, Quaid-i-Azam University, Islamabad 45320, Pakistan. Email: awaisse1@gmail.com

Madiha Imtiaz, Department of Botany, Bacha Khan University, Charsadda 24420, Khyber Pakhtunkhwa, Pakistan. Email: madiha92bkuc@gmail.com

Javed Iqbal, Department of Botany, Bacha Khan University, Charsadda 24420, Khyber Pakhtunkhwa, Pakistan. Email: javed89qau@gmail.com

Rashid Iqbal, Department of Agronomy, Faculty of Agriculture and Environment, The Islamia University of Bahawalpur 63100, Pakistan; Department of Life Sciences, Western Caspian University, Baku, Azerbaijan. Email: rashid.iqbal@iub.edu.pk

Waseem Ahmed Khattak, Department of Plant Sciences, Faculty of Biological Sciences, Quaid-i-Azam University, Islamabad 45320, Pakistan. Email: waseemkhattack@bs.qau.edu.pk

Pingwu Liu, College of Tropical Agriculture and Forestry, Hainan University, Danzhou, Hainan, China. Email: hnulpw@hainanu.edu.cn

Ghulam Murtaza, School of Agriculture, Yunnan University, Kunming, Yunnan 650504, China. Email: murtazabotanist@gmail.com

Naveed Mushtaq, Key Laboratory for Quality Regulation of Tropical Horticultural Crops of Hainan Province, School of Breeding and Multiplication (Sanya Institute of Breeding and Multiplication), Hainan University, Sanya 572025, China; Key Laboratory for Quality Regulation of Tropical Horticultural Crops of Hainan Province, School of Tropical Agriculture and Forestry, Hainan University, Haikou 570228, China. Email: naveedmushtaq@hainanu.edu.cn

Iqra Noor, School of Horticulture and Landscape Architecture, Yangzhou University, Yangzhou, Jiangsu 225009, China. Email: iqranoor@yzu.edu.cn

Salma Noureen, Department of Botany, Bacha Khan University, Charsadda 24420, Khyber Pakhtunkhwa, Pakistan. Email: salmanoreen738@gmail.com

Taskeen Rasheed, Department of Plant Sciences, Faculty of Biological Sciences, Quaid-i-Azam University, Islamabad 45320, Pakistan. Email: taskeenrashid18@gmail.com

Sadia Riaz, Department of Plant Sciences, Faculty of Biological Sciences, Quaid-i-Azam University, Islamabad 45320, Pakistan. Email: sadiariazkhan888@gmail.com

Hafiz Muhammad Rizwan, Shenzhen Key Laboratory of Food Nutrition and Health, College of Chemistry and Environmental Engineering, Shenzhen University, Shenzhen 518060, China. Email: chrizwan51@szu.edu.cn

Wajid Saeed, College of Tropical Agriculture and Forestry, Hainan University, Danzhou, Hainan, China. Email: wajidsaeed1055@gmail.com

Muhammad Muzamil Sattar, Department of Plant Sciences, Faculty of Biological Sciences, Quaid-i-Azam University, Islamabad 45320, Pakistan. Email: muzammilsattar55@gmail.com

Iffat Shaheen, Department of Agriculture, Bahauddin Zakariya University, Multan, Pakistan. Email: shaheen.iffat.malik@gmail.com

Hamza Sohail, School of Horticulture and Landscape Architecture, Yangzhou University, Yangzhou, Jiangsu 225009, China. Email: hamzasohail@yzu.edu.cn

Saquib Waheed, Department of Neurology, Yongchuan Hospital of Chongqing Medical University, Chongqing, China. Email: saquibwaheed@gmail.com

Muhammad Waseem, College of Tropical Agriculture and Forestry, Hainan University, Danzhou, Hainan, China. Email: m.waseem.botanist@gmail.com

Guangsheng Yang, College of Tropical Agriculture and Forestry, Hainan University, Danzhou, Hainan, China. Email: ysheg@hzau.edu.cn

Xiaodong Yang, School of Horticulture and Landscape Architecture, Yangzhou University, Yangzhou, Jiangsu 225009, China. Email: yxd@yzu.edu.cn

Tabassum Yaseen, Department of Botany, Bacha Khan University, Charsadda 24420, Khyber Pakhtunkhwa, Pakistan. Email: tabassumyaseen@bkuc.edu.pk

Sadia Zafar, Department of Botany, Division of Science and Technology, University of Education Lahore, Pakistan. Email: sadia.zafar@ue.edu.pk

Syeda Anber Zahra, Institute of Biomaterials and Biomedical, School of Food and Pharmacy, Shanghai Zhongqiao Vocational and Technical University, Shanghai, China. Email: sazahra@shzq.edu.cn

Farishta Zarshan, Department of Botany, Bacha Khan University, Charsadda 24420, Khyber Pakhtunkhwa, Pakistan. Email: farishta24bkuc@gmail.com

1 Introduction to Plant Stress Responses

Muhammad Anas[1,2], Waseem Ahmed Khattak[2], Afshan Farid[3], Banzeer Ahsan Abbasi[4*] and Javed Iqbal[5]

[1]Department of Soil Science & Plant Nutrition, University of Selcuk Campus, Konya 42079, Türkiye; [2]Department of Plant Sciences, Faculty of Biological Sciences, Quaid-i-Azam University, Islamabad 45320, Pakistan; [3]Department of Botany, PMAS Arid Agricultural University, Rawalpindi 46300, Pakistan; [4]Department of Botany, Rawalpindi Women University, Rawalpindi 46300, Pakistan; [5]Department of Botany, Bacha Khan University, Charsadda 24420, Pakistan

Abstract

Plants are constantly exposed to diverse abiotic and biotic stressors, including drought, salinity, extreme temperatures, heavy metals, pathogens, and herbivory, which compromise their growth and productivity. This chapter presents an integrated overview of the physiological, biochemical, molecular, and epigenetic mechanisms that plants employ to mitigate stress impacts. It highlights the role of osmotic and ionic homeostasis, reactive oxygen species (ROS) detoxification, and hormonal signaling involving abscisic acid (ABA), salicylic acid (SA), and jasmonic acid (JA) in stress adaptation. Key regulatory networks such as the mitogen-activated protein kinase (MAPK) cascade and calcium signaling, transcription factors (TFs) (DREB, WRKY, NAC, bZIP), and stress-responsive genes are discussed, emphasizing their role in modulating defense responses. The interplay between abiotic and biotic stress responses, mediated by shared signaling pathways and hormonal crosstalk, is also examined. Furthermore, the chapter explores the role of nanotechnology and biotechnology—such as nanoparticles, CRISPR–Cas9 genome editing, and "omics" technologies—in enhancing plant stress tolerance. It concludes with future perspectives in precision agriculture and sustainable crop improvement, underlining the need for integrated approaches to bolster plant resilience in a changing climate.

Keywords: plant stress responses, abiotic and biotic stress, antioxidant defense, hormonal signaling, nanobiotechnology

1.1 Introduction

1.1.1 Overview of plant stress responses

As stationary organisms, plants face a variety of natural and living stressors including water shortages, salt stressors, temperature extremes, chemical dangers, nutrient scarcity, attacks from pathogens and herbivores, and plant competition (Kalavalapalli *et al.*, 2024).

Under stressful environmental conditions plants use their developed defenses to sense and adapt to threats. The body uses structured biological methods to defend cells against stress while maintaining its internal stability and ensuring survival. The body deals with stress by

* Email: benazirahsanabbasi786@gmail.com

 Molecular Biology of Plant Stress Responses (Muhammad Waseem and Pingwu Liu eds)
DOI: 10.1079/9781800628823.0001

adjusting growth patterns that lower shoot height while boosting root development to secure needed resources and save water (Bhattacharya and Bhattacharya, 2021). Plants respond to drought by closing their stomata to slow water loss, while building longer roots to discover more soil water. Plants under salinity stress deal with osmotic imbalance by building up compatible solutes including proline and glycine betaine for cell protection. Under stress the body produces more antioxidants to protect cells from harmful reactive oxygen species (ROS), which build up when plants face stressful situations (Singh *et al.*, 2016). An excess of ROS, such as superoxide radicals (O_2^-), hydrogen peroxide (H_2O_2) and hydroxyl radicals (•OH), can cause structural damage to cells by reacting with lipid proteins and DNA, which can weaken the cell or cause cell death.

Under stress conditions, plants produce antioxidant enzymes including superoxide dismutase (SOD), catalase (CAT) and peroxidase (POD), plus non-enzymatic substances including ascorbic acid (Asc) (vitamin C), glutathione (GSH), and vitamin E to eliminate ROS from their cells (García-Caparrós *et al.*, 2021).

Stress response activities in cells depend on the coordination of several molecular signaling systems and DNA-transcribing elements that activate protein and chemical defense mechanisms. Through mitogen-activated protein kinase (MAPK) pathways and calcium signals, plants detect changes in their environment to begin appropriate responses. Hormones then help transmit these signals to various parts of the cell. The MAPK pathway reacts to multiple stresses, including drought and pathogen challenge, to switch on proper stress-responsive genes with help from various transcription factors (TFs). Cells use calcium-dependent protein kinases (CDPKs) and calmodulin (CaM) to read and respond to stress signals by adjusting functions of stomatal closure and the handling of ions and antioxidants (Ketehouli *et al.*, 2022). Hormones, including abscisic acid (ABA), jasmonic acid (JA), salicylic acid (SA), and ethylene (ET), run stress response networks in plants by guiding their actions.

The *Arabidopsis thaliana* plant uses ABA as a main regulator to handle drought stress by closing stomata while also growing roots and turning on genes that help plants deal with stress (Mukherjee *et al.*, 2023). JA and ET work primarily to protect plants from biological threats, but also support abiotic stress reaction by managing root development and stomatal closing under salt and drought conditions. SA operates as an immune system regulator of biotrophic pathogens that improves plant tolerance to abiotic stresses through both boosting antioxidant capacity and safeguarding against metal and ultraviolet (UV) radiation damage. The stress adaptation mechanism relies on the TFs DREB/CBF, NAC, bZIP, and WRKY, which recognize dynamic DNA sequences in stress-responsive gene promoters to affect expression patterns. Under abiotic conditions DREB/CBF activates target genes, while WRKY controls both abiotic and biotic defense mechanisms through regulation of pathogenesis-related (PR) protein genes and osmoprotectant-linked genes.

Environmental triggers cause quick epigenetic DNA changes that help plants respond better to stress. Under both abiotic and biotic stress, DNA methylation controls gene activity and histone changes to stress-sensitive regions impact gene expression. DNA methylation and histone modifications regulate stress responses, while non-coding RNAs, such as micro-RNAs (miRNAs) and long non-coding RNAs (lncRNAs), control the expression of genes that control stress signaling and hormone signaling pathways. Under oxidative stress miR398 controls plant defenses and miR393 balances JA response when herbivores endanger plants (Abajebal *et al.*, 2024). Through this systematic response system plants can defend themselves against many stress types, but resource limitations create opposing effects. Under severe water shortage plants would prefer conservation through tolerance to protection from pathogens, which makes them more vulnerable to pathogen attacks. Another pattern involves redirecting resources from growth to defend against herbivores. To create better stress tolerance in plants farmers must first fully understand how plants react to stress. Stress response studies help researchers discover ways to improve defense systems and protect farming output in our changing climate conditions (Shahzad *et al.*, 2021). We use this understanding to design plants that withstand various environmental pressures, so farming remains feasible under worsening conditions.

1.2 Types of Plant Stresses

1.2.1 Abiotic stress (drought, salinity, temperature (heat and cold), heavy metals, radiation (UV and ionizing), oxidative, nutrient deficiency and flooding)

Abiotic stress is the term used to describe non-living environmental factors that impact plant health in a negative way. From these stresses, we will include drought, salinity, temperature extremes, heavy metal toxicity, radiation, oxidative stress, and nutrient deficiency. Over time, abiotic stress response has created a wide spectrum of physiological, biochemical, and molecular response mechanisms in plants to combat abiotic stress and maintain homeostasis (Fig. 1.1).

Drought stress is one of the major constraints to plant growth and yield. Cell expansion is reduced, stomata are closed, photosynthesis is inhibited, and ROS are accumulated as a result of water deficit (Qi *et al.*, 2018). In response, plants activate drought-responsive genes, generate osmoprotectants (proline and glycine betaine), promote root growth, and synthesize ABA, which controls stomatal closure and stress signaling. Excessive salt concentrations in the soil cause osmotic stress, ionic toxicity, and nutrient imbalances (Arif *et al.*, 2020). A plant will accumulate sodium (Na^+) and chloride (Cl^-) ions, which results in oxidative stress and cell malfunction. Plants counteract salinity stress by using ion homeostasis mechanisms such as the Na^+/H^+ antiporter system, synthesis of compatible solutes, and upregulation of antioxidant enzymes (such as SOD, CAT, and POD).

Metabolic processes, membrane stability and enzyme activity are disrupted by extreme temperatures. Protein denaturation and oxidative damage occur at high temperatures, while low temperatures lead to a reduction in membrane fluidity and formation of ice crystals. Mitigation of temperature-induced damages is carried out by heat shock proteins (HSPs) and cold responsive genes (Abdullah *et al.*, 2022).

Heavy metals such as cadmium (Cd), lead (Pb), and arsenic (As), cause disruption of plant physiological processes from induction of oxidative stress, displacement of essential nutrient, and damaging of cell structures (Anas *et al.*, 2024a). Heavy metals are detoxified by plants using chelation, sequestration into vacuoles, and upregulation of metallothioneins and phytochelatins. Because ionizing radiation

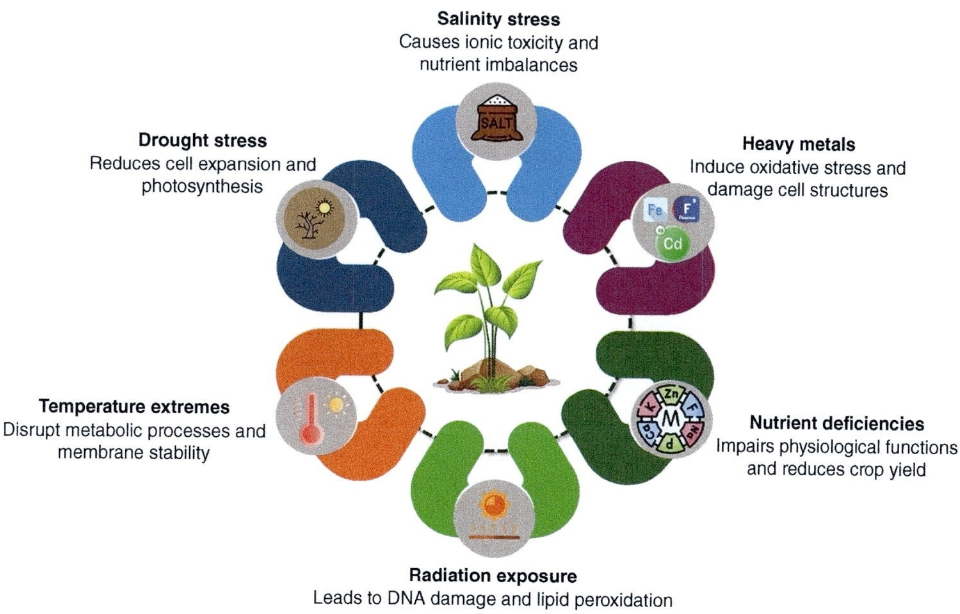

Fig. 1.1. Impacts of abiotic stresses on plant health. (Figure author's own.)

exposure can result in the production of ROS, DNA damage and lipid peroxidation can occur. DNA repair, accumulation of flavonoids and anthocyanin (UV-absorbing compounds), and the activation of the antioxidant defense system (AOS) protect plants from radiation damage (Mahdavian, 2024). The abiotic stresses usually cause an overproduction of ROS such as O_2^- radicals, H_2O_2, and OH• radicals. These can damage proteins, lipids, and nucleic acids. Antioxidant defense mechanisms used by plants to mitigate oxidative stress include making use of enzymatic antioxidants such as SOD, CAT, and ascorbate peroxidase (APX), and non-enzymatic antioxidants such as Asc, GSH, and carotenoids (Anas *et al.*, 2024c).

Essential macronutrients for growth of plants include nitrogen (N), phosphorus (P), potassium (K), calcium (Ca), magnesium (Mg), sulfur (S) and micronutrients including iron (Fe), zinc (Zn), manganese (Mn), copper (Cu), molybdenum (Mo) and boron (B) (Nadeem *et al.*, 2018). Deficiencies in these nutrients cause impaired physiological functions, chlorosis, necrosis and stand loss, along with a poor yield of the crop. Upon sensing these compounds, plants alter the structure of the root system, activate transporters to increase the abilities to uptake nutrients from the soil, and release root exudates to mobilize colloidal particles and nutrient compounds in the root environment. The decrease in soil oxygen availability due to excess water is known as hypoxia, which is followed by the changes in the metabolic pathways. To cope under low oxygen conditions, plants form aerenchyma tissue and enhance fermentation pathways while upregulating anaerobic metabolism genes for their energy production.

Plant growth and agricultural productivity are severely challenged by abiotic stresses (Kumari *et al.*, 2022; Anas *et al.*, 2025a). These stresses have selected plants to evolve complex sensing, responding and adaptive mechanisms. Understanding these responses at physiological, biochemical, and molecular levels will aid in the development of stress-tolerant crops by combining genetic engineering, breeding, and sustainable agricultural approaches (Fig. 1.1) (Bashir *et al.*, 2021).

1.2.2 Biotic stress (pathogens, insects, herbivory)

Living beings trigger biotic stress through their actions as pathogens or plant-eating creatures. Plants suffer major biological harm as stress triggers their growth reduction and hampers both their chemical processes and produces less. Microorganisms like bacteria and fungi enter plants to create diseases that harm their health (Anas *et al.*, 2025b). When plants detect pathogen-associated molecular patterns through pattern recognition receptors, they respond by triggering their defense system. An infected cell dies to keep pathogens from spreading throughout the plant. Plants develop lasting resistance through SA signaling, which activates systemic acquired resistance (SAR) processes (Roychowdhury *et al.*, 2024). Plants develop antimicrobial chemicals to stop pathogens from growing. Plants respond to insect damage by developing wounds and losing photosynthesizing areas while dying from plant pathogens transferred by the pests (Anas *et al.*, 2025b).

Plants defend themselves using trichomes and strong cuticles, along with walled cells that are hardened by lignin. Plants produce defensive compounds like alkaloids, terpenoids, and phenolics to protect themselves against feeding (Divekar *et al.*, 2022). The JA system triggers plants to make protease inhibitors and volatiles that push away or draw natural predators of herbivores.

Animals and insects damage plants through their natural feeding actions. In response, plants develop tolerance mechanisms such as compensatory growth and regrowth of damaged tissues, as well as defensive secondary metabolites, such as phenolics, tannins, and alkaloids, which deter herbivores.

Plants improve their cell wall strength and support their vascular tissue systems to shield themselves from further harm (Padayachee *et al.*, 2017; Anas *et al.*, 2023).

Plants struggle to survive and produce lower yields when faced with natural and living environment stress. Environmental elements trigger abiotic stress, while living beings create biotic stress for plants. Studying how plants respond to stress enables us to design better methods for plant breeding and protect crops better while preserving the environment.

1.3 Physiological and Biochemical Basis of Stress Responses

Stress in organisms is a broad concept that refers to a number of processes at tissue, physiological, and biochemical levels through which organisms are able to cope with unfavorable conditions (Anas *et al.*, 2025a). These are particularly relevant for homeostasis, which is the maintenance of appropriate internal conditions required for the proper functioning of the cells. Stress refers to pressure from factors such as drought, salinity, temperature fluctuation, pathogen invasion or any internal changes like an imbalance in potassium ions (K^+) and sodium ions (Na^+).

The three well-established aspects of stress include control of osmotic and ionic balance, oxidative stress and ROS regulation, and hormonal regulation in stress (Anas *et al.*, 2024a). Each of the afore-mentioned areas has an important function in assessing stress in organisms, as well as their ability to cope with it and heal from it.

1.3.1 Osmotic and ionic homeostasis

Osmotic and ionic balance is one of the most crucial aspects of stress since its communalities affect the abilities of plants and other organisms to survive when exposed to stressors like salinity, drought, or high temperatures. These stressors can cause imbalance of water and ions inside the cells, resulting in osmotic stress and subsequent toxicity, in case of improper regulation.

Osmotic and ionic balance gives stability to the cell membranes and enzymes and is important to various metabolic activities within the cell (Ataullakhanov *et al.*, 2022). Of all the stress stimuli, osmotic stress is characterized by the inequality of solute concentrations within the cytoplasm and in the external environment, which leads to the movement of water across the cell membrane. In hyperosmotic conditions, including unfavorable conditions such as high salinity or drought, water will move out of the cell, thus pulling it in on itself. At a hypoosmotic state the water moves into the cell eventually making the cell to equilibrate – a phenomenon that may cause the cell to burst (Wennerström and Oliveberg, 2022).

Osmotic pressure has, however, been offset through several ways by organisms to regulate the balance. Osmoprotectants, also known as compatible solutes, are substances that organisms accumulate to maintain osmotic balance. Some of the osmotically inactive solutes include proline, glycine betaine, sugars such as trehalose, and polyols such as mannitol. These solutes prevent protein denaturation and maintain cell, membrane, and organelle structure as well as preserve the membrane barrier function without influencing normal cell metabolism. For instance, under drought or salinity stress conditions, there is usually an increase in proline concentration in plants. Proline performs an act of molecular chaperone that stabilizes the proteins, and it is involved in removing any excess of ROS.

1.3.2 Ion transporters and channels

Ion transporters and channels are involved importantly in controlling ion balance. Under a salinity stress condition, the speed of transport of ions such as Na^+ and Cl^- into the cells can be increased, which affects enzyme action and metabolic reactions. To this end, organisms use ion transport proteins, such as the Na^+/H^+ antiporter where Na^+ is pumped out of the cell in exchange for H^+ (Palmgren, 2023). In plants, such molecule is also described in the salt overly sensitive (SOS) pathway, which is involved in the maintenance of Na^+ cation homeostasis.

The *SOS1* transporter, which is located next to the plasma membrane, transports Na^+ out of the cell, while *SOS2* and *SOS3* regulate the activity of the transporter. Aquaporins are small proteins that exist on cell membranes and enable the transport of water across the cell. Under osmotic stress it is imperative that the regulation of aquaporins decides on water flow and the pressure tension of the cell. For instance, in organisms with water shortage problems, aquaporin

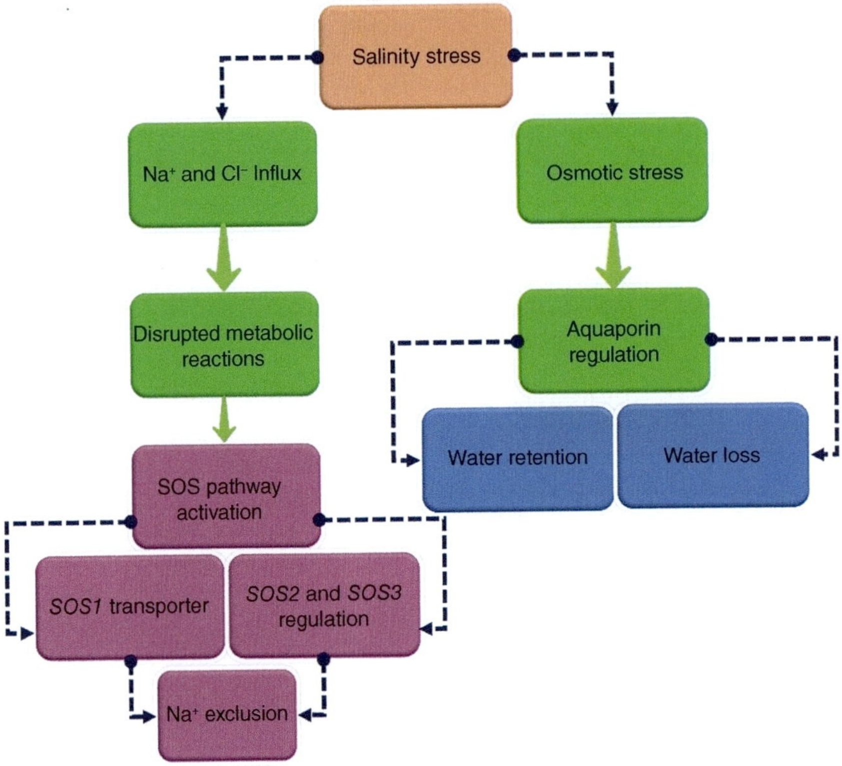

Fig. 1.2. Ion transport and aquaporin regulation in plants. (Figure author's own.)

reduces its activity to establish water-saving measures in the plant tissues (Fig. 1.2).

1.3.3 Role of osmotic and ionic homeostasis in stress tolerance

A high efficiency of osmotic and ionic regulation is known to be essential for the tolerance of stress. There are pre-existing universal mammal gene databases that are open access and easily accessible online, but there is no comparable source of human gene data since animals able to report the comprehension of several pressures influencing water and ion stability in their bodies can adapt better to the current climate change and other unfavorable conditions. For instance, in the case of halophytes (plants that are adapted to saline environments), Na^+ is trunked into the vacuoles to reduce its concentration in the cytoplasm. Likewise, the ability to tolerate drought conditions is made possible through the buildup of osmoprotectants besides the regulation of water transport (Saleem *et al.*, 2024).

1.3.4 Oxidative stress and reactive oxygen species (ROS) dynamics

Oxidation is thus one of the leading effects of environmental and physiological hassles (Anas *et al.*, 2024b). This is virtually losing the balance between the generation of oxidants and antioxidants when the species such as ROS surpass the capacity of antioxidants to neutralize the same. Some of these ROS include O_2^- radicals, H_2O_2, and OH• radicals, which can react with lipids, proteins, and DNA. However, they are also involved in signaling functions associated with stress and, therefore, the dynamic regulation of ROS has an extremely significant role to play in stress acclimatization. ROS are produced normally in cytochrome P450, xanthine oxidase, and complex I and II of the chloroplasts, mitochondria, and peroxisomes respectively. During stress conditions, the electron transport chains (ETCs) in the mitochondria and chloroplasts can disintegrate, leading to the production of ROS. The activation of NADPH oxidases and other ROS-producing enzymes further contributes to this process. Exposure to environmental stressors like UV radiation, heavy metals, and pollutants also leads to the generation of ROS. These free radicals or oxygen-derived species have detrimental effects on biological macromolecules.

To counteract the effects of ROS, organisms have developed an efficient AOS, which consists of enzymes as well as other structures for the removal of ROS and regulating redox balance.

1.3.5 Enzymatic and non-enzymatic antioxidants

Enzymatic and non-enzymatic antioxidants

- SOD converts the harmful O_2^- radicals into H_2O_2 and O_2.
- CAT, in the presence of a reducing agent, breaks down H_2O_2 and organic peroxides into water and O_2.
- APX is a key scavenger of H_2O_2 within the ascorbate-glutathione cycle, which detoxifies H_2O_2 in the chloroplasts and cytosol.
- Asc (Vitamin C) is a powerful non-enzymatic antioxidant that directly neutralizes ROS and helps to regenerate other antioxidants.
- GSH is a tripeptide that acts as a redox buffer and is a key participant in the ascorbate-glutathione cycle.
- Tocopherols (Vitamin E) are lipid-soluble antioxidants that are crucial for protecting cell membranes from oxidative damage.

ROS as signaling molecules

While an excessive amount of ROS can be damaging, they also serve as important signaling molecules that regulate stress responses. For instance, ROS can activate stress-responsive genes and pathways, such as the MAPK cascades. ROS also play a role in modulating stomatal closure, which helps plants reduce water loss during drought conditions.

Stress tolerance and antioxidant capacity

An organism's ability to tolerate stress is highly dependent on its capacity to manage oxidative stress. Organisms with a high antioxidant capacity can more effectively deal with the ROS produced under stress conditions. For example, transgenic plants engineered to express higher levels of antioxidant enzymes like SOD and APX often show a greater tolerance to abiotic stresses such as drought, salinity, and heavy metal toxicity.

1.3.6 Hormonal regulation in stress adaptation

Stress responses are centrally coordinated at the cellular, tissue, and organismal levels by hormones. They serve as signaling molecules involved in the regulation of gene expression, metabolic processes, and the physiological changes called for during the adaptation to stress. In plants, stress responses include ABA, JA, SA, ET and brassinosteroids (BRs), which are key hormones (Parwez *et al.*, 2022). Stress hormones, such as the hormones cortisol, adrenaline, and norepinephrine, are extremely important for animals' abilities to cope with stress situations.

1.3.7 Abscisic acid (ABA) in plant stress responses

In plants ABA plays the central and coordinating role in the mediation of responses to abiotic stresses like drought, salinity, and cold. ABA is therefore often known as the "stress hormone". ABA is synthesized in response to stress signals. These signals are more prevalent in conditions of water deficit. During biosynthesis, carotenoids are converted to ABA through the intermediates such as xanthoxin. Thus, ABA signaling includes receptor proteins (e.g. PYR/PYL/RCARs), which inhibit protein phosphatases (e.g. PP2Cs) and require activation of SnRK2 kinases. Phosphorylated by these kinases, TFs and ion channels stimulate stress-responsive gene expression and physiological changes. ABA causes stomatal closure to avoid water loss in drought. ABA upregulates stress genes, including most stress-responsive genes, genes encoding osmoprotectants, antioxidant enzymes, and dehydration protective proteins (Manna *et al.*, 2021). ABA also facilitates root growth for water and nutrient absorption.

JA and SA are two important hormones in the plant response to biotic stresses, like pathogen attack and herbivory. Defense responses against herbivores and necrotrophic pathogens are regulated by JA (Macioszek *et al.*, 2023). JA turns on the expression of defense genes such as those of protease inhibitors and secondary metabolites. There is also interaction between JA and other hormones, such as ET, in modulating stress responses. SA has mainly a defense against biotrophic pathogens. It provokes SAR, which is a lasting infection immunity. Activation of SA signaling engages non-expression of PR Genes 1 (NPR1), which is the major regulator of SAR (Mishra *et al.*, 2024).

1.3.8 Hormonal crosstalk in stress adaptation

Stress adaptation is dependent on crosstalk among hormones, which enables the integration of multiple signals and the fine-tuning of responses. For instance, ABA is typically involved in antagonizing ET, where ABA promotes stress tolerance, while ET promotes senescence and stress-induced damage. The interactions between JA and SA can be synergetic or antagonistic depending on the type of stress and on the situation. Stress hormones like cortisol, adrenaline, and norepinephrine take on major roles in the "fight or flight" response in animals. These hormones mediate the physiological changes needed to respond to acute stress, e.g. increased heart rate, increased blood glucose levels, and increased energy metabolism. Stress responses involve complex interplay of the mechanisms that, through the complex physiological and biochemical mechanisms that underlie them, govern the ability of the organism to maintain the homeostasis and adapt to adverse conditions. Ensuring stability of cellular environments, through both osmotic and ionic homeostasis, and protecting them against ROS-induced damage are important functions of oxidative stress management (García-Caparrós *et al.*, 2021; Hayat *et al.*, 2024). The details of these responses are coordinated at the organismal level by hormonal regulation where signals from different pathways are integrated to optimally adapt to stress. These mechanisms are useful for understanding how organisms cope with stress and present potential strategies for increasing stress tolerance in agriculture, medicine, and environmental conservation (Fig. 1.3).

1.4 Molecular Mechanisms of Stress Tolerance

Stress tolerance mechanisms as regulatory processes affecting all life forms can be understood as sets of signaling pathways, gene expression,

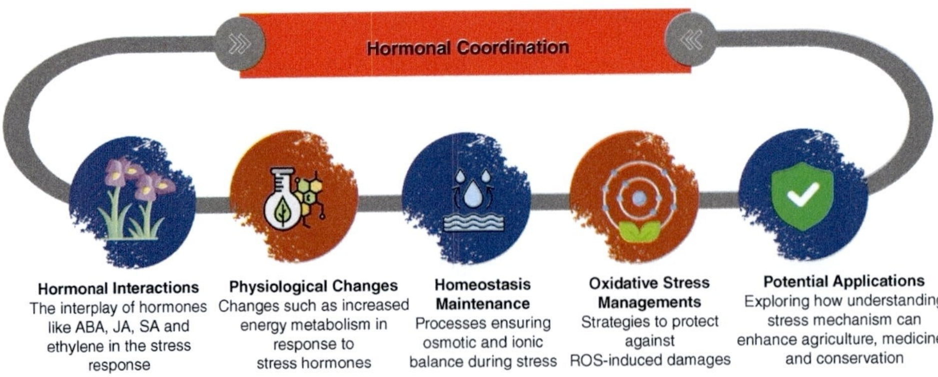

Fig. 1.3. Hormonal coordination in stress adaption. (Figure author's own.)

and epigenetic modifications that help organisms to sense and manage the effects of stress. These mechanisms are exercised for the ability to survive adverse effects such as drought, salt stress, extreme temperatures, and pathogen invasion. Stress tolerance at the molecular level underpins how organisms continue maintenance of homeostasis and potential ways to enhance stress resistance in crop plants and other organisms (Locato *et al.*, 2018).

Stress signaling pathways involve a set of signaling pathways through which stress signals that may originate from the environment are transmitted to the cell and utilitarian responses are generated. These pathways are present in almost all organisms and include activation of the receptor, signal transmission, and gene regulation pathways.

The MAPK pathway is a core process involved in response to stress stimulus. It is composed of three successively activated kinases: MAPKKK, MAPKK, and MAPK. Stress signals trigger the MAPKKK proteins to phosphorylate and activate the MAPKK proteins, which activates the MAPK proteins. The MAPK signal cascade can modulate activity of the TFs, enzymes, and other proteins in the stress signaling pathways (Yue and López, 2020). For example, in plants, the MAPK cascade mediates responses to abiotic stresses such as drought, salinity, and cold, as well as biotic stresses like pathogen attack.

Calcium ions (Ca^{2+}), per se, play the role of secondary messenger in stress signaling. Stress conditions automatically bring about an influx of Ca^{2+} into the cytoplasm resulting to activation of CDPKs and CaM. CDPKs and CaM work in a coordinated manner to regulate targets that include TFs and ion channels in stress responses. For instance, in plants, Ca^{2+} signaling is used in the closing of the stomata in response to drought and in the initiation of antioxidants in the case of oxidative stress (Liu *et al.*, 2022).

ROS are not only toxic moieties but especially ubiquitous stress mediators. They induce stress-responsive genes and signaling, as the MAPK cascade and the antioxidant adaptive response. ROS signaling is still tightly controlled to avoid oxidative stress on cells and tissues, but at the same time the stimulus is allowed to act as a signaling molecule for stress adaptation. For example, H_2O_2 has the ability to act as a signaling molecule to elicit the SAR in plants.

1.5 Hormonal Signaling and Integration of Stress Signaling Pathways

Some of the hormones that are involved in stress signaling include ABA, JA, and ET (Parwez *et al.*, 2022). For instance, ABA is involved in the management of drought stresses, while JA and ET have a role to play in defense against herbivores and pathogen attacks. Hormonal signaling is known to interconnect the pathways to regulate various stress responses. For instance, ABA and ET signaling networks play a synergistic/bipartite to become a pivotal for growth and stress in plants. Stress signaling pathways are linked so that the cells can receive several signals and manage a response to all of them. For instance, ROS signaling may activate the MAPK pathway, which in turn controls hormonal signaling (Raja *et al.*, 2017). Hormones are chemical signals that allow organisms to coordinate responses to various internal and external stimuli, including stressors.

1.5.1 Role of transcription factors (TFs) and stress-responsive genes

TFs are protein molecules capable of controlling gene activity when exposed to stress since these molecules determine the activation or inactivation of genes when stressed (Hussain *et al.*, 2021). TFs interact with specific DNA sequences, which are in the promoter regions of target genes to address stress stimuli. In the light of the above scene, the TFs involved in stress responses include DREB/CBF, NAC, bZIP, and WRKY families. DREBs are TFs that belong to the group of proteins involved in the regulation of dehydration response combined with C-repeat binding factors (CBFs) that take part in abiotic stress reactions including drought, salinity, and freezing stress. They interact with the DRE/CRT cis-element in the promoters of stress-responsive genes and stimulate the genes' activity.

Transgenic plants overexpressing DREB/CBF genes activate genes associated with the synthesis of osmoprotectants, antioxidant enzymes, and membrane maintenance. These NAC TFs are relevant to stress response such as drought, salinity, and oxidation stress. They control genes that produce proteins providing protection to the structures and maintaining internal environment of cells under stress. For instance, the NAC TF SNAC1 is responsible for improving drought tolerance in rice through controlling the closure of stomata and root development. Some of the well-known bZIP TFs involve AREB/ABF that mediate the ABA dependent stress response. They interact with the ABA-responsive element (ABRE) in the early part of the stress-responsive genes to control their expression. AREB/ABF TFs will always be significant to help in controlling genes that are involved with water retention and osmotic stress. In their study, Li *et al.* (2020) categorized all the identified WRKY TFs as biotic and abiotic stress-responsive proteins. The promoter sequences of genes that they control belong to the PR proteins and other stress-responsive proteins and antioxidant enzymes. Another function of WRKY TFs is their involvement in hormonal signaling that incorporates signals from ABA, JA, as well as SA.

1.6 Stress-responsive Genes

Stress-induced genes are considered as genes that produce proteins that shield them from adverse conditions such as stress-related damage and restore their biomolecules. These include:

- *Osmoprotectant biosynthesis genes.* These genes concern enzymes for the biosynthesis of proline, glycine betaine, and trehalose.
- *Antioxidant enzymes.* These enzymes include SOD, CAT, and APX for the removal of ROS
- *Stress-related proteins.* Stress-related proteins contain genes for late embryogenesis abundant (LEA) proteins and HSPs that protect the cells' structures during stress.

1.6.1 Epigenetic modifications in stress adaptation

Epigenetic modification is another type of modification of gene that can be passed on from one generation to the other but not through changes in the DNA sequence. MiRNAs are very important for stress adaptation since they are capable of mediating abrupt and reversible changes in the expression of multiple genes in response to stimulus (Prabhakar and Semenza, 2012).

1.6.2 DNA methylation, histone modifications and non-coding RNAs

DNA methylation refers to the process through which a methyl group is added to the cytosine base, in particular the quantitative cytosine base to which the methyl group is added most frequently in the CpG dimer. It usually downregulates gene expression through preventing the binding of the TFs or, in other cases, through recruitment of the repressive chromatin modifiers. The alt switch between genes with stress can lead to the activation or repression of stress-related genes due to alterations in DNA methylation.

Schmidt *et al.* (2020) established that drought stress results in the demethylation of genes for osmoprotectant production. Post-translational modifications including acetylation, methylation, and phosphorylation of histones, which change the chromatin conformation and have a role in the regulation of gene expression (Millán-Zambrano *et al.*, 2022). While acetylation is related to increased gene expression, methylation can either increase or decrease gene expression depending on the circumstance.

Histone modifications as a result of stress can change the way genes are expressed in stress-related genes. For instance, histone acetylation promotes the expression of DREB/CBF genes under cold stress condition.

There are other RNA molecules, for example, miRNA and lncRNA, which act in post-transcriptional regulation of genes expression, where miRNAs interact with target mRNAs leading to their degradation or translation suppression (Dykes and Emanueli, 2017). Schizophrenic responsive miRNA is implicated in stress tolerance through controlling of genes associated with stress stimulus, oxidative stress, and stress hormone signaling.

1.7 Plant Antioxidant Defense System

Cells need to stay within their normal redox range to survive, so they push for metabolic balance. Plants significantly contain compounds that stop oxidation from happening. Every kind of plant has some form of antioxidative protection, although this protection varies from one species to another. Lower levels of antioxidants slow down oxidation processes when present at concentrations below what needs to be oxidized. Plants possess mainly Asc and GSH as water-soluble antioxidants, and contain large quantities of tocopherols, tocotrienols, and carotenoids as liposoluble antioxidants.

The molecules eliminate ROS better when enzymes perform the reaction such as APX (EC 1.11.1.11), AscO, and other similar enzymes. Low weight antioxidants fight ROS by both cleaving them and by intercepting their path. Transition metal chelation blocks them from the Haber-Weiss or Fenton reactions, whereas direct action happens when antioxidants donate or accept electrons and catch free radicals before they can strike biological molecules. When electrons move between the compounds Asc, tocopherols, and phenolic antioxidants, the system stays stable through electrons spreading to and from different bonding areas. By nature, scavengers can enter the cells more easily than enzyme-based antioxidants and can position themselves close to biological cellular material that ROS damages (Zhang *et al.*, 2020).

Apart from major antioxidants, other biological molecules including amino acids sugars pigments, and secondary metabolites including flavonoids and terpenes, can function as antioxidants too. Secondary antioxidants can restore primary antioxidants such as Asc, which brings damaged α-tocopherol back to its active state. The α-tocopherol can then recover oxidized β-carotene (Gęgotek and Skrzydlewska, 2022). The antioxidants both reduce newly produced free radicals inside the lipid membranes to protect them from becoming oxidized. The joint activity of the AOS allows it to preserve cell membranes.

Plants possess large levels of Asc stored in chloroplasts and significantly lower amounts of GSH as its vital counterpart. The precise POD enzyme allows H_2O_2 to react quickly while regeneration systems restore their active state using powerful reductases. Most plant compartments that have weak redox recycling systems show high levels of oxidized antioxidants since their well-equipped regions use up over 95% of the original antioxidants. Environmental studies suggest these antioxidants cooperate in the water-water and Asc-GSH cycle to control H_2O_2 breakdown and chloroplast excitatory energy preservation. CAT serves a major purpose by breaking down H_2O_2 materials through peroxisome systems (Anwar *et al.*, 2024).

1.8 Water-water and Asc-GSH Cycles

Molecular oxygen (O_2) enters the water-water cycle at PS I where the enzyme turns ground state O_2 into a superoxide radical (O_2^-). The cell commodity O_2^- gets turned into water as SOD instantly works on it.

Under some conditions oxygen free radicals can change directly into H_2O_2 and O_2 without enzyme action. The H_2O_2 elimination through APX triggers the oxidation of two Asc molecules as its dedicated electron partners. MDA• becomes short-lived in this reaction before it naturally evolves into Asc or DHA while MDAR uses NADPH to restore MDA• back to Asc. MDAR operates in all internal cell spaces including chloroplasts, the cell membrane, and mitochondria as well as the cytosol and peroxisomes. DHA needs two electrons from GSH to transform back into Asc through the function of DHA reductases (DHAR, EC 1.8.5.1). The enzymes that use GSH in DHAR activity exist in chloroplasts, mitochondria, and peroxisomes. The enzyme GR uses NADPH electrons to convert GSSG back into GSH which completes the regeneration cycle of Asc and GSH (Anas *et al.*, 2024c).

Client cells throughout the body contain the Asc-GSH cycle active in their cytosol, mitochondria, chloroplasts, and peroxisomes. APX and SOD exist among multiple isoforms at both stroma and thylakoid membrane locations but chloroplast GR and DHAR are found in the stroma. Researchers now show how peroxiredoxin (PRX) helps maintain suitable H_2O_2 levels in plant chloroplasts together with the Asc-GSH cycle.

PRX reduces H_2O_2 along with organic peroxides through the POD family and team with thioredoxin (TRX) to detoxify ROS inside chloroplasts. PRX proteins destroy H_2O_2 by reacting with thiol groups. The Asc-GSH cycle works best with H_2O_2, but chloroplast APX shows enhanced performance against H_2O_2 even though PRX targets lipid peroxides and/or reactive nitrogen species (RNS).

1.9 Enzymatic Antioxidants

1.9.1 Superoxide dismutase (SOD)

The dismutation of O_2^- radicals proceeds at high speed through SOD, which decreases harmful •OH radical formation through metal reaction pathways. The natural dismutation speed of SOD is 10,000 times faster than the spontaneous breaking down of O_2^- radicals. The enzyme is active in all living cells and cellular areas that experience damage from oxidative stress (Murphy *et al.*, 2022). The plant cell hosts three SOD metalloenzymes that work through different metal ions at their cores. Cu/Zn-SOD stands as the most common plant SOD enzyme that resides within chloroplast stroma, cytosol, peroxisomes, and cell wall space. Mn-SOD appears in both mitochondrial and peroxisomal locations as well as in parts of the internal cell environment while Fe-SOD production remains more limited. This specialized SOD enzyme exists mainly inside the chloroplast stroma of plant specimens. Each SOD isoenzyme reacts differently to H_2O_2 and potassium cyanide (KCN) treatment (Campos-Shimada *et al.*, 2020). Genes that make SOD need environmental protection, and plants show more tolerance to stress when these enzymes work harder according to Sharma *et al.* (2020). SOD defends cells at the first line by breaking down O_2^- into H_2O_2 and other protective biochemical systems, then neutralizing the H_2O_2 against ROS damage. Plants with reduced chloroplast Cu/Zn-SOD function grow slowly and produce smaller plant cells with damaged photosynthetic zones.

1.9.2 Catalase and peroxidase

The level of H_2O_2 within cells is maintained through a system of enzymes especially CAT and POD, which control ROS presence. The high concentration of CAT appears inside peroxisomes while scientists have not identified this enzyme in chloroplasts yet. H_2O_2 forms in peroxisomes when plants break down fatty acids and perform photorespiration. The CAT enzyme breaks down two H_2O_2 molecules into H_2O and O_2 by oxidizing four heme component Fe^{2+} ions. CAT works best by transforming two H_2O_2 molecules into H_2O and O_2 through oxidation of Fe^{2+} ions without using any other reducing agent. High oxide molecules react with the Fe^{2+} ion of heme to trigger oxidation, then restores the ion back to Fe^{2+} by binding with H_2O_2. CAT reacts with millimolar H_2O_2 levels mainly during times of stress because its Michaelis constant (Km) value lies in this concentration range. CAT demonstrates a weak capacity to break down chemical peroxides. POD can remove small amounts of H_2O_2 because it needs higher H_2O_2 levels for reactions compared to CAT. Using GSH and Asc in POD-driven reactions costs more cellular energy for H_2O_2 removal since two small antioxidant molecules are needed for each H_2O_2 molecule. POS breaks down into three groups based on its aminoalkanoic acid arrangement (He *et al.*, 2022).

The APX belongs to the highest enzyme category, standing apart from the class III POD family, which includes horseradish peroxidase (HRP). The function of APX in plant cells is well understood, as the plant produces this enzyme in five specialized cellular areas. A deficiency in thylakoid-bound APX (tAPX) makes it hard or plants to survive, while plants that overexpress tAPX develop resistance to methyl viologen (paraquat).

Findings from a specific study by Collado-Arenal *et al.* (2024) showed that when plants were engineered to express cytosolic APX genes in their chloroplasts, they exhibited better resistance to drought and salt stress. Conversely, the corresponding knock-out mutants suffered a decreased scavenging ability and had increased H_2O_2 levels in their leaves. The APX enzyme varieties do not efficiently decrease lipid hydroperoxides. A unique characteristic of chloroplast APX is its rapid molecular breakdown when Asc is absent. When exposed to less than 20 mmol/l Asc and H_2O_2 reaction, chloroplast APX shows a rapid enzyme decline with a half-life of just 30 seconds. In contrast, cytoplasmic and peroxisomal isoforms inactivate more slowly over the course of an hour. Researchers view chloroplast Asc depletion and chloroplast APX denaturation as potential areas for increasing photosynthetic efficiency under stressful conditions.

All plant organs possess just one type of POD enzyme for H_2O_2 scavenging in their apoplastic and vacuolar spaces. The enzymes catalyze the H_2O_2 reduction of various substances through an electron reaction process that generates active molecules from phenolic compounds. This H_2O_2 dependent process is referred to as the peroxidase/phenolics/ascorbate (PPA) system.

This enzyme family reacts to many different environmental pressures through essential communication between plant antioxidants. Glutathione peroxidase (GPx) turns lipid peroxides and H_2O_2 into harmless substances through using both GSH and TRX as their main reductive agents. Together the enzymes CAT and GPx, as well as SOD, reinforce their O_2^- radical removal capacity.

1.10 Non-enzymatic Antioxidants

1.10.1 Ascorbate

The electron-donating ability of ascorbate makes it an exceptionally powerful antioxidant substance that can operate in multiple enzymatic and non-enzymatic reactions (Zandi and Schnug, 2022). The leaves contain an elevated amount of ascorbate compared to GSH, with most of it detected in this compartment. The monoanionic form of ascorbate mainly exists as its reduced state (90%) along with DHA (10%) in chloroplasts under physiological conditions. Ascorbate transfers its electrons, generating semidehydroascorbate after one electron donation. After giving its second electron, it becomes dehydroascorbate (DHA).

Dehydroascorbate reductase (DHAR) carries out the regeneration of ascorbate by utilizing two GSH molecules as its source of reduction. The cellular compartments of the cell wall, along with all subcellular organelles, host ascorbate apart from vacuoles where this substance exists in minimal levels. The majority of ascorbate located in cytoplasmic space differs from general soluble antioxidants since it maintains high concentrations in apoplasts, which serve as the main protective barrier against dangerous external prooxidants. Within chloroplasts ascorbate enables the activity of violaxanthin de-epoxidase to produce xanthophylls that directly quench excessive energy on PS II. The plant cell contains ascorbate as its principal antioxidant substance, which occurs throughout all subcellular regions and achieves its highest concentrations of 2–25 mmol/l or more in chloroplast stromal compartments (Wang *et al.*, 2023). The intracellular amounts reach millimolar strength while the cytosol contains 20 mmol/l and chloroplast stroma contains 20–300 mmol/l. The direct action of ascorbate enables capture of •OH, O_2^-, and $^1O_2$ in addition to the ascorbate peroxidase reaction, which reduces H_2O_2 to water. The reduced ascorbate functions as a prooxidant by lowering Fe, Cu, and Mn ion concentrations, enabling these ions to revisit redox reactions.

1.10.2 Glutathione (GSH)

The most abundant low relative molecular mass thiol within the cell is γ-glutamyl-cysteinyl glycine and scientists have located considerable amounts of it in cytosol, chloroplast, endoplasmic reticulum, vacuoles, and mitochondria (Rachet, 2023). Plant cells contain GSH as a redox buffer specifically for Asc functions and more cells of all types display this crucial redox property (Anas *et al.*, 2024c). The reduced form of GSH confers significant biological functions as a primary sulfur storage depot while serving critical roles in cellular growth and development, together with sulfur transport, signal transduction, nucleic acid and protein synthesis, phytochelatins-based metal chelation, xenobiotic detoxification, and stress-related gene expression. GSH occurs in two production locations as cells have enzymes that create it in chloroplasts and within the cytosol (Foyer and Hanke, 2022). The substance chelates Cu^{2+} ions thus blocking their participation in the Haber-Weiss reaction.

The redox balance of the cell relies on both reduced GSH along with its counterpart oxidized form, GSSG. The cysteine chemical group located at the core of the molecule enables GSH to retain high reduction potential. GSH functions as an antioxidant because it possesses low relative molecular mass to eliminate H_2O_2 while simultaneously interacting with •OH radicals, $^1O_2$, and O_2^- radicals without enzymes. As the primary antioxidant function of GSH involves regenerating the potent hydrophilic antioxidant Asc through the Asc-GSH cycle. GSH ensures complete regeneration of the reduced form of Asc through DHAR activity. The reducing ability of GSH works non-enzymatically to reduce DHA when both pH reaches above 7 and GSH reaches concentrations over 1 mmol/l (Anas *et al.*, 2024d).

1.11 Carotenoids

Lutein together with β-carotene, xanthophyll, zeaxanthin and lycopene belong to the carotenoid class that performs lipid membrane protection by detoxifying ROS through their efficient neutralization of the LOO• radical (Dumanović *et al.*, 2021). Carotenoids react with LOO• radicals to produce lipid hydroperoxide (LOOH) along with a carotenoid radical that can be restored by tocopherol. Both radicals can be subsequently reduced by Asc. Within plant cells these antioxidants protect plant structures by absorbing three chlorophyll (Chl) molecules, one O_2 molecule and 3Chl, $^1O_2$ as well as excited chlorophyll (Chl∗). β-carotene demonstrates better efficiency in capturing O_2 compared to the capturing ability of α-tocopherol according to Kern (2005) and Sharma *et al.* (2012). Carotenoids contain double bonds arranged in a conjugated system, which makes it effortless for excited molecules to absorb energy while releasing this energy as heat. The molecule zeaxanthin functions as part of non-photochemical excess energy quenching at PS II.

1.12 Tocopherols and Tocotrienols

Szewczyk *et al.* (2021) conducted research into the role of the cell membrane components tocopherols and tocotrienols. They found that these two components tocopherols manifest antioxidant capabilities and various other mechanisms that do not require antioxidant effects. A total of four compounds form the tocopherol and tocotrienol group including α, β, γ, and δ types. Photosynthetic organisms synthesize this group of lipophilic antioxidants while it exists exclusively within the green photosynthetic plant tissues. The antiradical kinetics of tocopherol depends on the donating electron properties of its chromanol ring. The antioxidants defend membrane elements by both physically ensnaring and chemically neutralizing $^1O_2$ in chloroplasts, which supports PS II functioning and protects its structural stability. The reaction of $^1O_2$ neutralization occurs at high efficiency as a single α-tocopherol molecule successfully eliminates up to 220 $^1O_2$ molecules according to laboratory results. Their reaction pace stands two times slower than β-carotene while they lack the capability to seize both •OH and alkoxy radicals (RO•) within the body. Asc as well as GSH and ubiquinone participate in restoring oxidized tocopherol in biological systems (Szewczyk *et al.*, 2021).

The order of antioxidant activity observed in the body (α > β > γ > δ) matches the sequence defined by phylogenetic polarity combined with the number of methyl substitutions on the phenolic ring manifold (Molino *et al.*, 2023). The best antioxidant effectiveness belongs to α-tocopherol because of its three methyl substitutions. Relevant research has shown that α-tocotrienol displays stronger membrane-based antioxidant power than α-tocopherol. The chloroplast membrane has α-isomer of tocopherol that provides optimal protection against photooxidative damage to chloroplasts. Inside plant leaves α-tocopherol functions as the primary form of vitamin E, while the seed contains mostly γ-tocopherol. The potential for vitamin E to generate lipid peroxyl alkyl and alkoxy radicals during polyunsaturated fatty acids oxidation allows it to block the propagation sequence of lipid layer auto-oxidation (Anas *et al.*, 2025a). Vitamin E gets converted into a tocopherol radical following its hydrogen donation to the radical, yet the stabilized radical remains too unreactive to induce membrane peroxidation independently, thus showing conductive characteristics of superior antioxidants. Ultra-high amounts of tocopherol function as prooxidants, together with transition metal ions and lipid peroxides.

1.13 Phenolic Compounds

The plant cell contains multiple varieties of secondary substances (flavonoids, tannins, hydroxycinnamate esters, lignin, etc.) known as phenols. The chemical structure of phenols enables them to trap radicals because of their aromatic ring combined with –OH or –OCH_3 substituents. Phenols exhibit strong electron donation and hydrogen donation capacity because they easily stabilize the formed phenol radical while achieving better antioxidant performance than Asc and α-tocopherol in vitro systems (Dumanović *et al.*, 2021). Phenols with o-dihydroxy groups in their makeup create metal ion complexes that stop ROS formation during the Haber-Weiss reaction. The biochemical compound has two functions: capturing the radical $^1O_2$ and stopping lipid peroxidation by neutralizing lipid alkoxy radicals. Antioxidant properties of phenols also work by changing how lipids are packaged which reduces membrane fluidity to affect the peroxidative process (Saha *et al.*, 2020). The modifications in the system both restrict free radical movement and decrease the reaction process. The research confirms that phenolic compounds serve a role in the capture cascade of H_2O_2. Under stress conditions phenolic compounds build up in plants while specific anthocyanins and flavanols delivered fourfold more antioxidant capacity than Asc.

In vacuoles H_2O_2 passively diffuses while POD reduces it alongside the use of phenols as primary electron donors for this hydrogen-peroxide-peroxidase system. MDA• radicals, together with Asc, possess the ability to reduce phenoxy radicals. The regenerative pathway of Asc inside the cytosol enables the vacuoles transport of Asc back to allow the PPA system to capture H_2O_2 within the vacuole. The protein mechanism exists only in plant tissues to help plants tolerate damaging oxidative stress conditions. MDAR demonstrates its ability to change the phenoxy radical similar to quercetin radical into a phenol compound (Nambiar, 2022).

1.14 Crosstalk Between Abiotic and Biotic Stress Responses

Organisms connect abiotic and biotic stress responses through a complex set of molecular and physiological and biochemical processes that allow them to handle several environmental stressors, including drought and high salt levels along with extreme temperatures and pathogen and herbivore threats at the same time or in succession.

The communication between various stress triggers relies on combined signaling pathways with shared transcriptional networks alongside hormonal and epigenetic systems that help organisms connect and rank their responses to various stressors (Singhal *et al.*, 2021). ROS molecules play dual functions in stress responses because they work simultaneously as harmful agents and signaling molecules. ROS exist in two states during plant stress as abiotic stress produces them as metabolic process byproducts, but biotic stress generates them through NADPH oxidase activation for pathogen defense (Anas Khan *et al.*, 2025; Anas Khattak *et al.*, 2025).

ROS function as dual-use molecules to establish stress signaling pathways since they both activate defense-related genes and trigger the MAPK cascade, which serves as a universal signaling module for responding to abiotic and biotic stress. The MAPK pathway, consisting of MPK3, MPK4, and MPK6, operates as an activation system in response to drought and salinity and PAMPs to control stress defense and tolerance gene expression (Sharma *et al.*, 2020). The signaling process of Ca^{2+} demonstrates crosstalk functionality because this messenger molecule activates both abiotic and biotic stress responses through the regulation of CDPKs and CaM activities. The stress adaptation balance between growth management functions derives from hormonal signaling networks that integrate ABA with JA, SA and ET. ABA signaling originally linked to abiotic stress performances like drought and salt stress, now shows involvement in biotic stress defense by controlling defense-related genes while connecting to signaling pathways involving JA and SA. Plants become more defenseless to specific pathogens when ABA reduces SA defense mechanisms, while ABA strengthens the JA defense system against herbivores.

Anti-stress hormones JA and ET, which usually protect against biotic threats, demonstrate abiotic stress influence through their action to control root development and maintain stomatal functionality within drought and saline conditions. Plants utilize SA as a primary pathogenic defense regulator to boost their resistance against abiotic stress by strengthening antioxidants for heavy metal defense and protection against UV radiation-induced oxidative stress. The interaction of hormones results in two main concerns because activating one signaling network typically suppresses the other, which causes specific defense mechanisms to take precedence over alternative responses (Figueroa-Macías *et al.*, 2021). Plants facing drought situations tend to direct their priority towards maintaining water conservation through ABA-based mechanisms instead of SA-dependent defense thus making them vulnerable to pathogen attacks.

TFs function as the vital elements of stress-responsive communication because they activate genes for multidimensional stress defense mechanisms. The W-box regulatory sequence in gene promoters attracts WRKY TFs which mediate the activation of PR protein along with antioxidant enzyme and osmoprotectant genes (Anas *et al.*, 2024a; Anas *et al.*, 2024c). The NAC TFs regulate genes for cellular protection and homeostasis control when plants face drought, salt stress or pathogen attacks, and bZIP TFs specifically like AREB/ABF control ABA-mediated stress responses as well as affect biotic stress responses by interacting with JA and SA signaling networks.

Gene expression changes due to environmental stimuli become rapid and reversible through epigenetic modifications that include DNA methylation and histone modifications and the impact of non-coding RNAs (Kumar *et al.*, 2020). DNA methylation can either activate or silence genes that respond to stress under recurrent abiotic and biotic conditions through similar pathways as histone modifications like acetylation and methylation control chromatin structure to regulate gene expressions.

Post-transcriptional regulation by microRNAs (miRNAs), together with long non-coding RNAs (lncRNAs), controls gene expression of significant stress signaling pathways and antioxidant defense elements and hormonal responses. The cellular defense mechanism miR398 controls antioxidant defenses under conditions of oxygen-based damage, yet miR393 manages the relations between JA signaling during attacks from herbivorous organisms. These integrated molecular mechanisms enable organisms to manage definite responses towards various stressors however they involve cost-benefit trade-offs emerging from how stress responses compete and require allocation of resources. The need to tolerate dry conditions causes plants to sacrifice their protection from pathogens thus making them more vulnerable to disease injuries, while management of herbivorous threats might lead to stunted growth due to resource reallocation. The relationship of the stressful interactions between abiotic and biotic factors needs complete examination because it unlocks vital secrets about stress adaptation, while creating strategies to improve tolerance in agricultural crops. The discovery of shared signaling networks, transcriptional activities, hormonal systems, and epigenetic patterns governing stress communication enables researchers to develop methods for strengthening stress resistance and securing food production in a changing climate. The understanding gained allows researchers to create crops that are able to withstand multiple environmental stress factors, so agriculture becomes sustainable under worsening climate conditions (Anas *et al.*, 2025a) (Fig. 1.4).

1.15 Role of Nanotechnology and Biotechnology in Enhancing Stress Tolerance

Nanotechnology and biotechnology have come a long way as means for strengthening plant and organisms stress tolerance, and they are promising solutions that cancel out the negative impacts of abiotic and biotic stressors (Azameti and Imoro, 2023). They make use of the distinct properties of nanomaterials and the accuracy of genetic engineering to enable crops to withstand extreme stress, maximize recoverable resources, and increase agricultural productivity. Researchers can integrate nanotechnology and biotechnology to develop superior strategies to combat the problems of increasing climate change, population, and environmental degradation.

1.15.1 Nanotechnology in stress tolerance

Manipulation of matter at 1–100 nm scale is termed nanotechnology, and this has been applied to the development of novel applications in the areas of agriculture and environment. Nanoparticles (NPs) and nanomaterials, as such, have unique physical, chemical, and biological properties that are very useful in imparting stress tolerance (Zulfiqar and Ashraf, 2021).

1.15.2 Nanofertilizers and nanopesticides

Nanofertilizers are better because there are nano encapsulated nutrients that increase the nutrient use efficiency and reduce environmental losses (Babu *et al.*, 2022). Zinc oxide (ZnO)

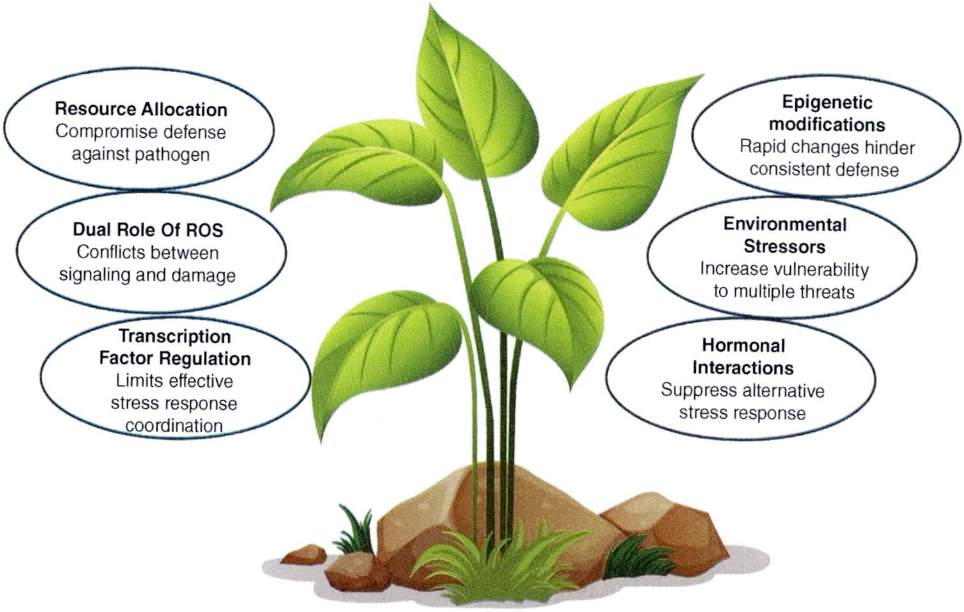

Fig. 1.4. Competing stress response in plants. (Figure author's own.)

and iron oxide (Fe_2O_3) nanoparticles improve nutrient uptake and stress relieving. Silver (Ag) and copper (Cu) nanoparticles are among the nanopesticides that provide targeted delivery of active ingredients and reduce the use of chemical inputs and hence, their impact on environment. The potential of these nanomaterials is not only to combat climate change but also to prevent pathogen attack in the plants through their antimicrobial properties (Anas *et al.*, 2025c).

1.15.3 Nanosensors for stress detection

The ability to continuously monitor environmental conditions and plant health, and for early detection of stress is afforded by nanosensors. For instance, sensors based on carbon nanotubes can sense variations in soil moisture, pH, and nutrient levels, allowing for precision agriculture. Timely interventions like irrigation or nutrient application can be made to avoid negative effects of the stress and enhance crop performance when stress is detected early.

1.15.4 Nanomaterials for oxidative stress management

Cerium oxide (CeO_2) and titanium dioxide (TiO_2) nanoparticles that have an antioxidant function in "scavenging" of ROS compounds that induce oxidative stress in plants under the conditions of drought, salinity, and heavy metals. These nanomaterials increase antioxidant enzymes' activities including SOD and CAT, strengthening stress tolerance and cellular homeostasis.

1.16 Biotechnology in Stress Tolerance

1.16.1 Genetic engineering, molecular breeding, omics technologies

Developing stress-tolerant crops and organisms using genetic engineering, molecular breeding and driving "omics" technologies is a form of biotechnology (Anas *et al.*, 2025d). They also make it possible to precisely manipulate the involved genes and pathways that mediate the stress responses. However, the introduction or modification of stress-responsive genes can be achieved through genetic engineering to enhance tolerance. For instance, overexpressed genes for osmoprotectants (e.g. proline, glycine betaine) or antioxidant enzymes (e.g. SOD, APX) help in developing drought and salinity tolerance.

Several foreign DNA free abiotic stress tolerance strategies based on CRISPR–Cas9 genome editing are applied to target these stress-related genes, for instance, DREB/CBF TFs for improved abiotic stress tolerance. Molecular breeding is a combination of traditional breeding with the help of identifiable molecular markers for selecting the stress tolerant traits (Ashraf and Foolad, 2013).

Accelerated development of stress resilient crop varieties is due to identification and introgression of desirable alleles by marker assisted selection (MAS). For example, MAS has been used for example to select for alleles associated with root architecture and water use efficiency to develop drought-tolerant rice and wheat varieties. Obtaining comprehensive information at the molecular level of stress response involves omics technologies, for example, genomics, transcriptomics, proteomics, and metabolomics (Anas *et al.*, 2025d).

The approaches described identify key genes, proteins, and metabolites involved in stress adaption. For example, transcriptomic analysis allows for stress-responsive genes and pathways to be identified, which then can be used to engineer crops that are more tolerant to multiple stressors.

1.16.2 Integration of nanotechnology and biotechnology

Synergistic benefits to augment stress tolerance can be achieved by the integration of nanotechnology and biotechnology. For example, a nanotechnology can deliver genes like DNA or RNA to plant cell for modern genetic engineering with micro precision (Squire *et al.*, 2023). The efficiency of CRISPR–Cas9 genome editing is enhanced for the development of stress tolerant crops via nanoparticle mediated gene delivery. In addition, nanomaterials can be used as the delivery agent of biostimulants; plant growth promoting rhizobacteria (PGPR) in our case, which can increase the stress resiliency of crops and their nutrient uptake. Abiotic and biotic stresses impact plant growth and development, having adverse effects on agricultural production. Nanotechnology and biotechnology, however, are involved in increasing stress tolerance by offering unique solutions that prevent the impact of abiotic and biotic stressors. These methods allow such things as the creation of stress resilient crops, resource optimization and increasing agricultural productivity (Srivastav *et al.*, 2021). The integration of nanotechnology and biotechnology allows researchers to counter challenges of climate change and ensure food security for an ever-increasing global population.

1.17 Perspectives in Plant Stress Research

The integration of advanced technologies and multidisciplinary approaches is also a future perspective in plant stress research dealing with the challenges of climate change and environmental variability. As temperatures increase worldwide, water gets scarce, and soil salinity rises, researchers are breeding crops that can endure more than one stressor at a time. High throughput omics technologies—genomics, transcriptomics, proteomics, and metabolomics—are revolutionizing our knowledge of plant stress responses at the molecular, cellular, and whole plant levels. Precise identification of stress-responsive genes and pathways allows for specific genetic modifications that result in increased tolerance. Finally, the emerging imaging techniques such as computed tomography, magnetic resonance imaging, and positron emission tomography are eliminating requirements of invasiveness in stress detection methods and are providing non-invasive approaches to study plant physiology across the scales from the cellular level to the field level. Real time stress monitoring in agricultural landscapes through combining remote sensing technologies with artificial intelligence (AI) is investigated to develop early warning systems for optimizing the use of resources and reducing crop losses. In addition, MAS and genomic prediction have started being used by breeding programs for development of genotypes bred for stress adapted genotypes with higher yields under adverse conditions.

Other researchers are exploring epigenetic mechanisms and small RNA-mediated regulation as an avenue to increase plant resilience, but not via changes to core genetic structures.

Future studies will examine the difficulty of combined stress scenarios of drought with heat or salinity and how, under these scenarios, plant growth is affected in a dose dependent way and whether the effect results in synergistic or antagonistic responses. This collaboration of plant scientists, physicists, computer scientists, and agronomists will advance precision agriculture in driving innovation and sustainable food production for a growing global population.

1.18 Conclusion

Stress response in plants provides corner stone information on understanding physiological, biochemical, and molecular basis that helps them to counter different adverse states. The plant has complex mechanisms of adapting to abiotic and biotic stress factors that include osmotic and ionic balance, oxidative stress, and hormonal signaling. These responses are regulated by signaling pathways, transcription preparations, and epigenetic regulation, thus enabling plants to recognize, counter, and restore to normalcy any harm. It only complicates the signaling mechanisms further as the plant has to interpret and consolidate signals from both abiotic and biotic stressors for it to survive. Biotechnology, as well as nanotechnology, has brought about an improvement in stress tolerance, genetic modification or advancement of molecular breeding schemes, and nanocarrier delivery. These technologies not only enhance the stress biology knowledge but are also helpful for the designing of stress-tolerant crops to fulfill the food demand required in the context of climate change and land degradation. Studying stress responses in more detail provides an opportunity for a further breakthrough in the development of effective approaches to solving the problems of sustainable agriculture and feeding the world's population in conditions of an active conquest of space.

References

Abajebal, N., Barnabas, J. and Abebe, T. (2024) Impact of RNA enzymes and RNA protein complexes in unleashing RNA interference for manyfold crop improvements: A review. *Agricultural Reviews* 45(4), 421–435.

Abdullah, S.N.A., Azzeme, A.M. and Yousefi, K. (2022) Fine-tuning cold stress response through regulated cellular abundance and mechanistic actions of transcription factors. *Frontiers in Plant Science* 13, 850216.

Anas, M., Saeed, M., Elahi, M., Naeem, K., Shafique, M.A. and Qurashi, U.M (2023) Histological and ionomics assessment to elucidate tolerance mechanisms of nickel-tolerant and sensitive cultivars of bread wheat (Triticum aestivum L.). *Plant Stress* 10, 100277. DOI: 10.1016/j.stress.2023.100277.

Anas, M., Bibi, S., Farwa, U., Khan, A. and Quraishi, U.M. (2024a) Manganese oxide nanoparticles elicit cadmium tolerance in wheat seedlings by ionomic and phenomic adjustment, regulation of AsA–GSH pathway, cellular thickness recovery, and antioxidant modulation. *Environmental Science: Nano* 11(11), 4588–4605.

Anas, M., Bashir, M.S., Saleem, K., Noor, A. and Quraishi, U.M. (2024b) Adaptive mechanisms of wheat cultivars to lead toxicity through enhanced oxidative defense, ionomic redistribution, and anatomical modifications. *Journal of Plant Physiology* 303, 154370. DOI: 10.1016/j.jplph.2024.154370.

Anas, M., Saeed, M., Naeem, K., Shafique, M. A., and Quraishi, U. M. (2024c) Anatomical and ionomics investigation of bread wheat (Triticum aestivum L.) to decipher tolerance mechanisms under arsenic stress. *Journal of Plant Growth Regulation* 1–17.

Anas, M., Falak, A., Khan, A., Khattak, W. A., Nisa, S. G., Aslam, Q. *et al.* (2024d) Therapeutic potential and agricultural benefits of curcumin: a comprehensive review of health and sustainability applications. *Journal of Umm Al-Qura University for Applied Sciences*, 1–16.

Anas, M., Khan, A., Falak, A., Shinwari, Z.K., Mohamed, H.E.A., Hkiri, K. *et al.* (2025a) Harnessing biomedical applications of ascorbic acid, berberine, and glutathione-capped quantum dots through multifunctional physico-chemical, biochemical, and anti-microbial assays. *BioNanoScience* 15(2), 259–271.

Anas, M., Falak, A., Hassan, S., Khattak, W.A., Saleem, M.H., Khan, K.A. *et al.* (2025b) Microbial interactions and bacterial responses to metal stress in plants: Mechanisms, adaptations, and applications for sustainable agriculture. *Journal of Crop Health* 77, 36. DOI: 10.1007/s10343-025-01108-w.

Anas, M., Khattak, W.A., Alrawiq, N, Jaremko, M., Quraishi, U.M., Alrawiq, H.S. *et al.* (2025c) Mitigating heavy metal pollution in agriculture: A multi-omics and nanotechnology approach to safeguard global wheat production. *Journal of Hazardous Materials Advances* 17, 100584. DOI: 10.1016/j.hazadv.2024.100584

Anas, M., Khalid, A., Saleem, M.H., Ali Khan, K., Ahmed Khattak, W. and Fahad, S. (2025d) Symbiotic synergy: Unveiling plant-microbe interactions in stress adaptation. *Journal of Crop Health* 77(1), 1–21.

Anwar, S., Alrumaihi, F., Sarwar, T., Babiker, A.Y., Khan, A.A., Prabhu, S.V. and Rahmani, A.H. (2024) Exploring therapeutic potential of catalase: Strategies in disease prevention and management. *Biomolecules* 14(6), 697.

Arif, Y., Singh, P., Siddiqui, H., Bajguz, A. and Hayat, S. (2020) Salinity induced physiological and biochemical changes in plants: An omic approach towards salt stress tolerance. *Plant Physiology and Biochemistry* 156, 64–77.

Ashraf, M. and Foolad, M.R. (2013) Crop breeding for salt tolerance in the era of molecular markers and marker-assisted selection. *Plant Breeding* 132(1), 10–20.

Ataullakhanov, F.I., Martinov, M.V., Shi, Q. and Vitvitsky, V.M. (2022) Significance of two transmembrane ion gradients for human erythrocyte volume stabilization. *PLoS one* 17(12), e0272675.

Azameti, M.K. and Imoro, A.W.M. (2023) Nanotechnology: A promising field in enhancing abiotic stress tolerance in plants. *Crop Design* 2(2), 100037.

Babu, S., Singh, R., Yadav, D., Rathore, S.S., Raj, R. *et al.* (2022) Nanofertilizers for agricultural and environmental sustainability. *Chemosphere* 292, 133451.

Bashir, S.S., Hussain, A., Hussain, S.J., Wani, O.A., Zahid Nabi, S. *et al.* (2021) Plant drought stress tolerance: Understanding its physiological, biochemical and molecular mechanisms. *Biotechnology & Biotechnological Equipment* 35(1), 1912–1925.

Bhattacharya, A. and Bhattacharya, A. (2021) Effect of soil water deficit on growth and development of plants: A review. *Soil Water Deficit and Physiological Issues in Plants* 393–488.

Campos-Shimada, L.B., Hideo Gilglioni, E., Fernandes Garcia, R., Rizato Martins-Maciel, E., Luiza Ishii-Iwamoto, E. and Luzia Salgueiro-Pagadigorria, C. (2020) Superoxide dismutase: A review and a modified protocol for activities measurements in rat livers. *Archives of Physiology and Biochemistry* 126(4), 292–299.

Collado-Arenal, A.M., Exposito-Rodriguez, M., Mullineaux, P.M., Olmedilla, A., Romero-Puertas, M.C. and Sandalio, L.M. (2024) Cadmium exposure induced light/dark-and time-dependent redox changes at subcellular level in Arabidopsis plants. *Journal of Hazardous Materials* 477, 135164.

Divekar, P.A., Narayana, S., Divekar, B.A., Kumar, R., Gadratagi, B.G. *et al.* (2022) Plant secondary metabolites as defense tools against herbivores for sustainable crop protection. *International Journal of Molecular Sciences* 23(5), 2690.

Dumanović, J., Nepovimova, E., Natić, M., Kuča, K. and Jaćević, V. (2021) The significance of reactive oxygen species and antioxidant defense system in plants: A concise overview. *Frontiers in Plant Science* 11, 552969.

Dykes, I.M. and Emanueli, C. (2017) Transcriptional and post-transcriptional gene regulation by long non-coding RNA. *Genomics, Proteomics & Bioinformatics* 15(3), 177–186.

Figueroa-Macías, J.P., García, Y.C., Núñez, M., Díaz, K., Olea, A.F. and Espinoza, L. (2021) Plant growth-defense trade-offs: Molecular processes leading to physiological changes. *International Journal of Molecular Sciences* 22(2), 693.

Foyer, C.H. and Hanke, G. (2022) ROS production and signalling in chloroplasts: Cornerstones and evolving concepts. *The Plant Journal* 111(3), 642–661.

García-Caparrós, P., De Filippis, L., Gul, A., Hasanuzzaman, M., Ozturk, M., Altay, V. and Lao, M.T. (2021) Oxidative stress and antioxidant metabolism under adverse environmental conditions: A review. *The Botanical Review* 87, 421–466.

Gęgotek, A. and Skrzydlewska, E. (2022) Antioxidative and anti-inflammatory activity of ascorbic acid. *Antioxidants* 11(10), 1993.

Hayat, A., Anas, M., Shaheen, Z., Falak, A. and Quraishi, U.M. (2024) Comparative morpho-physiological traits, antioxidant defense and nutritional profiling under Cd stress of japonica-indica elite rice (Oryza sativa L.) cultivars. *Journal of Crop Science and Biotechnology* 27(2), 175–186.

He, X., Li, X., Zhao, J., Mu, C., Mi, H., Hu, J. and Shi, F. (2022) Improving the catalytic and mechanical performance of alginate catalyst through functionalization by aminopolycarboxylic acids. *Journal of Colloid and Interface Science* 628, 1058–1066.

Hussain, Q., Asim, M., Zhang, R., Khan, R., Farooq, S. and Wu, J. (2021) Transcription factors interact with ABA through gene expression and signaling pathways to mitigate drought and salinity stress. *Biomolecules* 11(8), 1159.

Kalavalapalli, V.S., Aishwarya, B. and Prasuna, R.G. (2024) Unveiling stress tolerance mechanisms in photosynthetic cyanobacterial symbionts: A comprehensive review. *Stress Biology in Photosynthetic Organisms: Molecular Insights and Cellular Responses* 121–146.

Kern, J. C. and Kehrer, J. P. (2005) Free radicals and apoptosis: relationships with glutathione, thioredoxin, and the BCL family of proteins. *Front Biosci*, 10, 1727–1738.

Ketehouli, T., Quoc, V.H.N., Dong, J., Do, H., Li, X. and Wang, F. (2022) Overview of the roles of calcium sensors in plants' response to osmotic stress signalling. *Functional Plant Biology* 49(7), 589–599.

Kumar, S., Gonzalez, E. A., Rameshwar, P. and Etchegaray, J. P. (2020). Non-coding RNAs as mediators of epigenetic changes in malignancies. *Cancers* 12(12), 3657.

Kumari, V.V., Banerjee, P., Verma, V.C., Sukumaran, S., Chandran, M.A.S. *et al.* (2022) Plant nutrition: An effective way to alleviate abiotic stress in agricultural crops. *International Journal of Molecular Sciences* 23(15), 8519.

Li, W., Pang, S., Lu, Z. and Jin, B. (2020) Function and mechanism of WRKY transcription factors in abiotic stress responses of plants. *Plants* 9(11), 1515.

Liu, H., Song, S., Zhang, H., Li, Y., Niu, L., Zhang, J. and Wang, W. (2022) Signaling transduction of ABA, ROS, and Ca2+ in plant stomatal closure in response to drought. *International Journal of Molecular Sciences* 23(23), 14824.

Locato, V., Cimini, S. and De Gara, L. (2018) ROS and redox balance as multifaceted players of cross-tolerance: epigenetic and retrograde control of gene expression. *Journal of Experimental Botany* 69(14), 3373–3391.

Macioszek, V.K., Jęcz, T., Ciereszko, I. and Kononowicz, A.K. (2023) Jasmonic acid as a mediator in plant response to necrotrophic fungi. *Cells* 12(7), 1027.

Mahdavian, K. (2024) Effects of ultraviolet radiation on plants and their protective mechanisms. *Russian Journal of Plant Physiology* 71(6), 184.

Manna, M., Thakur, T., Chirom, O., Mandlik, R., Deshmukh, R. and Salvi, P. (2021) Transcription factors as key molecular target to strengthen the drought stress tolerance in plants. *Physiologia Plantarum* 172(2), 847–868.

Millán-Zambrano, G., Burton, A., Bannister, A.J. and Schneider, R. (2022) Histone post-translational modifications—cause and consequence of genome function. *Nature Reviews Genetics* 23(9), 563–580.

Mishra, S., Roychowdhury, R., Ray, S., Hada, A., Kumar, A. *et al.* (2024) Salicylic acid (SA)-mediated plant immunity against biotic stresses: An insight on molecular components and signaling mechanism. *Plant Stress*, 100427.

Molino, S., Francino, M.P. and Henares, J.Á.R. (2023) Why is it important to understand the nature and chemistry of tannins to exploit their potential as nutraceuticals? *Food Research International* 173, 113329.

Mondal, S. (2023) Heavy metal stress-induced activation of mitogen-activated protein kinase signalling cascade in plants. *Plant Molecular Biology Reporter* 41(1), 15–26.

Mukherjee, A., Dwivedi, S., Bhagavatula, L. and Datta, S. (2023) Integration of light and ABA signaling pathways to combat drought stress in plants. *Plant Cell Reports* 42(5), 829–841.

Murphy, M.P., Bayir, H., Belousov, V., Chang, C.J., Davies, K.J. *et al.* (2022) Guidelines for measuring reactive oxygen species and oxidative damage in cells and in vivo. *Nature Metabolism* 4(6), 651–662.

Nadeem, F., Hanif, M.A., Majeed, M.I. and Mushtaq, Z. (2018) Role of macronutrients and micronutrients in the growth and development of plants and prevention of deleterious plant diseases-a comprehensive review. *International Journal of Chemical and Biochemical Sciences* 13, 31–52.

Nambiar, S.S. (2022) Antioxidant activity of polyphenols in combating atherosclerosis. *International Journal of Research and Analytical Reviews* 9, 905–924.

Padayachee, A., Day, L., Howell, K. and Gidley, M.J. (2017) Complexity and health functionality of plant cell wall fibers from fruits and vegetables. *Critical Reviews in Food Science and Nutrition* 57(1), 59–81.

Palmgren, M. (2023) Evolution of the sodium pump. *Biochimica et Biophysica Acta (BBA)-Molecular Cell Research* 1870(7), 119511.

Parcheta, M., Ś wisłocka, R., Orzechowska, S., Akimowicz, M., Choińska, R. and Lewandowski, W. (2021) Recent developments in effective antioxidants: The structure and antioxidant properties. *Materials* 14(8), 1984.

Parwez, R., Aftab, T., Gill, S.S. and Naeem, M. (2022) Abscisic acid signaling and crosstalk with phytohormones in regulation of environmental stress responses. *Environmental and Experimental Botany* 199, 104885.

Prabhakar, N.R. and Semenza, G.L. (2012) Adaptive and maladaptive cardiorespiratory responses to continuous and intermittent hypoxia mediated by hypoxia-inducible factors 1 and 2. *Physiological Reviews* 92(3), 967–1003.

Qi, J., Song, C.P., Wang, B., Zhou, J., Kangasjärvi, J., Zhu, J.K. and Gong, Z. (2018) Reactive oxygen species signaling and stomatal movement in plant responses to drought stress and pathogen attack. *Journal of Integrative Plant Biology* 60(9), 805–826.

Rachet, A. (2023) Characterizing Transport Modalities of Glutathione Towards the Mammalian Endoplasmic Reticulum, and Chemical Screenings for the Discovery of Novel Compounds Modulating its Metabolism (Doctoral dissertation, Université Paris-Saclay).

Raja, V., Majeed, U., Kang, H., Andrabi, K.I. and John, R. (2017) Abiotic stress: Interplay between ROS, hormones and MAPKs. *Environmental and Experimental Botany* 137, 142–157.

Roychowdhury, R., Mishra, S., Anand, G., Dalal, D., Gupta, R., Kumar, A. and Gupta, R. (2024) Decoding the molecular mechanism underlying salicylic acid (SA)-mediated plant immunity: An integrated overview from its biosynthesis to the mode of action. *Physiologia Plantarum* 176(3), e14399.

Saha, S., Panieri, E., Suzen, S. and Saso, L. (2020) The interaction of flavonols with membrane components: Potential effect on antioxidant activity. *The Journal of Membrane Biology* 253, 57–71.

Saleem, M.H., Zafar, S., Javed, S., Anas, M., Ahmed, T. *et al.* (2024) Modulatory effects of glutamic acid on growth, photosynthetic pigments, and stress responses in olive plants subjected to cadmium stress. *Journal of King Saud University-Science* 36(11), 103540.

Schmidt, J., Claussen, J., Wörlein, N., Eggert, A., Fleury, D., Garnett, T., and Gerth, S. (2020). Drought and heat stress tolerance screening in wheat using computed tomography. *Plant Methods*, 16(1), 15.

Shahzad, A., Ullah, S., Dar, A.A., Sardar, M.F., Mehmood, T. *et al.* (2021) Nexus on climate change: Agriculture and possible solution to cope with future climate change stresses. *Environmental Science and Pollution Research* 28, 14211–14232.

Sharma, D., Verma, N., Pandey, C., Verma, D., Bhagat, P.K. *et al.* (2020) MAP kinase as regulators for stress responses in plants: An overview. *Protein Kinases and Stress Signaling in Plants: Functional Genomic Perspective*, 369–392.

Singh, R., Singh, S., Parihar, P., Mishra, R.K., Tripathi, D.K. *et al.* (2016) Reactive oxygen species (ROS): beneficial companions of plants' developmental processes. *Frontiers in Plant Science* 7, 1299.

Singhal, R.K., Saha, D., Skalicky, M., Mishra, U.N., Chauhan, J. *et al.* (2021) Crucial cell signaling compounds crosstalk and integrative multi-omics techniques for salinity stress tolerance in plants. *Frontiers in Plant Science* 12, 670369.

Squire, H.J., Tomatz, S., Voke, E., González-Grandío, E. and Landry, M. (2023) The emerging role of nanotechnology in plant genetic engineering. *Nature Reviews Bioengineering* 1(5), 314–328.

Srivastav, A.L., Dhyani, R., Ranjan, M., Madhav, S. and Sillanpää, M. (2021) Climate-resilient strategies for sustainable management of water resources and agriculture. *Environmental Science and Pollution Research* 28(31), 41576–41595.

Szewczyk, K., Chojnacka, A. and Górnicka, M. (2021) Tocopherols and tocotrienols—bioactive dietary compounds; What is certain, what is doubt? *International Journal of Molecular Sciences* 22(12), 6222.

Wang, L., Yang, T., Pan, Y., Shi, L., Jin, Y. and Huang, X. (2023) The metabolism of reactive oxygen species and their effects on lipid biosynthesis of microalgae. *International Journal of Molecular Sciences* 24(13), 11041.

Wennerström, H. and Oliveberg, M. (2022) On the osmotic pressure of cells. *QRB Discovery* 3, e12.

Yue, J. and López, J.M. (2020) Understanding MAPK signaling pathways in apoptosis. *International Journal of Molecular Sciences* 21(7), 2346.

Zandi, P. and Schnug, E. (2022) Reactive oxygen species, antioxidant responses and implications from a microbial modulation perspective. *Biology* 11(2), 155.

Zhang, J., Fu, Y., Yang, P., Liu, X., Li, Y. and Gu, Z. (2020) ROS scavenging biopolymers for anti-inflammatory diseases: classification and formulation. *Advanced Materials Interfaces* 7(16), 2000632.

Zulfiqar, F. and Ashraf, M. (2021) Nanoparticles potentially mediate salt stress tolerance in plants. *Plant Physiology and Biochemistry* 160, 257–268.

2 Plants' Responses to Abiotic Stress: Signal Transduction and Perception Mechanism

Waseem Ahmed Khattak[1], Ifra Batool[1], Muhammad Anas[1,2], Banzeer Ahsan Abbasi[3] and Javed Iqbal[4*]

[1]Department of Plant Sciences, Faculty of Biological Sciences, Quaid-i-Azam University, Islamabad 45320, Pakistan; [2]Department of Soil Science & Plant Nutrition, University of Selcuk Campus, Konya 42079, Türkiye; [3]Department of Botany, Rawalpindi Women University, Rawalpindi 46300, Pakistan; [4]Department of Botany, Bacha Khan University, Charsadda 24420, Pakistan

Abstract

A natural outcome faced by plants is the ever-increasing range of various environmental stresses including drought, salinity, temperature extremes, and pathogen attacks, all of which can have very deleterious effects on growth, development, and agricultural productivity. To face these challenges, plants have developed complex stress perception and signal transduction mechanisms to quickly trigger the right signals in response to stresses imposed. In this chapter, the molecular and physiological aspects of plant stress adaptation are studied including the roles of membrane receptors, signaling cascades, and transcriptional regulation. Stress perception is initiated through membrane bound receptors, such as receptor-like kinases (RLKs) and ion channels, which activate key signaling pathways such as calcium signaling, generation of reactive oxygen species (ROS), and mitogen-activated protein kinase (MAPK) cascades. Expression of these genes is regulated by transcription factors (TFs) including WRKY, MYB and NAC, which regulate stress responses such as osmotic adjustment, antioxidant defense, and hormonal regulation. This chapter also shows how the perception of stress integrates into the sphere of plant growth and development by the means of hormonal crosstalk comprising abscisic acid (ABA), ethylene (ET), cytokinins (CKs) and auxins. Stress-induced secondary metabolites with far reaching consequences to human health due to their antioxidant and therapeutic properties are also discussed. The study underscores the need for more interdisciplinary research in embracing plant biology and in the human health sciences, such as that involved in functional foods, nutraceuticals, and biopharmaceuticals. This includes further application of genome editing, RNA interference (RNAi), and omics approaches to improve stress tolerance of crops. Knowledge of these mechanisms is essential to creating resilient crop varieties that contribute to food security, and to leveraging plant derived compounds as medicinal agents during the age of climate change and global environmental challenges.

Keywords: plant stress adaptation, signal transduction, reactive oxygen species (ROS), hormonal regulation, crop resilience

* Email: javed89qau@gmail.com

DOI: 10.1079/9781800628823.0002

2.1 Introduction

Climate adversities including salinity, water deficit/flooding, low or high temperature, metal/metalloids, and nutrient stress, are basically severe threats to global agriculture since they limit capability of crop plants to express their genetic potentiality and decrease the yield globally (Choudhary and Meena, 2024; Nehra *et al.*, 2024). These are some of the ways through which the above-mentioned abiotic stresses are worsened by the changing climatic conditions. Salinity and our drought are reckoned to impact more than 10% of the agricultural cultivated land. This results in an approximate average loss of 50% crop yield loss worldwide (Soni *et al.*, 2023).

Stresses are polygenic factors that lessen the growth and development of the plant, and consequently the produce, in a big way. As a result of these environmental insults, plants have developed mechanisms that contain the effect by harnessing their tolerance potential through numerous molecular and cellular processes. In general, the defense mechanism is the formation of stress response into genes and their products, which are functional or regulatory to achieve direct stress tolerance or regulatory through signal transduction pathway (Xiong and Ishitani, 2006). This defense mechanism, along with stress hormones including abscisic acid (ABA), plays a great role in the stress response of plants (Tuteja, 2007).

However, plants also have other ways of reacting to different stress factors – both ABA independent and ABA dependent ways. That is why the molecular and physiological features concerning the functionality of plants under stress, for instance, stress signal perception, transducers, stress-related transcription mechanisms, target genes and stress metabolites, should be investigated intimately (Anumalla *et al.*, 2016). That is also why the use of genetic engineering to combat the load of the given and other abiotic stress factors can be regarded as a perspective strategy.

This chapter focuses on stress signaling and stress responses in plants to provide a thorough understanding of abiotic stress signaling (Nykiel *et al.*, 2023).

2.2 Significance of Stress Responses in Plant Adaptation

Plants exist in an environment that is always under pressure from factors that are destructive and unfavorable to their existence and productivity. Therefore, their capacity to reason towards the stressors explains their growth, development, and prospects of the future agriculture.

Stress responses in plants have a significant function in the modification of ability in the living plant systems to manage harsh environments by changing inherent constitution, chemicals and molecular processes (Aborderin *et al.*, 2024). Stress factors like drought, salinity, temperature stress, and high ion toxicities can interfere with spatial and temporal regulation and homeostasis of cells, and turn off photosynthesis, interfere with metabolism, and induce oxidative damage. To avoid such effects, plants employ complex defense responses such as the synthesis of osmoprotectants, including proline and glycine betaine, the availability of reactive oxygen species (ROS) and activation of genes that are responsive to stress. As biotic stresses, pathogens, insects, and herbivores also affect the plant's immune response through pattern recognition receptors (PRRs), hypersensitive response (HR), and systemic-acquired resistance (SAR) mediated through salicylic acid (SA) and jasmonic acid (JA) (Roychowdhury *et al.*, 2025).

The ability to modulate these responses, thereby enabling the plant to optimize the above factors, and adaptation determines plant survival in such environments. Stress signaling is another important factor in stress adaptation as plants can choose which signal to respond to depending on the extent of the stress. Thus, ABA, ethylene (ET), gibberellin (GA), and cytokinin (CK) are the hormones involved in the regulation of adaptive mechanisms like stomata, root, and shoot formation, as well as synthesis of secondary metabolites. Besides structural changes, DNA methylation and histone acetylation permit the "memory" of stress in plants such that primal responses are increased and more efficient the second time around (Sun *et al.*, 2022). Knowledge of these adaptive mechanisms can prove useful in developing crops with higher resistance to biotic and abiotic stresses by conventional breeding, genetic engineering, or by means of biotechnology techniques (Tomar, 2021). That is why working to improve stress tolerance would enable genetics to be used to grow crops that could perform well under unfavorable environmental conditions when it comes to producing crops in terms of yield and quality. Addressing plant stress in relation to climate change can help in ending hunger and malnutrition since the occurrence of climate change will affect plant growth.

2.3 Relevance of Plant Stress Mechanisms to Human Health and Organisms

Stress responses can be referred to as one of the most important processes in the life of plants, as well as in the mutual relations between people and the environment. Plants face numerous biotic and abiotic stresses such as water scarcity, somatic stress, high and low temperatures, salinity and environmentally toxic substances such as heavy metal and pathogenic organisms. In response to these stressors, plants generally produce a mechanism that impacts physiological, biochemical, and molecular levels of the plant, such as antioxidant defenses, second metabolisms, and signaling pathways (Isah, 2019). These mechanisms are associated with human health directly since the majority of the bioactive compounds generated when under stress are used as functional foods, drugs, and nutraceuticals. For instance, stress-elicited secondary products, including flavonoids, alkaloids, and phenolics, have antioxidants, anti-inflammatory and anticancer attributes that make them relevant as dietary source and medicinally essential components (Roy *et al.*, 2022). Among the ramifications of the mechanism of stress in plants to human health, antioxidants are of tremendous importance. In response to environmental stresses, plants produce ROS as signals; nonetheless, the generation of high levels of ROS may impair the cells (Manna *et al.*, 2019). To combat this, the plants use antioxidant enzymes such as superoxide dismutase (SOD), catalase (CAT), and ascorbate peroxidase (APX), as well as non-enzymic antioxidants like vitamin C, vitamin E, and glutathione (GSH). These antioxidants also aid in the protection of cells from plants, thus having a significant health impact when incorporated in plant diets. Chronic diseases such as oxidative stress-related cardiovascular diseases, neurodegenerative diseases, and cancers, are some of the diseases that affect humans. Hence, improving stress tolerance in plants through agronomic and biotechnological means can also help restore vitamin content in foods and improve human health by providing additional natural sources of antioxidants (Fig. 2.1).

Furthermore, stress responses have great relevance to issues of crop productivity and food security for human consumers. These migratory stress conditions consist of heat stress, drought, and soil contamination. This will influence crop yield and nutrient density. Therefore, knowledge of plant defense strategies facilitates the creation of crop varieties capable of yielding under conditions of stress. They help to feed the world consistently and avail food nutrients that are free from chemically produced fertilizer, which has more likelihood of causing serious effects to the body since it contains chemicals and pesticides that harm the lives of human beings.

Moreover, different forms of plant stress have also been shown to play a role in the biosynthesis of useful products with anti-microbial activities. A great number of stress-induced products have been used in medical field for synthesizing various drugs, including taxol derived from yew trees for cancer diseases and artemisinin derived from the *Artemisia annua* plant used in malaria treatment (Malabadi *et al.*, 2024). These bioactive compounds evidence the fact that the plant stress response contains therapy and secondary metabolites that can affect other organisms such as insects, herbivores, and microbes (Muhammad *et al.*, 2023), for instance, inhibitors or plant secondary metabolites that discourage herbivory and structures including thorns or waxy cuticle. Some of these chemical defenses, like alkaloids and terpenoids, affect the pests by being toxic or deterring them, while on the other hand they have some medical uses to human beings. In terms of microbes, stress responses change interaction with beneficial and pathogenic microbes affecting soil properties and plant-microbe relationships (Binyamin *et al.*, 2019). From these interactions, it is possible to utilize the beneficial microbes in such areas as plant growth promotion, bioremediation, and agriculture. Stress that is evident in

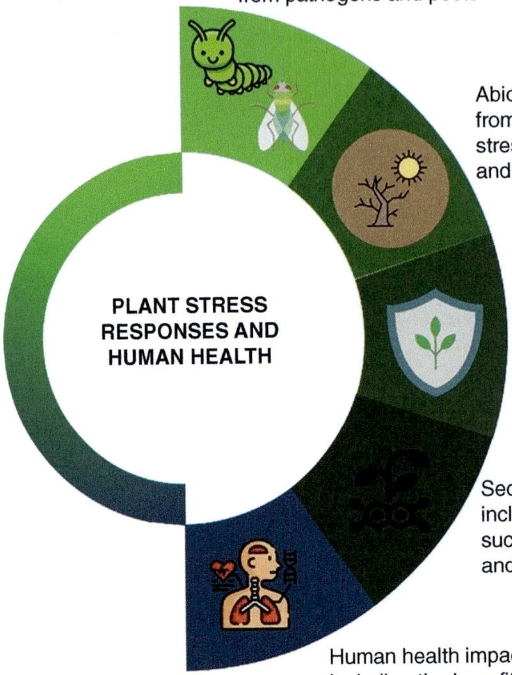

Fig. 2.1. Plant stress responses and human health. (Figure author's own.)

plants has a direct relation with human health as well as other living organisms. This is in view of the numerous attributes they possess, such as production of bioactive compounds, improvement of crop traits, food production and security, and conserving the environment in the face of global change. The gains to be made by studying plant stress responses include the possible health benefits associated with the metabolites that plants produce.

2.4 Abiotic Stress (Drought, Salinity, Temperature Extremes)

Plants need to adapt physiologically to moderate the negative effects of abiotic stress factors, which include high temperatures along with salinity and drought conditions. Through ABA production from stress signals a selection of physiological and developmental processes activate resulting in necessary adaptations to ensure survival (Campalans *et al.*, 1999; Laid *et al.*, 2021).

The abilities of plants to survive and adapt to their environment represent crucial elements in their evolution because physiological adaptations remain vital for how well the plant can survive and remain resilient. The atmosphere and soil experience a state of water deficiency defining drought conditions. During drought stress plants perform multiple physiological and biochemical reactions within their system. The combination of these responses leads to stomatal closure, followed by a drop of transpiration rate, decreased photosynthetic and cellular growth, reduced osmotic pressure, and heightened respiratory activity. Furthermore, plants modify themselves through changes at both the cellular and molecular structure. The accumulation of mannitol, proline, sorbitol, trehalose, and fructans, alongside ABA and stress-tolerant protein synthesis, occurs in plants (Hossain *et al.*, 2022).

Microarray analysis studies found various genes that respond to stress conditions in both *Arabidopsis* and rice. A full-length cDNA microarray revealed 299 genes that become active in *Arabidopsis* when exposed to drought conditions through the examination of 1300 genes. More than 50% of the drought-inducible genes showed induction patterns after exposure to high salinity and/or ABA treatment (Roychoudhury and Paul, 2012). The analysis of 1700 independent cDNAs in rice generated results with 73 stress inducible genes. Drought-stress specific genes represented around 40% of the total induced genes that also responded to cold treatment. The genes that are expressed during high salt treatment along with ABA exposure also respond to drought conditions. The signaling interactions between different stress conditions including drought, cold, and high salinity, and ABA treatment reached extremely significant levels. Drought-responsive genes exist in two biological groups that differentiate their functions. The genes encoding functional proteins listed in this group stabilize cellular homeostasis through membrane protection as well as protein stabilization along with heat shock proteins (HSPs), late embryogenesis abundant (LEA) proteins and lipid transfer proteins (LTPs) (Bhardwaj *et al.*, 2013). CIPK16 along with PP2C represent the second group of drought-responsive genes because they perform as signal molecules during stress events (Guo *et al.*, 2009). The drought tolerance mechanism becomes understandable from studying different functional and regulatory protein-encoding genes that show their expression patterns. The integration of SuperSAGE with high-throughput Roche 454 sequencing tech delivered valuable information about the drought-triggered root transcriptome in chickpea. This technique proved beneficial to address the issue of absent EST and genomic data in the non-model crop plant chickpea. Scientists obtained 17,493 unique transcripts from both drought-stressed and non-stressed control roots which contained genes that participate in stress-perception, signaling, and transcription initiation (Verma *et al.*, 2013).

Abiotic stress due to salinity is a menace in dry regions in as far as plant's yielding capacity and standard is concerned. In this, it generates hazardous complications for some plants by generation of ROS, malnutrition, respiratory burst, ions imbalance, toxicity in genes, etc. As a result, the electron transport chain (ETC) pathway gets disturbed, and there is excess production of ROS.

To ward off these symptoms in the plants, we must take protective measures against hypersalinity. Each year salinity affects a large area of the land around the world, and the size of that area is growing day by day. Because of this, salinity becomes the main environmental stress that is very constraining for plant production. The plants become unlikely to take up water due to high salinity. It quickly results in slowing down the growth rate with an accompanying set of metabolic changes, which are the same as the changes due to water stress. Excessive salt entering the plant is seen as salt specific impact. In older transpiring leaves, the amount of salt is raised above the toxic levels. Finally, the premature senescence and lower photosynthetic leaf area reduce the photosynthetic leaf area of the plant, which cannot sustain the growth of the plant.

With respect to extreme temperature stress, some proteins such as COLD1, ELF3, TWA1 and HSPs can only sense cold stress or heat stress. PhyB, however, can sense both cold stress and heat stress. In rice, transmembrane protein COLD1 was identified as a potential cold sensor protein that interacts with G-protein α subunit 1 (RGA1) in stimulating Ca^{2+} channels and increasing cold resistance of rice (Ma *et al.*, 2015). ELF3 negatively regulates elevated temperature tolerance and is a thermosensor in *Arabidopsis*. ELF3 is a polyglutamine (polyQ) containing gene containing a polyQ repeat embedded in a predicted prion domain (PrD). Higher temperatures induce formation of speckles of ELF3-GFP within minutes, in a PrD dependent manner. *In vitro*, the PrD of ELF3 showed reversible formation of liquid droplets as a direct biophysical response conferred by the PrD and depending on the temperature. By a phase transition, temperature can quickly shift ELF3 between active and inactive states. However, TWA1 has been shown to be a temperature sensor required for basal and acquired thermotolerance in *Arabidopsis* only recently. TWA1 is a transcriptional co-regulator. TWA1 accumulates in nuclear subdomains at elevated temperatures, changes its conformation, and physically associates with JASMONATE-ASSOCIATED MYC-LIKE (JAM) transcription factors

(TFs) and TOPLESS (TPL) and TOPLESS-RELATED (TPR) proteins to activate transcription of the heat shock transcription factor (HSF) A2 (HSFA2) and HSPs (Bohn *et al.*, 2024).

It has been reported that HSPs participate in perception of high temperature. Protein denaturation and misfolded proteins participate in protein aggregation, which is detected by HSPs as a result of heat stress. The heat stress responses are activated when HSPs bind to aggregated proteins resulting in the release of HSFs (Pirkkala *et al.*, 2001). The canonical photoreceptor phyB, which also appears to function as a temperature sensor, is necessary for plant responses to both warm and cold temperatures by virtue of its temperature dependent Pr/Pfr interconversion from the active Pfr (a far-red light absorbing form) to the inactive Pr state (a red-light absorbing form). The cold stabilization of phyB is induced through the accumulation of key transcription factor chain (CBF) that binds PHYTOCHROME-INTERACTING FACTOR 3 (PIF3) to antagonize mutually assured destruction of PIF3–phyB. Feeding *Arabidopsis* with cold stabilized phyB makes it more tolerant to freeze. Activation of the downstream responses is facilitated by reversion of the phyB Pfr to Pr under elevated temperatures to release the inhibition of the Pfr dimer on PIF4 and PIF7.

2.5 Global Impact of Abiotic Stress on Agriculture and Ecosystems

Salinization expansion leads to a worsening impact on both crop production and agricultural sustainability, as well as environmental stability. Salt-affected soils worldwide cover more than 833 million hectares based on FAO data, and arid and semi-arid climatic zones contain over two thirds of these affected areas (Negacz *et al.*, 2022). The increase in salinization of cultivable lands happens at a rate of 10% annually. A soil can be classified as saline when its saturated hydraulic conductivity reaches or exceeds 4.0 dS·m^{-1} (≈40 mM NaCl) thus creating significant negative effects on plant development. Plants receive salt-tolerant or salt-sensitive categorization from researchers through their growth conditions under stress challenges that simultaneously depend on plant species along with genetic types and how old they are as well as stress exposure length and intensity (Seeda *et al.*, 2022).

Plant roots suffer from an initial water deficit because they encounter increased salt ions, which also create high osmotic pressure near the root system. The first reaction takes place during the first few minutes of exposure. The water uptake ability tends to recover completely or partially throughout time before intracellular metabolic functions normalize. Toxic ions access plant cells by membrane transporters and gradually build up inside shoots and leaves, thus disrupting plant normal metabolism. The consecutive ion accumulation manifests as leaf deformities alongside etiolation symptoms, as well as wilting and the ultimate death of plants that may become evident throughout several months. The multiple impacts of salt stress cause osmotic stress together with ion toxicity and nutrient deficiency and their concurrent effects to produce various physical, physiological, biochemical, metabolic, and genetic alterations in plants (Seeda *et al.*, 2022). Study of plant adaptation mechanisms under salt stress requires deep examination because this knowledge holds essential value for understanding plant responses to high salt environments. The article gathers information about plant salt resilience across four research domains including morphology, physiology, biochemistry, and molecular investigation. Multiple plant mechanisms operate in salt stress through organ-specific effects and physiological and hormonal regulation and ion homeostasis and sensory signaling pathways and genetic regulations (Fu and Yang, 2023). The acquired information holds significant importance for plant cultivation research and fruit breeding programs in the future domain.

Plant existence depends on environments regularly shifting between conditions that normally stress their development and growth process. Plant growth faces two types of adverse conditions, including pathogen invasion and herbivore predation under biotic stress and drought, together with heat exposure or cold exposure, nutrient shortages and high levels of aluminum (Al), arsenate, and cadmium (Cd) found in the soil as abiotic stress. The natural distribution patterns of plants in their environment are mostly influenced by drought and salt stress and temperature extremes, which simultaneously constrain agricultural productivity while endangering global food security (Majeed and Muhammad, 2019). Climate change acts as an exponent that worsens abiotic stress effects because scientists predict more frequent occurrences of extreme weather will happen. Primary biological questions concern how plants detect stress indications while adapting to harsh environmental conditions. Enhancing plant resistance to environmental stress is essential because poor-stress resistant crops utilize excessive water and fertilizers which lead to environmental deterioration. Distinction between primary stress signals that originate from drought or salt stress is essential when determining the secondary stress signals that stem from insufficient water or excessive salt (Zhu, 2002). The main drought-triggered effect on plant cells is hyperosmotic stress, which scientists commonly call osmotic stress due to rare occurrences of hypo-osmotic conditions. Salt stress creates dual effects against cells by spreading osmotic pressure as well as toxic ions. The series of reactions that follow drought and salt stress in plants create multiple compounding stress effects, which affect cell membranes and proteins as well as nucleic acids while also causing metabolic dysfunction along with oxidative stress. The cellular responses to primary stress signals remain separate from those that trigger mainly from secondary signals. Drought stress delivers distinct signaling pathways that combine partly with those produced by salt stress. Plants experience many adaptive responses because the phytohormone ABA accumulates through hyperosmotic signals produced by drought and salt stress (Ma and Qin, 2014).

Plants experience WRKY TF activation in response to different environmental stress conditions originating from both living and non-living agents. Different biological processes within plants need their operation. Researchers have extensively studied the working mechanisms of these proteins through developmental and hormonal and abiotic variable tests across *Arabidopsis*, rice, potatoes, and parsley plant species. When *Arabidopsis* experiences cold stress, it depends on CBF and DREB1A TFs to enhance plant tolerance against oxidative stresses (Akhtar *et al.*, 2012). The study shows that melatonin plays a substantial role in manipulating the transcriptional activity of TFs. It has been established that the *Arabidopsis* ZAT6 and HSFA1 TFs directly manage signal pathways specific to heat- and cold-resistance responses.

2.6 Physiological and Molecular Effects of Stress on Plants

Plants change the expression of stress-response and tolerance genes as a natural response to abiotic stress conditions. Different research techniques have been employed for studying gene activation during this process. AREB1/ABF2, AREB2/ABF4 and ABF3 respond to all three stressors of dehydration, high salt levels and ABA treatment (Liu *et al.*, 2014; Yoshida *et al.*, 2014). According to this data, plant reactions to abiotic stress are reliant on these proteins, together with ABA. Among all transcription factors, the WRKY factors represent leading regulators that control plant hormone signal transduction systems during biotic and abiotic stress. The NAC-type TF OsNAC6, which is specifically found in plants, maintains control over genes that activate biological and physical stress responses. The manipulation and regulation of plants by the signaling compound JA remains beneficial for plant growth under stress encountered from abiotic environmental changes (Awan *et al.*, 2017). Modern research indicates that HSFs function as controllers to activate stress-responsive genes with HSPs. Plants experience different abiotic stressors through the regulatory actions of these HSFs. The endoplasmic reticulum (ER) functions as a medium through which plants receive continuous stimuli for interpretation of both biotic and abiotic signals.

OsNAC5 functions as a plant TF, along with its abiotic stress response capabilities to control stress-induced gene expression and enhance stress tolerance in the overexpressing rice plant (Takasaki *et al.*, 2010). Scientific evidence shows that stabilizing EIN3/EIL1 proteins in *Arabidopsis* efficiently helps plants survive salt stress conditions, thus demonstrating the essential role of ET signaling for abiotic stress adaptation. The significant regulatory power of B-BOX genes makes them effective at helping plants adapt to stress through development and growth and stress reaction control. Different molecular pathways and mechanisms activate abiotic stress signaling by modulating TF activity hormonal gene expression and stress regulation processes inside plant biological models (Nykiel *et al.*, 2023). The upregulation of functional genes in response to this signaling encompasses various molecular pathways and mechanisms. Plants exhibit compound and coordinate reactions to abiotic stress events that become obvious in these examples.

The physiological changes in plant models that result from abiotic stress events happen at three different levels involving molecules and biochemical components and developmental operations. Plant adaptation serves as a vital mechanism, which enables their development under harsh environmental situations. Investigations demonstrate different biological processes through which plants adjust to abiotic stress, whereas alterations in development and photosynthetic mechanics and stomata behavior represent some examples along with changes in plant physiological, biochemical, and molecular processes. Picturesque plants adjust their photosynthetic systems while improving membrane fat distribution and managing calcium movement to adapt to stress environments, as well as producing antioxidant and osmoprotectant compounds. Plants adopt stress through several interconnected approaches that utilize multiple molecular, physiological, and biochemical procedures for their adaptation processes.

Plants need hormones to adapt their growth procedures against abiotic stress factors. The fundamental regulator of physiological pathways involving vegetative tissues uses ABA to respond to water stress and developmental events and growth physical triggers (Murti and Upreti, 2007). The stress adaptation through ABA shows significant interaction with zeatin (ZA), which belongs to the CK hormone group. The biological and physiological changes happening to plants under environmental stress conditions like oxidative stress, drought, and heat and cold require a complete examination of complex molecular signaling pathways and biological processes. Devireddy *et al.* (2021) expands the knowledge base regarding plant hormones and ROS involvement in temperature response mechanisms. The research shows that plant hormones use thermotolerance mechanisms through ROS production along with NADPH oxidase activation and changing redox signaling, along with various cellular and physiological processes.

2.7 Mechanisms of Stress Perception in Plants

The perception of stress triggers initial signaling mechanisms within plants of *Arabidopsis* and rice through ROS production, calcium signaling and mitogen-activated protein kinase (MAPK) activation for successful stress adaptation. The signaling pathways function to solve biological and environmental stress conditions in plants. The plant system initiates multiple molecular processes after detecting stress stimuli leading to activation of defense protocols along with adaptive measures throughout the plant organization. Republished research data verifies ROS signaling plays an elementary role in plant stress responses. ROS molecules control physiological functions of dehydration and oxidative stress responses (Manna *et al.*, 2019). Plants use ROS as their mechanism to detect dehydration as well as salt-induced stress. Plants use ROS molecules to control their physiological reactions toward oxidative stress and dehydration, while these molecules serve as signaling substances. Evidence reveals that although ROS can inflict damage at times they act as essential signal transduction molecules through which plants respond to pathogen invasions, environmental stresses, programmed cell death, and various developmental triggers (Baxter *et al.*, 2013).

Literature shows ROS perform an important role during many critical biological activities such as cellular growth and cell division. ROS function as essential signaling molecules that enable plant acclimatization to process various abiotic environmental stress stimuli. Survival chances for plants increase when researchers regulate genes regulate stress defense pathways and maintain ROS production under a specific threshold. Different physiological processes like transpiration and hormonal balance require the essential functions of ROS and nitric oxide (NO) in addition to their roles in germination and stomatal gaseous exchange, as well as fruit maturation and blossoming, defense mechanisms, and programmed cell death (Wani *et al.*, 2021). Plants receive distant stress warnings through

ROS formation because of environmental stress conditions. For plants to sustain normal ROS balance and execute various physiological functions both compounds need to interact openly (Fig 2.2).

The signaling mechanisms of plant cells include calcium signaling, which stands as one of the top significant pathways of this type. The research ability regarding cell signaling patterns has been transformed by plant cellular signaling pathways and stress response functionality. MAPK cascades get activated in plants after they encounter stress from their environment. The stress response mechanisms get regulated through these cascades at the same time that these cascades control core body temperature in situations involving cold and dry conditions. Plant MAPK signaling cascades function as the mechanism that controls stress responses when plants experience environmental threats such as drought and cold (Lin *et al.*, 2021). Plants use MAPK signaling to defend their internal heat levels when exposed to extreme temperatures and to respond to cold stress. Plants employ calcium signaling as a versatile signaling mechanism in response to drought stress or limited soil water availability. Plants primarily use three calcium sensors to monitor drought-stress signals, which demonstrate the vital role of sensing stress reactions.

Plants enhance their immune system when supplemented with calcium while simultaneously establishing relationships with plant microbial organisms. The functioning of plant calcium signaling depends on three regulatory mechanisms that include calcium-dependent protein kinases (CDPKs), calcium-modules (CaMs) and calcium block proteins (CBLs) (Tuteja and Mahajan, 2007). Many proteins including metabolic enzymes along with TFs require CaMs to bind calcium molecules, which then cause their structural alterations. CDPKs belong to the protein kinase family, which has two functional domains consisting of calmodulin-like elements and protein kinase activity elements. When the CDPK molecule binds calcium, it activates itself to phosphorylate signaling cascade proteins. Binding to CBL-interacting protein kinase (CIPK) enzymes activates them through the calcium-dependent interaction of the CBL protein sensor. Through the CBL–CIPK combination, important proteins involved in development and stress receive regulatory control.

The human body starts early signaling processes after stress detection through various signaling pathways. The stress perception activates three major pathways: ROS signalling, calcium signalling, and MAPK cascades. All plants use complex entangled signaling mechanisms to determine their reactions towards multiple stress factors. Plants use signaling pathways to reduce the harmful effects that are caused by environmental stress. Plants use elaborate processes that act along various routes as they react to various types of stress. Plants survive and adapt to rough ecological situations through linked signaling pathways, which form a complete network of reactions. Phytohormones, particularly JA and ABA, control how plants respond to abiotic stress (El Sabagh *et al.*, 2022). As protective signals in plants, these hormones produce plant defense through their interactions with multiple

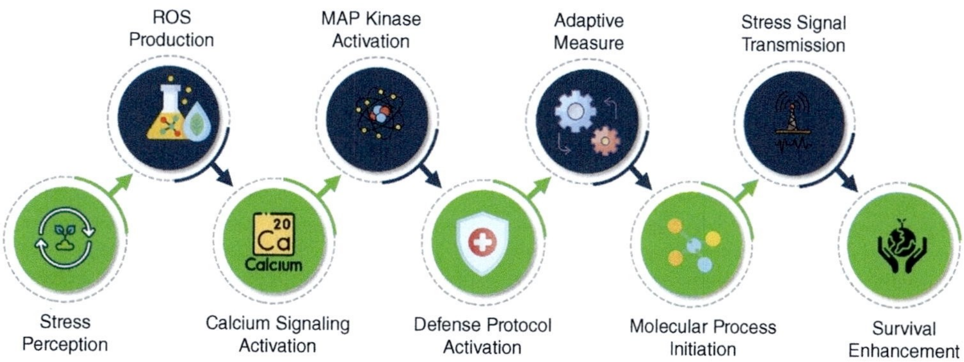

Fig. 2.2. Plant stress response sequence. (Figure author's own.)

signaling pathways. The protective abilities of polyamines during abiotic stress responses make it possible to improve plant stress tolerance by manipulating their chemical pathways (Gill and Tuteja, 2010). Plants receive protective benefits from stress, together with signaling through polyamines and nitrogen, which regulates their response to stressful conditions. Plants achieve better defense mechanisms and adaptation to environmental threats through the sequential actions of the ABA, MAPK and ROS signaling pathways (Danquah *et al.*, 2013). Multiple studies have demonstrated the necessity of systemic signaling in plants, especially when they need to endure abiotic combinations of stress. Plants show distinct systemic signaling reactions to various stressors, which produce stress-specific accumulation of transcripts and metabolites and aid adaptation to stress (Table 2.1 and Fig. 2.2).

2.8 Role of Membrane Receptors in Stress Perception

Early stress signal perception in plants depends heavily on membrane receptors, which function as the fundamental point of connection between external stimuli and the cellular signaling pathway processes. Stress receptors positioned on plasma membranes detect multiple stress signals originating from drought and salinity stress, together with temperature variations, pathogen intrusions, and mechanical damages. Stress perception depends on three essential receptor families, which include receptor-like kinases (RLKs), receptor-like proteins (RLPs), and ion-channel-linked receptors (Jamieson *et al.*, 2018; Zhiqi *et al.*, 2025). Early stress recognition by RLKs relies on their three-part structure, which includes an extracellular portion and a transmembrane segment linked to an intracellular kinase domain. When ligands bind to RLKs the receptors start phosphorylating cascades that activate signaling pathways to give rise to stress adaptation mechanisms. The class of membrane receptors known as G-protein-coupled receptors (GPCRs) uses hormones and environment-specific signals to sense external stimuli. Through interactions between GPCRs and G-proteins these compounds adjust signaling within cells, which controls stress adaptation in cells by changing gene expression. Stress perception depends on ion channels particularly calcium-permeable channels because they control ion equilibrium while triggering calcium-based signaling mechanisms. Salt exposure activates both sodium transporters and CDPKs, which bring about osmotic adjustment and stress tolerance mechanisms (Latz *et al.*, 2012). Through stress signal integration, membrane receptors allow plants to deliver fast and appropriate responses to unfavorable environmental factors. Stress-responsive genes need these receptors to become active while protective metabolites are manufactured and growth processes get adjusted. The study of membrane receptors in stress perception leads to better stress-tolerant crop development, which makes plants more resistant to weather fluctuations thus creating sustainable agriculture systems.

2.9 Early Detection of Stress Signals and Transduction Cascades

Plants require quick detection of their stress signals to launch prompt and suitable responses when facing environmental threats. Specialized sensors and stress receptors detect external stress factors including drought stress and high salt concentrations, together with heavy metal contamination and pathogen entry. As soon as a stress signal reaches perception the cell activates multiple intracellular transduction cascades to transmit signals across membranes until they reach the nucleus for initiating suitable responses (Xiong and Ishitani, 2006). The recognition of early stress requires calcium signaling mechanisms as a fundamental process. Stress signals cause calcium-permeable channels to become active so the cytosolic calcium levels rise briefly. The distinct patterns of calcium levels in cells act as a code recognized by calcium-binding proteins, including CaMs, CDPKs, and CBLs, that stimulate their respective kinases to modify genetic expression along with metabolic adaptations. The stress signaling pathways contain hydrogen peroxide (H_2O_2) as one type of ROS that functions as a secondary messenger. ROS signaling at controlled levels causes activation of stress-responsive genes and starts antioxidant

Table 2.1. Signal transduction pathways involved in plant responses to abiotic stress, highlighting key components, hormonal and transcriptional regulation, physiological outcomes, and supporting references. (Table author's own.)

Stress signal pathway	Key components and processes	Hormonal/transcriptional involvement	Physiological outcome	References
Ion homeostasis and cellular stress responses	ABA-induced stomatal closure, ROS generation, activation of genes like GRI, CAT1, Cu/Zn-SOD, APX1	ABA, CK, JA, SA, ET	Regulation of stomatal aperture, oxidative balance, gene activation	Miao *et al.* (2006); Choudhary *et al.* (2007)
ROS regulation under drought and ABA	Xanthine dehydrogenase (XDH) and aldehyde oxidase (AO) genes involved in H_2O_2 and O_2^- production during stress	ABA	ABA mutants show impaired ROS response, confirming ABA's role in ROS signaling	Miao *et al.* (2006)
Cytokinin and senescence interaction	IPT gene under SARK promoter enhances CK synthesis; improves water efficiency	CK, ABA	Leaf senescence delayed,drought tolerance enhanced	Choudhary *et al.* (2007)
Hormonal crosstalk in guard cell signaling	H_2O_2 production in guard cells via NADPH oxidase pathway; MJ promotes ROS	ABA, MJ, SA, JA, ET	Stomatal closure, ROS accumulation, hydraulic conductivity regulation	Thirugnanasambantham *et al.* (2014); Tang *et al.* (2024)
ROS production sites and triggers	Chloroplasts, mitochondria, peroxisomes, ER, plasma membrane generate ROS under stress and light	Redox regulation, photorespiration	Impaired photosynthesis, disrupted homeostasis	Ray *et al.* (2012)
Hormonal regulation under stress	ABA triggers antioxidant system and osmoprotectants; CKs delay senescence; ET and auxin contribute to stress recovery	ABA, ET, CKs, Auxin	Drought resistance, growth regulation, hormonal equilibrium	Ahammed and Yu (2016); Reguera *et al.* (2013)
Transcriptional regulation	Activation of NF-Y, WRKY, bZIP, NAC, DREB1/2, CBFs; MAPK cascades and calcium-signaling pathways	ABA-dependent and ABA-independent pathways	Expression of osmotic, detoxification, and damage repair genes	Narusaka *et al.* (2003); Zhang *et al.* (2023); Khan *et al.* (2024)

defense networks. The stress transduction pathway includes MAPK molecules, which act as amplifiers due to their sequential phosphorylation patterns (Cargnello and Roux, 2011). TFs controlled by MAPKs lead to the activation of stress responses by triggering osmoprotectants manufacturing and HSP synthesis and stress hormone production including ABA, SA and JA. The signaling molecules of these hormones direct plants to activate stress-related response programs, which encompass stomatal closure together with metabolic alterations and immune defense development. Alerting plants about environmental stress and the subsequent signal transduction process becomes essential for their ability to overcome adverse conditions. Stress-resistant crops emerge via genetic together with biotechnological methods to enhance these mechanisms thus guarantee better crop productivity and improved climate change resilience (Fig. 2.3).

2.10 Integration of Stress Perception with Growth and Development

Plants need to manage stress-related activities along with their developmental pathways to achieve maximum survival and output performance. The activation of stress perception signals initiates defense processes that disrupt standard growth rules, thus stress adaptation integration with developmental patterns stands as key for plant physiological research. Phytohormones, including ABA as well as GA, auxins and CK alongside brassinosteroids (BRs), control the integration between stress tolerance and developmental programs in plants (El Sabagh *et al.*, 2022). The stress mediator ABA controls drought and salt responses through its regulation of stomatal regulation, osmotolerance development, and gene activity. The plant growth decreases because ABA concentrations exceeding certain thresholds cause inhibition of cell division and elongation processes. Like JA and ET defense regulators that govern biotic stress defense, the two agents suppress growth procedures by restricting root growth and affecting leaf development. The hormone GA encourages plant growth through cell division and elongation, but stress conditions control its levels to avoid wasting energy on unrestrained growth. Planetary resource allocation priorities stem from a molecular regulatory process between stress-responsive TFs WRKYs, MYBs, NACs, and growth-related regulators. Plants experiencing gentle stress conditions select tolerance mechanisms that preserve their health while continuing development. During severe stress periods survival mechanisms take control, which results in both growth reduction and resource-saving measures. Plants employ epigenetic modifications such as DNA methylation alongside histone modifications to achieve stress adaptation-development equilibrium, which enables them to learn from past stress exposures through a process of memory retention for enhanced stress response in future stress situations (Miryeganeh, 2021). Study of plant integration between stress perception and growth and development processes brings important value for crop enhancement efforts. Scientists use their understanding of hormonal signaling pathways and stress-responsive genes to develop farming staples with stress resilience which preserves yield quality even when faced with harsh environmental stressors in order to enhance food security worldwide.

2.11 Signal Transduction Pathways in Response to Abiotic Stress

2.11.1 Ion homeostasis and cellular stress responses

Plants use hormones to control the development of senescence and abscission while they employ these hormones for acclimation responses including stomatal closure and changed hydraulic conductivity to protect against drought and salinity stress. The research demonstrated that ABA leads to elevated H_2O_2 levels in maize leaves and embryos along with seedlings and simultaneously activates genes related to the ROS network including glutathione reductase (GR), CAT1, cytosolic Cu/Zn, SOD, APX1, GR, and APX, which function in maize leaf tissues. Part of the NADPH oxidase mechanism supports the stomata closure processes activated by ABA. Plants become more sensitive to drought through an increased

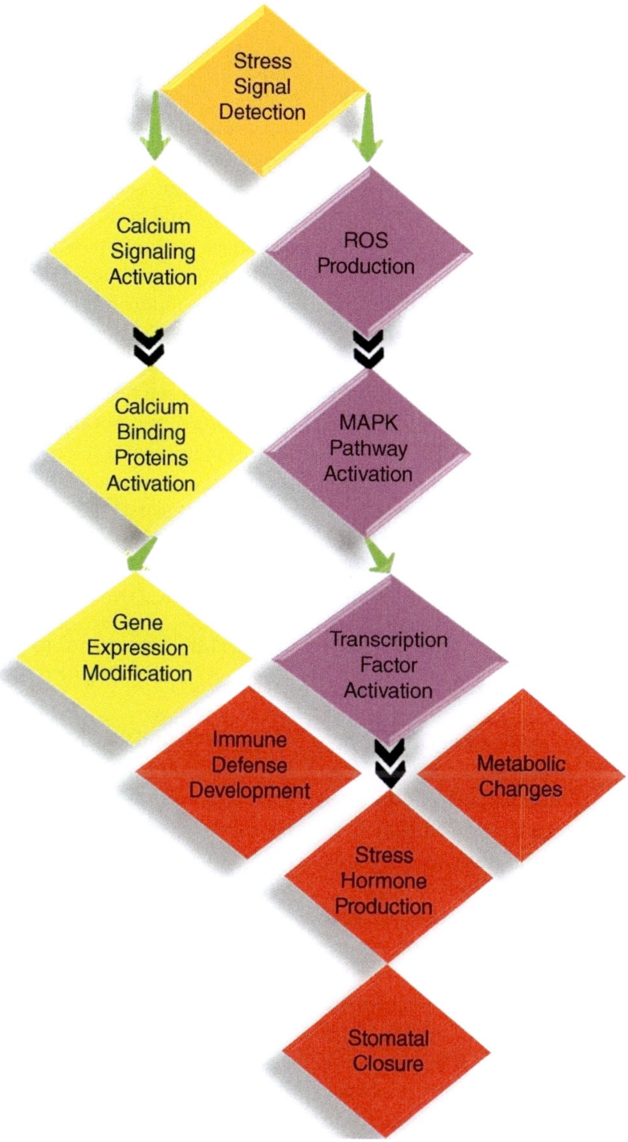

Fig. 2.3. Plant stress signal transduction pathway. (Figure author's own.)

H_2O_2 production in guard cells when AtGPX3 acts at deficient levels in mitochondria. In *Arabidopsis* GPX3 plants the ABA-responsive gene expression plus activation of calcium channels by ABA showed disruption, thus indicating redox state regulation serves as a crucial factor for plants to respond to ABA during drought (Miao *et al.*, 2006). Studies prove that both drought and ABA treatment boost XDH and cytosolic AO activity which then results in the production of H_2O_2 and O_2. The ABA-deficient mutants of tobacco plants as well as tomato and *Arabidopsis* plants fail to exhibit AO or XDH activities. The investigation demonstrated that drought stress generates ROS through XDH and AO genes following ABA treatment thus proving that ROS accumulation inside plants depends on ABA.

Stress conditions such as drought result in stomatal closure, which triggers CK level reduction and boosts the sensitivity of shoots to ABA. Plant canopy size decreases through leaf abscission combined with senescence processes because the stress alters CK and ABA levels in the plant system. Improved water efficiency and enhanced drought tolerance occurred in transgenic tobacco plants that contained the IPT gene under regulation of the SARK promoter. This gene functions as the enzyme responsible for the rate determining step in CK synthesis. Enhanced ROS metabolism gene activity occurred in the AsA-GSH cycle gene-transgenic plants. The defense mechanisms as well as systematically acquired resistance induction within plants is regulated through JA and ET in combination with SA (Choudhary *et al.*, 2007). The promoters that control expression of Zat7 Zat12 WRKY25 and APX1 genes have combined SA, JA, ET-sensitive elements, and other regulatory parts, which indicate that these hormones signaling mechanisms function beyond pathogen defense systems (Rizhsky *et al.*, 2004).

The root hydraulic conductivity (LPR) decreased in addition to *Arabidopsis* roots accumulating higher levels of ROS after NaCl and SA treatment. A research study showed that ROS must form to enable SA regulation of LPR while H_2O_2 elimination led to an incomplete prevention of this inhibitory effect. The increased activity of alternative oxidase (AOX) potentially allows SA to control mitochondrial ROS production without using cytochrome pathway technology at the UQ pool. The closure of stomata through methyl jasmonate (MJ) application becomes less effective when analyzing the AtrbohD/F double mutant. The hormone MJ enables guard cells to produce more H_2O_2. This increased production leads to stomata closure. Plant resistance to several stresses increases because of ethylene response factor (ERF) proteins, which contain the binding abilities of CE1-cis, GCC-box and DRE-/CRT- elements within the ERF/AP2 protein family (Thirugnanasambantham *et al.*, 2014; Tang *et al.*, 2024). Plants with the transgenic ERF3 and ERF factors show enhanced resistance to salt stress and drought conditions when grown in tobacco organisms. These factors promote gene expression of reducing ROS detoxifying enzymes.

2.11.2 Role of ROS in stress signaling

Plants naturally generate ROS as hazardous metabolic byproducts during their aerobic metabolic activities in both stress-free and stressful conditions. ROS appear in high quantities because of redox reactions or excessive energetic conditions and multi-stress conditions that result in O_2 reduction while causing photosynthesis impedance and increased photorespiration and cellular homeostasis disruptions (Ray *et al.*, 2012). The activation of ROS requires energy sources and excess electron discharge before various cell organelles such as chloroplasts, peroxisomes, membranes, cell wall, mitochondria, apoplasts, and the ER transports electrons. During the presence of light both peroxisomes and chloroplasts create ROS but under darkness mitochondria serves as the major ROS producer.

2.11.3 Hormonal regulation: ABA, ET, CK, and auxins

The growth and survival of plants under stressful situations depend strongly on hormonal regulation mechanisms that control their response to adverse conditions. Plants use the hormones ABA and ET, together with CKs and auxins, for coordinating numerous stress adaptation mechanisms that occur during environmental stress events (Ahammed and Yu, 2016). ABA functions as a central regulator, which controls the responses to abiotic stress and specifically manages conditions of drought and salinity. The stress response induces the elevation of ABA levels, which triggers stomatal movement to limit water evaporation and enable antioxidant activation and osmoprotectant storage. ABA brings positive change to stress tolerance by altering gene expression yet produces negative effects on growth by limiting cell expansion and division. The main association of ET pertains to biotic stress responses while it contributes to abiotic stress adaptation mechanisms. The hormone manages the aging process of leaves, together with regulating fruit maturation and influencing tissues when they encounter mechanical forces. Stress-induced changes in ET signaling combine with ABA signaling to manage stomatal responses along with restricting stress-related growth impairment. CK signaling promotes cell division processes and prevents aging-related tissue deterioration. When stress copurification of CK levels decrease to preserve energy, however external administration of CK enhances drought along with salt tolerance by sustaining antioxidant systems and helping keep photosynthetic functions (Reguera *et al.*, 2013). The growth, along with development of roots, falls under auxin regulatory functions. The stress causes modifications within the auxin distribution system, which generates changes in root structure. Under severe conditions the plant survival depends on auxin compounds to maintain equilibrium with ABA and ET for stress adaptation and growth management.

2.11.4 Transcriptional regulation and stress-induced gene expression

Stress-induced signaling creates genome-wide transgenomic reprogramming, thus activating mechanisms including osmoregulation detoxification and stress-damage repair in addition to its protective functions for maintaining ion and water homeostasis through rapid regulation. Multiple types of TFs present in various plant species were identified today through genome-wide investigation including NF-Y WOX WRKY bZIP and NAC (Zhang *et al.*, 2023; Khan *et al.*, 2024). Plants use particular key TFs for their responses against abiotic stress. The transcription patterns for stress-run environments affect signaling pathways through TF control. Multiple abiotic stress factors including cold stress and hypertonic stress, together with hypersalinity stress, activate similar transcriptional responses in plants. Most stress response genes control the production of proteins, which function within MAPK cascades along with calcium signaling, ROS and apoptosis systems, and proteolytic activities throughout numerous cellular areas.

When faced with drought or high salinity stress conditions, ABA synthesis occurs and controls stress responses through several linked phosphorylation signaling pathways. Numerous stress-responsive genes receive independent activation through ABA pathways alongside the hundreds of genes controlled by drought, high salt, and cold stress through ABA signaling. ACGTGGTC, TCGTGG and ACGTGG/TC function as promoter-specific cis-elements for ABA-regulated genes, whereas TACCGACAT operates as the promoter element of genes that respond to dehydration and ABA-independent cold or drought conditions (Narusaka *et al.*, 2003). TF CBF (also known as DREB1), together with DREB2, find their target genes through binding to DRE. Osmotic stress and high salinity primarily activate the expression of CBF4 and DREB2 genes, yet CBF1/2/3 genes specifically respond to cold conditions for downstream gene regulations, and this process occurs independently from ABA regulation (Table 2.1).

2.12 Genome Editing and Gene Stacking for Drought Tolerance

2.12.1 Genome editing

Genome editing is a method of precise biotechnology to edit targeted DNA sequences of an organism's genome. One of the most well-known is CRISPR-Cas9, enabling molecular "scissors" (Cas9 enzyme) to be guided by a small RNA (sRNA) molecule to a specific DNA sequence to cut it (McFadden and Laohasiripanya, 2022). Next, the cell's natural repair machinery fixes the cut and can, by introducing desired changes, help the attempt to take root. Genome editing facilitates scientists to modify the genes that are responsible for water use efficiency, osmotic adjustment and stress responses in drought tolerance. For instance, researchers can modify genes involved in controlling stomatal closure, which conserves water, or trigger stress-responsive attributes.

2.12.2 Gene stacking

Gene stacking refers to a series of genes incorporated into a single organism to produce a single organism that has more than one gene involved in its phenotype. Accumulation of many beneficial traits in a crop variety is done by the use of this technique. Gene stacking is observed in drought tolerance, which stacks genes related to mechanisms to save water, stress signaling, and to enhance growth under water deficit (Shailani et al., 2020). Genome editing and gene stacking can be used by scientists to develop crop varieties with increased drought tolerance. This involves stacking engineered or naturally existing genes, which are regarded to enhance the plant's ability to fight off drought stress with the genes that were edited for the desired traits.

2.13 RNA Interference and Gene Expression Manipulation

A natural cellular mechanism of RNA interference (RNAi) is essential to regulate gene expression. Using sRNA molecules, this is a process through which the production of a particular protein is silenced or degraded, resulting in the prevention of the production of the corresponding protein. Two classes of sRNA molecules are involved in RNAi and include microRNAs (miRNAs) and small interfering RNAs (siRNAs).

siRNAs are short double-stranded RNA molecules of about 20–25 nucleotides in length that are introduced into the cell. The process of "dicing" is used to produce siRNAs from longer RNA molecules. However, one of the siRNA strands, known as the guide strand, unites its complementary base with part of the target mRNA sequence. Normally the passenger strand is degraded. Once the target mRNA is bound with the guide strand, it forms a protein complex called RNA-induced silencing complex (RISC). It cleaves this target mRNA such that it cannot be translated into a protein.

miRNAs are endogenous, naturally occurring single-stranded RNA molecules of about 21–23 nucleotide long that are transcribed from specific genes and are produced through dicing similar to siRNAs. miRNAs also act like siRNAs. The miRNA comprises two copies, and like siRNAs, one strand of the miRNA binds to target mRNA with a complementary pattern of base pairing. Nevertheless, many miRNAs have imperfect sequence complementarity to their targets and regulate multiple mRNAs that are partially complementary to a miRNA. Typically, miRNA binds to the target mRNA and causes a decrease in translation (mRNA to protein) or causes degradation of the mRNA. RNAi plays many important roles in the cell, in development, immunity and in fighting against viruses. Scientists have also used the RNAi mechanism for gene silencing experiments to study the function of genes and for the development of therapeutic methods for diseases associated with overexpression or mutants' genes (Ghosal et al., 2011).

There are two prevailing methods for gene expression manipulation: RNAi and gene editing mediated by CRISPR–Cas9. CRISPR–Cas9 is a revolutionary genome editing technology that enables scientists to precisely cut and modify specific DNA sequences. This system can disrupt a target gene's function by introducing mutations, allowing researchers to determine the gene's role in cellular processes. In addition to gene disruption, CRISPR–Cas9 can insert new sequences—such as functional elements or fluorescent tags—into the genome, facilitating the study and tracking of gene activity.

Beyond altering the DNA sequence, CRISPR–Cas9 can also regulate gene expression by fusing the Cas9 protein with transcriptional activator or repressor domains. This approach allows targeted modulation of transcription without permanent DNA changes. The technology can also be adapted to modify epigenetic marks, such as DNA methylation or histone modifications, thereby influencing gene expression patterns on a broader scale (Pulecio et al., 2017).

These capabilities mean that CRISPR–Cas9 could one day correct disease-causing mutations, restoring normal gene function. Together, RNAi and CRISPR–Cas9 have transformed molecular biology and genetics, enabled precise control of gene activity and offered powerful strategies for understanding complex biological systems and developing novel treatments for genetic disorders (Fig. 2.4).

2.14 Application of Omics Technologies for Drought-related Gene Discovery

Omics technologies were applied to drought-related gene discovery by applying techniques of high throughput to comprehensively measure genes, metabolites, and proteins in an organism to determine which genes are associated with genes that assist in drought response and tolerance. The use of this approach gives a global

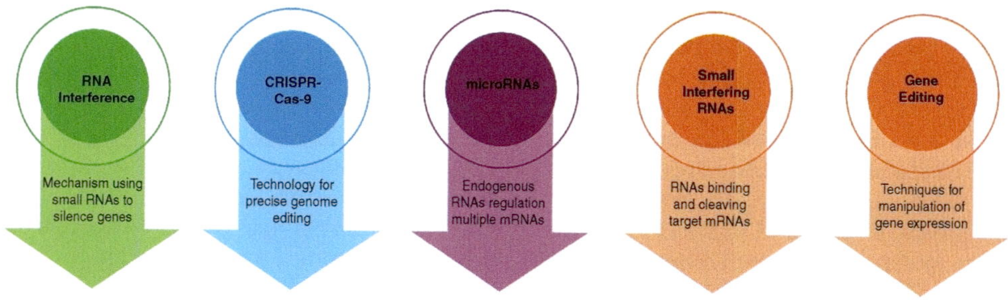

Fig. 2.4. Unravelling gene editing and regulation. (Figure author's own.)

picture of the molecular mechanisms behind drought adaptation. They include the genomics (study of an organism's entire genome), transcriptomics (analysis of gene expression patterns), proteomics (study of all proteins in a cell), and metabolomics (study of all small molecules/metabolites in a cell). These components can then be analyzed on a large scale to identify key molecules and pathways that are associated with drought response. Extensive use of omics technologies has been employed to identify genes, proteins, and metabolites involved in plant response to drought stress. These techniques allow identification of regulatory networks, signaling pathways, and of important molecular actors implicated during drought tolerance.

2.15 Interdisciplinary Research: Bridging Plant Biology and Human Health

Research is taking off that combines plant biology with human health, revealing these plants' deep connections with human health through their stress response and bioactive compounds. Secondary metabolites, such as polyphenols, flavonoids, alkaloids and terpenoids, produced by plants function as natural defense broaden against environmental stress (Ashraf *et al.*, 2018). Thereafter, these compounds possess extensive pharmacological benefits, e.g. antioxidant, anti-inflammatory, and anticancer, leading to their high usability applications in human health. In the meantime, the bioactive compounds, such as curcumin, resveratrol, and catechins are synthesized through metabolic pathways stimulated by stress in plants and have been studied extensively for their prevention of chronic diseases including cancer, diabetes, and cardiovascular disorders (Chen *et al.*, 2022). In addition, plant stress physiology has led to advances in the field of functional foods and nutraceuticals, which provide plant-based natural solutions to many human health problems. In a nutshell, apart from nutrition, plant biology is useful in medical prospects like biopharmaceuticals, where the stress-induced plant metabolites are used as precursors to make drugs. Their applications in targeted drug delivery and antimicrobial therapies have also expanded as part of research on plant-derived nanoparticles. Interdisciplinary research that bridges plant biology and the human health sciences provides opportunities to develop novel disease prevention solutions, as well as sustainable medicine and global food security, thereby reinforcing the connection between plant resilience and human wellbeing.

2.16 Future Challenges and Opportunities in Stress Adaptation Studies

All plants living in their natural environment must deal with multiple types of pressure. A controlled environment became necessary to reach optimal plant growth limits. Research demonstrates that the trade-off arises mainly from biological mechanisms that slow down plant growth when stress signaling networks become active. Regulatory systems for growth regulation and stress response management link up throughout different parts of the network. Future research needs to identify the main connections between stress responses and growth while finding ways to enhance resistance without harming yield (Jing et al., 2024). Scientists need to investigate further ways the plants react to natural stresses at the molecular level. Scientists still face hurdles while searching for plants that detect stress signals. Better gene editing technologies and genetic manipulation methods help scientists find stress sensors despite genetic challenges. Research into and understanding of organelles' role in stress perception and response, along with studying dispersed stress perception systems, will help us comprehend how plants perceive and handle stress, but we still need to uncover how damaged organelle signals connect with each other. The process of how plants respond to stress needs better understanding because research methods and genetic deletions still face barriers to discovering all parts of the pressure response network. Plants face many types of environmental pressure because they cannot move, which means scientists need to study how distinct defensive networks influence each other. Research opportunities into the effects of both natural and living stress factors remain unknown. Our understanding of how plants react to many stresses needs further research from different science fields to benefit agriculture.

2.17 Conclusion

The study of how plants sense and process stress information help us strengthen plant health and safety while securing food supply and enhancing wellness. As stationary organisms, plants have developed effective ways to detect and adjust to different forms of pressure in their environment. The plants use membrane sensing systems to send messages through signaling pathways that modify DNA for better survival under stressful conditions. Stress perception and hormonal signaling together supervise how plants handle stress and use their resources to survive. Plant stress research uncovers ROS as signaling elements to direct stress response activation. A low level of ROS acts as a message to turn on antioxidants, while excessive ROS buildup starts damaging plant cells.

The pathways of MAPK signal transduction along with involvement of WRKY and NAC TFs support plant stress adaptation by directing gene expression and metabolic activities. Research demand continues to grow because plant stress biology operates through intricate systems that need further understanding. Researchers from different fields prove that the way in which plants respond to stress affects human health. Plants develop defense chemicals like flavonoids, alkaloids, and polyphenols to resist stress and these substances show antioxidant, anti-inflammatory, and anticancer benefits in humans. Research into phytochemical production during plant stress has led to better healthcare provisions through functional food development and natural treatments.

Researchers have succeeded in delivering drugs precisely and fighting microorganisms through the use of plant-based nanoparticles. Expert research on plants should focus on finding their stress receptors along with connecting various defense systems and making crops tolerate many types of stress through genetic engineering.

Scientists use CRISPR-based genome editing systems plus RNAi and multi-omics tools to detect better ways plants handle stress and survive under harsh conditions. Examining how plants handle abiotic and biotic stress simultaneously will create new understanding about their natural adaptation to changing environments. Research on plants under stress drives significant progress in both safe farming and natural environment protection and provides medical advances for humans. Through combined plant biology and applied research scientists develop effective solutions to produce better crops under climate change and identify plant substances for therapeutic use. Strong teamwork among different scientific fields will help us better use plant stress research to protect Earth's ecosystems and communities.

References

Aboderin, O.S., Oyekunle, M., Bankole, F.A. and Olaoye, G. (2024) Combining ability for grain yield and low-N tolerance of intermediate/late maturing maize (Zea mays) inbred lines. *Journal of Crop Improvement* 38(6), 731–751.

Ahammed, G.J. and Yu, J.Q. (2016) Plant Hormones under Challenging Environmental Factors. DOI: 10.1007/978-94-017-7758-2.

Akhtar, M., Jaiswal, A., Taj, G., Jaiswal, J.P., Qureshi, M.I. and Singh, N.K. (2012) DREB1/CBF transcription factors: Their structure, function and role in abiotic stress tolerance in plants. *Journal of Genetics* 91, 385–395. DOI: 10.1007/s12041-012-0201-3.

Anumalla, M., Roychowdhury, R., Geda, C., Bharathkumar, S., Dash, G. and Tavva, M.D. (2016) Mechanism of stress signal transduction and involvement of stress inducible transcription factors and genes in response to abiotic stresses in plants. *International Journal of Recent Scientific Research* 7(8), 12754–12771.

Ashraf, M., Riaz, M. and Arif, M. (2018) Environmental Stress and Secondary Metabolites in Plants: An Overview.

Awan, F., Khurshid, Y. and Mehmood, A. (2017) Plant growth regulators and their role in abiotic stress management. *The International Journal of Innovative Research in Biosciences* 1, 9–22.

Baxter, A., Mittler, R. and Suzuki, N. (2013) ROS as key players in plant stress signaling. *Journal of Experimental Botany* 65(5), 1229–1240. DOI: 10.1093/jxb/ert375.

Bhardwaj, R., Sharma, I., Kanwar, M., Handa, N., Sharma, R. *et al.* (2013) Lea proteins and their role in salt stress tolerance, pp. 79–112.

Binyamin, R., Nadeem, S., Akhtar, S., Yahya Khan, M. and Anjum, R. (2019) Beneficial and pathogenic plant-microbe interactions: A review. *Soil and Environment* 38, xx–xx. DOI: 10.25252/SE/19/71659.

Bohn, L., Huang, J., Weidig, S., Yang, Z., Heidersberger, C. *et al.* (2024) The temperature sensor TWA1 is required for thermotolerance in Arabidopsis. *Nature* 629(8014), 1126–1132. DOI: 10.1038/s41586-024-07424-x.

Campalans, A., Messeguer, R., Goday, A. and Pagès, M. (1999) Plant responses to drought, from ABA signal transduction events to the action of the induced proteins. *Plant Physiology and Biochemistry* 37(5), 327–340. DOI: 10.1016/S0981-9428(99)80039-4.

Cargnello, M. and Roux, P.P. (2011) Activation and function of the MAPKs and their substrates, the MAPK-activated protein kinases. *Microbiology and Molecular Biology Reviews* 75(1), 50–83. DOI: 10.1128/mmbr.00031-10.

Chen, X., Pan, S., Li, F., Xu, X. and Xing, H. (2022) Plant-derived bioactive compounds and potential health benefits: Involvement of the gut microbiota and its metabolic activity. *Biomolecules* 12, 1871. DOI: 10.3390/biom12121871.

Choudhary, R. and Meena, B. (2024) *Challenges in the Green World: Abiotic Stress and Plants.* pp. 16–24.

Choudhary, D., Prakash, A. and Johri, B. (2007) Induced systemic resistance (ISR) in plants: Mechanism of action. *Indian Journal of Microbiology* 47, 289–297. DOI: 10.1007/s12088-007-0054-2.

Danquah, A., de Zélicourt, A., Colcombet, J. and Hirt, H. (2013) The role of ABA and MAPK signaling pathways in plant abiotic stress responses. *Biotechnology Advances* 32(1), 40–52. DOI: 10.1016/j.biotechadv.2013.09.006.

Devireddy, A.R., Zandalinas, S.I., Fichman, Y. and Mittler, R. (2021). Integration of reactive oxygen species and hormone signaling during abiotic stress. *The Plant Journal* 105(2), 459–476. https://doi.org/10.1111/tpj.15010

El Sabagh, A., Islam, M.S., Hossain, A., Iqbal, M.A., Mubeen, M. *et al.* (2022) Phytohormones as growth regulators during abiotic stress tolerance in plants. *Frontiers in Agronomy* 765068. DOI: 10.3389/fagro.2022.765068.

Fu, H. and Yang, Y. (2023) How plants tolerate salt stress. *Current Issues in Molecular Biology* 45, 5914–5934. DOI: 10.3390/cimb45070374.

Ghosal, A., Kabir, A. and Mandal, A. (2011) RNA interference and its therapeutic potential. *Central European Journal of Medicine* 6, 137–147. DOI: 10.2478/s11536-011-0005-5.

Gill, S. and Tuteja, N. (2010) Polyamines and abiotic stress tolerance in plants. *Plant Signaling and Behavior* 5, 26–33. DOI: 10.4161/psb.5.1.10291.

Guo, P., Baum, M., Grando, S., Ceccarelli, S., Li, R. *et al.* (2009) Differentially expressed genes between drought-tolerant and drought-sensitive barley genotypes in response to drought stress during the reproductive stage. *Journal of Experimental Botany* 60, 3531–3544. DOI: 10.1093/jxb/erp194.

Hossain, A., Pamanick, B., Venugopalan, V.K., Ibrahimova, U., Rahman, M.A. *et al.* (2022) Chapter 1 – Emerging roles of plant growth regulators for plants adaptation to abiotic stress–induced oxidative stress. In: Aftab, T. and Naeem, M. (eds), *Emerging Plant Growth Regulators in Agriculture*. Academic Press, pp. 1–72. DOI: 10.1016/B978-0-323-91005-7.00010-2.

Isah, T. (2019) Stress and defense responses in plant secondary metabolites production. *Biological Research* 52, 39 DOI: 10.1186/s40659-019-0246-3.

Jamieson, P., Shan, L. and He, P. (2018) Plant cell surface molecular cypher: Receptor-like proteins and their roles in immunity and development. *Plant Science* 274, 242–251. DOI: 10.1016/j.plantsci.2018.05.030.

Jing, Z., Liu, N., Zhang, Z. and Hou, X. (2024) Research progress on plant responses to stress combinations in the context of climate change. *Plants* 13, 469. DOI: 10.3390/plants13040469.

Khan, F.S., Goher, F., Hu, C. and Zhang, J.-Z. (2024) WUSCHEL-related homeobox (WOX) transcription factors: Key regulators in combating abiotic stresses in plants. *Horticulture Advances* 2(1), 2. DOI: 10.1007/s44281-023-00023-2.

Laid, B., Saibi, W. and Brini, F. (2021) Role of ABA in overcoming environmental stress:sensing, signaling and crosstalk. *Annals of Agricultural and Crop Sciences* 6, 1070. DOI: 10.26420/annagriccropsci.2021.1070.

Latz, A., Mehlmer, N., Zapf, S., Mueller, T., Wurzinger, B. *et al.* (2012) Salt stress triggers phosphorylation of the arabidopsis vacuolar K+ channel TPK1 by Calcium-Dependent Protein Kinases (CDPKs). *Molecular Plant* 6. DOI: 10.1093/mp/sss158.

Lin, L., Wu, J., Jiang, M. and Wang, Y. (2021) Plant mitogen-activated protein kinase cascades in environmental stresses. *International Journal of Molecular Sciences* 22(4), 1543. DOI: 10.3390/ijms22041543.

Liu, N., Ding, Y., Fromm, M. and Avramova, Z. (2014) Different gene-specific mechanisms determine the 'revised-response' memory transcription patterns of a subset of A. thaliana dehydration stress responding genes. *Nucleic Acids Research* 42. DOI: 10.1093/nar/gku220.

Ma, Y. and Qin, F. (2014) ABA Regulation of Plant Responses to Drought and Salt Stresses. pp. 315–336. DOI: 10.1007/978-94-017-9424-4_16.

Ma, Y., Dai, X., Xu, Y., Luo, W., Zheng, X. *et al.* (2015) COLD1 confers chilling tolerance in rice. *Cell* 160(6), 1209–1221. DOI: 10.1016/j.cell.2015.01.046.

Majeed, A. and Muhammad, Z. (2019) *Salinity: A Major Agricultural Problem – Causes, Impacts on Crop Productivity and Management Strategies.* pp. 83–99. DOI: 10.1007/978-3-030-06118-0_3.

Malabadi, R., Kolkar, K., Sultan, S., Chalannavar, R., Baijnath, H. and Castaño Coronado, K. (2024) Role of plant-derived medicine for controlling cancer. *International Journal of Science and Research Archive* 11, 2502–2539.

Manna, M., Achary, V.M. and Reddy, M. (2019) ROS Signaling and its Role in Plants, pp. 361–388. DOI: 10.1007/978-981-13-8922-1_14.

McFadden, M. and Laohasiripanya, N. (2022) CRISPR-Cas9, the genetic scissor. *International Journal of Advanced Research* 10, 946–950. DOI: 10.21474/IJAR01/14297.

Miao, Y., Lv, D., Wang, P., Wang, X.C., Chen, J., Miao, C. and Song, C.P. (2006) An Arabidopsis glutathione peroxidase functions as both a redox transducer and a scavenger in abscisic acid and drought stress responses. *Plant Cell* 18(10), 2749–2766. DOI: 10.1105/tpc.106.044230.

Miryeganeh, M. (2021) Plants' epigenetic mechanisms and abiotic stress. *Genes* 12(8), 1106. DOI: 10.3390/genes12081106.

Muhammad, M., Basit, A., Wahab, A., Li, W.-J., Shah, S. and Mohamed, H. (2023) *Response mechanism of plant stresses to secondary metabolites production.* pp. 469–492. DOI: 10.1016/B978-0-323-95241-5.00012-5.

Murti, G. and Upreti, K.K. (2007) Plant growth regulators in water stress tolerance. *Journal of Horticultural Sciences* 2, 73–93. DOI: 10.24154/jhs.v2i2.611.

Narusaka, Y., Nakashima, K., Shinwari, Z.K., Sakuma, Y., Furihata, T. *et al.* (2003) Interaction between two cis-acting elements, ABRE and DRE, in ABA-dependent expression of Arabidopsis rd29A gene in response to dehydration and high-salinity stresses. *Plant J* 34(2), 137–148. DOI: 10.1046/j.1365-313x.2003.01708.x.

Negacz, K., Malek, Ž., Vos, A. and Vellinga, P. (2022) Saline soils worldwide: Identifying the most promising areas for saline agriculture. *Journal of Arid Environments* 203, 104775. DOI: 10.1016/j.jaridenv.2022.104775

Nehra, A., Kalwan, G., Gill, R., Nehra, K., Agarwala, N. *et al.* (2024) *Status of impact of abiotic stresses on global agriculture*. pp. 1–21. DOI: 10.1016/B978-0-443-18500-7.00001-6.

Nykiel, M., Gietler, M., Fidler, J. and Labudda, M. (2023) Abiotic stress signaling and responses in plants. *Plants* 12, 3405. DOI: 10.3390/plants12193405.

Pirkkala, L., Nykänen, P. and Sistonen, L. (2001) Roles of the heat shock transcription factors in regulation of the heat shock response and beyond. *FASEB Journal: Official Publication of the Federation of American Societies for Experimental Biology* 15, 1118–1131.

Pulecio, J., Verma, N., Mejía-Ramírez, E., Huangfu, D. and Raya, A. (2017) CRISPR/Cas9-based engineering of the epigenome. *Cell Stem Cell* 21(4), 431–447. DOI: 10.1016/j.stem.2017.09.006.

Ray, P.D., Huang, B.W. and Tsuji, Y. (2012) Reactive oxygen species (ROS) homeostasis and redox regulation in cellular signaling. *Cell Signal* 24(5), 981–990. DOI: 10.1016/j.cellsig.2012.01.008.

Reguera, M., Peleg, Z., Abdel-Tawab, Y., Tumimbang, E., Delatorre, C. and Blumwald, E. (2013) Stress-induced CK synthesis increases drought tolerance through the coordinated regulation of carbon and nitrogen assimilation in rice. *Plant Physiology* 163, 1609–1622. DOI: 10.1104/13.227702.

Rizhsky, L., Davletova, S., Liang, H. and Mittler, R. (2004) The zinc finger protein Zat12 is required for cytosolic ascorbate peroxidase 1 expression during oxidative stress in Arabidopsis. *Journal of Biological Chemistry* 279(12), 11736–11743.

Roy, A., Khan, A., Ahmad, I., Alghamdi, S., Rajab, B.S. *et al.* (2022) Flavonoids a bioactive compound from medicinal plants and its therapeutic applications. *BioMed Research International* 2022, 5445291. DOI: 10.1155/2022/5445291.

Roychoudhury, A. and Paul, A. (2012) *Abscisic Acid-Inducible Genes during Salinity and Drought Stress.* pp. 1–78.

Roychowdhury, R., Hada, A., Biswas, S., Mishra, S., Prusty, M.R. *et al.* (2025) Jasmonic acid (JA) in plant immune response: Unravelling complex molecular mechanisms and networking of defence signalling against pathogens. *Journal of Plant Growth Regulation* 44(1), 89–114. DOI: 10.1007/s00344-024-11264-4.

Seeda, A., Abou El-Nour, E.-Z., Abdallah, M., El-Bassiouny, H. and El-Monem, A. (2022) Impacts of salinity stress on plants and their tolerance strategies: A review. *Middle East Journal of Applied Sciences* 12, 282–400. DOI: 10.36632/mejas/2022.12.3.27.

Shailani, A., Joshi, R., Sneh, Singla-Pareek, L. and Pareek, A. (2020) Stacking for future: Pyramiding genes to improve drought and salinity tolerance in rice. *Physiologia Plantarum*, 172(2), 1352–1362. DOI: 10.1111/ppl.13270.

Soni, P., Basak, N., Rai, A., Sundha, P., Chandra, P. and Yadav, R. (2023) Occurrence of Salinity and Drought Stresses: Status, Impact, and Management, pp. 1–28. DOI: 10.1007/978-981-99-4669-3_1.

Sun, M., Yang, Z., Liu, L. and Duan, L. (2022) DNA methylation in plant responses and adaption to abiotic stresses. *International Journal of Molecular Sciences* 23. DOI: 10.3390/ijms23136910.

Takasaki, H., Maruyama, K., Kidokoro, S., Ito, Y., Fujita, Y., Shinozaki, K., Yamaguchi-Shinozaki, K. and Nakashima, K. (2010) The abiotic stress-responsive NAC-Type transcription factor OsNAC5 regulates stress-inducible genes and stress tolerance in rice. *Molecular Genetics and Genomics: MGG* 284, 173–183. DOI: 10.1007/s00438-010-0557-0.

Tang, Q., Wei, S., Zheng, X., Tu, P. and Tao, F. (2024) APETALA2/ethylene-responsive factors in higher plant and their roles in regulation of plant stress response. *Critical Reviews in Biotechnology* 44, 1–19. DOI: 10.1080/07388551.2023.2299769.

Thirugnanasambantham, K., Sekar, D., Subramanian, S., Karikalan, K., Senguttuvan, M. and Hairul-Islam, V. (2014) Role of ethylene response transcription factor (ERF) and its regulation in response to stress encountered by plants. *Plant Molecular Biology Reporter* 33(3), 347–357. DOI: 10.1007/s11105-014-0799-9.

Tomar, R.S. (2021) Biotechnology and Genetic Engineering Tools for Abiotic Stress Management.

Tuteja, N. (2007) Abscisic acid and abiotic stress signaling. *Plant Signaling and Behavior* 2(3), 135–138. DOI: 10.4161/psb.2.3.4156.

Tuteja, N. and Mahajan, S. (2007) Calcium signaling network in plants: An overview. *Plant Signaling and Behavior* 2, 79–85. DOI: 10.4161/psb.2.2.4176.

Verma, S., Nizam, S. and Verma, P. (2013) Biotic and Abiotic Stress Signaling in Plants. Vol. 1, pp. 25–49. DOI: 10.1007/978-1-4614-6372-6_2.

Wani, K.I., Naeem, M., Castroverde, C.D.M., Kalaji, H.M., Albaqami, M. and Aftab, T. (2021) Molecular mechanisms of Nitric Oxide (NO) Signaling and reactive oxygen species (ROS) Homeostasis during abiotic stresses in plants. *International Journal of Molecular Sciences* 22(17), 9656. DOI: 10.3390/ijms22179656.

Xiong, L. and Ishitani, M. (2006) Stress Signal Transduction: Components, Pathways and Network Integration, pp. 3–29. DOI: 10.1007/1-4020-4389-9_1.

Yoshida, T., Fujita, Y., Maruyama, K., Mogami, J., Todaka, D., Shinozaki, K. and Yamaguchi-Shinozaki, K. (2014) Four Arabidopsis AREB/ABF transcription factors function predominantly in gene expression downstream of SnRK2 kinases in abscisic-acid signaling in response to osmotic stress. *Plant, Cell and Environment* 38(1), 35–49. DOI: 10.1111/pce.12351.

Zhang, Y., Xu, J., Li, R., Ge, Y., Li, Y. and Li, R. (2023) Plants' response to abiotic stress: Mechanisms and strategies. *International Journal of Molecular Sciences* 24, 10915. DOI: 10.3390/ijms241310915.

Zhiqi, H., Tingyi, W., Dongdong, C., Lan, S., Guangheng, Z., Qian, Q. and Li, Z. (2025) Leucine-rich repeat protein family regulates stress tolerance and development in plants. *Rice Science* 32(1) 32–43. DOI: 10.1016/j.rsci.2024.12.003.

Zhu, J.-K. (2002) Salt and drought stress signal transduction in plants. *Annual Review of Plant Biology* 53, 247–273. DOI: 10.1146/annurev.arplant.53.091401.143329.

3 Gene Expression Regulation of Plants Under Stress Conditions

Hajira Bibi[1], Taskeen Rasheed[2], Javed Iqbal[3], Banzeer Ahsan Abbasi[4]*, Muhammad Anas[5,2] and Syeda Anber Zahra[6]

[1]Department of Biochemistry, Faculty of Biological Sciences, Quaid-i-Azam University, Islamabad 45320, Pakistan; [2]Department of Plant Sciences, Faculty of Biological Sciences, Quaid-i-Azam University, Islamabad 45320, Pakistan; [3]Department of Botany, Bacha Khan University, Charsadda 24420, Khyber Pakhtunkhwa, Pakistan; [4]Department of Botany, Rawalpindi Women University, Rawalpindi 46300, Pakistan; [5]Department of Soil Science and Plant Nutrition, University of Selcuk Campus, Konya 42079, Turkiye; [6]Institute of Biomaterials and Biomedical, School of Food and Pharmacy, Shanghai Zhongqiao Vocational and Technical University, Shanghai, China

Abstract

This chapter provides detailed analysis of complicated mechanisms that control gene expression in plants during stressful circumstances. Transcription regulation depends on the DNA sequence interactions between cis-regulatory components of stress-related genes and trans-acting elements or transcription factors (TFs) for gene expression regulation under cold, drought and high salinity environmental stress. Three types of cis-acting elements, comprising abscisic acid-responsive elements (ABAREs), dehydration-responsive elements (DREs) and low-temperature-responsive elements (LTREs), function together to help plants execute stress responses that match environmental stimuli types and intensities. The discussion in this chapter examines the significance of gene expression post-transcriptional regulation through small non-coding RNAs under different stress environments in plants. The expression of non-coding RNAs helps plants manage their responses to various types of stress situations. The author shows in this chapter how epigenetic mechanisms control plant gene expression in stressful environmental situations. Gene expression regulation of plants under stress conditions depends on two main epigenetic approaches, which are DNA methylation and histone modifications. The chapter progresses to explore translation regulation mechanisms of plants used for gene expression control in various stress environments. Environmental stress triggers the translation process that converts mRNAs into functional proteins in specific ways. Plants use this mechanism to biologically create ribosomes, initiate translation and specifically draw stress-induced mRNAs toward polysomes to rapidly generate adaptive resistant behaviours against environmental stress. This chapter extensively explores plant biochemical reaction frameworks that regulate how plants handle different types of environmental hardships, which demonstrates their advanced techniques for surviving challenging environmental conditions.

Keywords: gene expression, transcription regulation, environmental stress, non-coding RNAs

* Email: benazirahsanabbasi786@gmail.com

DOI: 10.1079/9781800628823.0003

3.1 Introduction

Plants regularly experience numerous environmental stress factors, which include drought, high and low temperatures and pathogen attacks, as well as salinity. Environmental stresses impair development and growth as well as reproductive functions of plants.

Plants that lack mobility need stress adaptation mechanisms because of their static lifestyle (Saini *et al.*, 2012; Zhuang *et al.*, 2014; Shaw, 2018; Ahmad *et al.*, 2022; Ahmad *et al.*, 2024). Plants utilize a detailed set of regulatory pathways to reprogram gene expression based on the different environmental stress conditions, which allows them to survive challenging environmental situations. This chapter explains the complicated mechanisms that control plant gene expression throughout diverse environmental stress exposures.

Plant stress gene expression requires essential analysis of cis-acting elements (specific DNA sequences) in gene promotor regions (Marques *et al.*, 2017). Different transcription factors (TFs) work as trans-active elements to find specific DNA sequences on regulatory regions of genes through dehydration-responsive elements (DREs) (Nakashima *et al.*, 2014), abscisic acid-responsive elements (ABAREs) (Reddy *et al.*, 2013), and low-temperature-responsive elements (LTREs) (Reddy *et al.*, 2013). This interaction drives expression of stress-responsive genes under drought or heat stress, as well as high salinity and cold stress (Haider *et al.*, 2021; Ali *et al.*, 2022).

The complex gene expression regulatory system functions through unique configurations and element set-ups in promoter regions of stress-responsive genes, which exactly controls plant responses to environmental stress factors.

Studies have proven that epigenetic modifications, especially DNA methylation and histone modifications, work together with TFs to control plant gene expression under different environmental stress situations by enabling the accessibility and expression of stress-responsive genes. The modification process ultimately controls transcription factor interactions combined with their target DNA sequences located within stress-responsive gene promoters (Bobadilla and Berr, 2016; Haider *et al.*, 2022).

Plants use multiple post-transcriptional regulatory methods for gene expression modulation under various stress scenarios, and these methods include non-coding RNA-based mechanisms that control stress-responsive genes. Plant organisms utilize patterned translational regulatory mechanisms to control the expression of stress-response genes across different environmental stress conditions after transcription and post-transcriptional regulation. The process of translation during the central dogma plays an essential control role for gene expression because it interprets mRNA sequences to create protein amino acid sequences. Plants acquired multiple sophisticated systems to optimize their stress response through mRNA translation regulation during both biotic and abiotic conditions. Gene expression regulation in plants depends primarily on three main translational-regulatory mechanisms where plants control ribosomal biogenesis, while simultaneously controlling the timing of translational initiation and how stress-responsive mRNAs use polysomes. The plant maintains rapid proteome turnover through translational regulatory mechanisms when facing environmental changes and stress conditions, thus achieving improved stress tolerance and survival and adaptation capabilities (Asghar *et al.*, 2024; Ijaz *et al.*, 2024). These gene expression regulatory mechanisms in plants under different stress conditions have been summarized in Table 3.1 and Fig. 3.1.

3.2 Transcriptional Regulation of Gene Expression in Plants Under Stress

The main process of genetic expression regulation starts with transcription while requiring both cis-acting elements and trans-acting elements for their interaction. The promoter areas of genes contain specific sequences referred to as cis-acting elements, which affect how genes respond to environmental stress conditions (Lata *et al.*, 2011). Stress-responsive genes contain promoter sequences that include DREs among other cis-acting elements. DREs represent cis-acting stress-responsive elements that exist inside promoter regions of drought and high salinity and

Table 3.1. Mechanisms of gene expression regulation in plants under stress with different regulator elements under different stress conditions. (Table author's own.)

Gene expression regulation	Regulatory elements	Stress	Example	Model plant	Effect
Transcriptional regulation	Cis-acting elements (DNA sequences) Trans-acting elements (TFs)	Biotic and abiotic	DREs ABAREs LTREs	*Arabidopsis thaliana*	Improved stress tolerance Lata *et al*. (2011)
Post-transcriptional regulation	MicroRNAs (miRNAs) Small-interfering RNAs (siRNAs)	Drought	miR393 targets auxin receptor transcripts (TIR1, AFB2, and AFB3) miR396 targets growth-regulating factor transcripts	*Arabidopsis thaliana*	Improved drought tolerance Tan *et al*. (2020); Jiang *et al*. (2022)
Translational regulation	Eukaryotic transcription initiation factors RNA secondary structures RNA-binding proteins (RBPs) RNA modifications Stress granules (SGs) and processing bodies (PBs)	Heat and cold stress	eIF2α phosphorylation 5' untranslated region of PIF7 mRNA controls its translation in response to temperature RBPs binding to EIN3-BINDING F-BOX 1 and EBF2 mRNAs N6-methyladenosine (m6A) and mRNA stability	*Arabidopsis thaliana*	Reduces global protein synthesis; Improved temperature tolerance Ethylene (ET)-mediated stress responses Managing mRNA storage Degradation Translation suppression
Post-translational regulation	Molecular chaperones and chaperonins (Protein folding regulation)	Heat	HSPs (HSP70, HSP90, and small HSPs) bind to unfolded or partially folded proteins and promote their correct three-dimensional structure (Mukhopadhyay *et al*., 2024)		
Epigenetic modifications	DNA methyltransferases (DNA methylation) Histone acetyltransferases (Histone acetylation) Histone methyl transferase (Histone methylation)	Drought	DNA demethylation at the RD29A promoter The TF DREB2A, a key regulator of drought response, exhibits increased acetylation under drought conditions Histone H3 trimethylation at lysine 4 (H3K4me3), H3K9me2/3 and H3K27me3	*Arabidopsis thaliana*	Improved drought tolerance (Liu *et al*., 2022b; Yadav *et al*., 2024) Stress-responsive gene expression

(Continued)

Table 3.1. Continued.

Gene expression regulation	Regulatory elements	Stress	Example	Model plant	Effect
Signaling pathways	Abscisic acid (ABA) pathway ABAREs Jasmonic acid (JA) signaling pathway (JA-responsive TFs) ET signaling pathway (ET-responsive factor)	Drought High salinity Wounding Pathogen attack Abiotic stress	AREB/ABF interacting with the promoters of these stress-responsive genes like RD29A and RAB18 MYC2 TF upregulation by JA accumulation Ethylene-response factor (ERF) DREB2A, acting as a master regulator of drought response; ET-induced expression of RBPs, regulating the stability, localization, and translation of stress-responsive mRNAs (Duarte-Conde *et al.*, 2022)		Activation of stress-responsive genes; Improved osmotic stress tolerance
Mitogen-activated protein kinase (MAPK) cascades	MAPK modules	Environmental stresses, including pathogen attack, cold, and oxidative stress	MPK6 has been demonstrated to phospho-activate the drought stress-responsive transcription factor, DREB2A to activate drought-responsive genes in *Arabidopsis* (Zhang and Zhang, 2022)	*Arabidopsis thaliana*	Improved stress tolerance

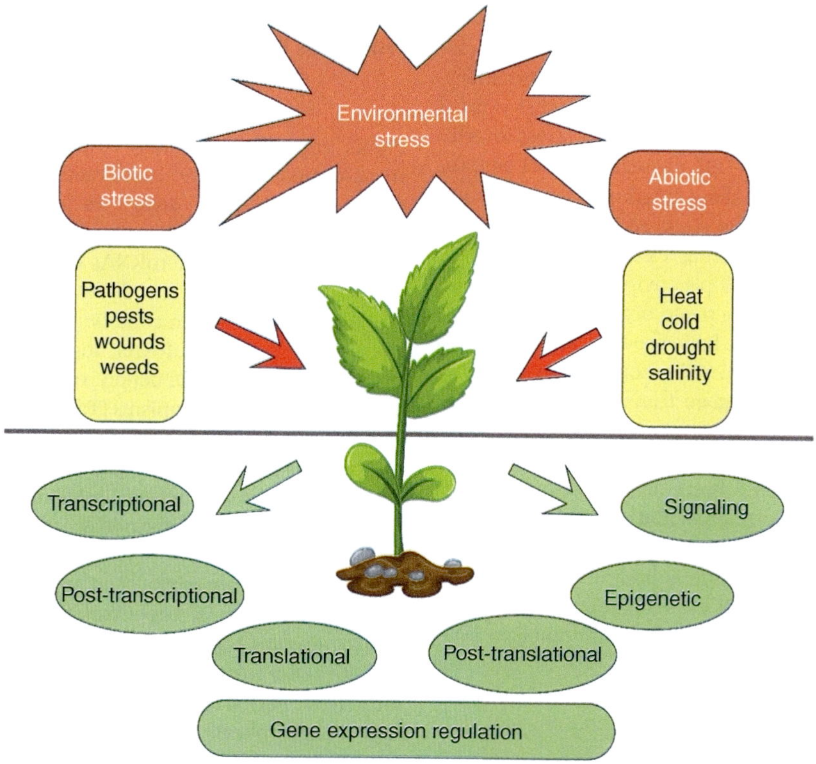

Fig. 3.1. Modes of gene expression regulation in plants under different stress conditions. (Figure author's own.)

cold-temperature responsive genes (Mizoi *et al.*, 2012, Ijaz *et al.*, 2024). Stress-responsive genes contain DREs as well as ABAREs and LTREs, which trans-acting elements use for regulation of stress-responsive gene expression across different stress situations.

The specific proteins referred to as DRE-binding TFs (DREBs) function as the AP2/ERF family member proteins. The AP2/ERF family of proteins demonstrates vital gene expression regulatory functions for plant stress-responsive genes and particularly controls genes therefore controlling drought stress reactions among other responses. Through the binding of trans-acting elements to specific promoter-sequence DNA motifs (cis-acting elements) these proteins regulate different stress responses in a plant organism (Mizoi *et al.*, 2011).

Genes that respond to stress get their expression regulated through interactive feedback between cis- and trans-acting elements that both respond to epigenetic modifications of DNA methylation and histone modifications. DNA methylation together with histone modifications modify the stress-responsive gene chromatin structure as well as its accessibility (Rehman and Tanti, 2020; Liu *et al.*, 2022b). The chromatin structure changes because of these modifications making them influence transcription factor (trans-acting element) binding to cis-acting elements. DNA methylation together with histone deacetylation creates extremely compact chromatin structures that block trans-acting elements (TFs) from reaching promoter region cis-acting elements. Histone demethylation, together with histone acetylation, creates greater accessibility of trans-acting elements to the cis-acting elements, which allows them to bind and interact more frequently and upregulates stress-responsive gene expression (Candela-Ferre *et al.*, 2024).

Plants channel their stress responses through particular cis-acting elements located within the promoter, which helps them adjust their gene expression patterns in response to different environmental stress stimuli (Yamaguchi-Shinozaki and Shinozaki, 2005; Hernandez-Garcia and Finer, 2013). Plants express various levels of gene transcripts based on stress severity because multiple DRE elements within the promoter sequence allow different gene expression outcomes. Different stress-responsive genes contain several cis-acting elements in their promoters where multiple environmental messages can converge to activate organized plant stress responses. During both drought and salt stress conditions AREB1 TFs together with other ABREs influence gene transcription of abscisic acid (ABA) responsive genes (Joo *et al.*, 2020).

The precise activation of plant stress mechanisms requires both cis-acting factors and trans-acting elements that function under the influence of cis-acting element arrangements and DNA modifications including methylation and histone modifications and DNA demethylation processes (Fig. 3.2). The complete DNA methylation and histone modification mechanisms receive further explanation in "Epigenetic Regulation of Plant Stress Response."

3.3 Post-transcriptional Regulation of Gene Expression in Plants Under Stress

Plants control their stress response through stress-responsive gene expression, while the post-transcriptional level includes small non-coding RNAs such as small interfering RNAs (siRNAs) and microRNAs (miRNAs). The small RNAs function crucially within plants to manage stress response conditions by selecting stress-related gene mRNAs for degradation through selective destruction or translation-based expression mechanisms (Tan *et al.*, 2020). The drought tolerance mechanism in *Arabidopsis* improves due to miR393 guiding the degradation of Auxin receptors AFB2 AEB3 and TIR1 mRNA transcripts. Laboratory observations have shown that the miRNA 396 targets growth factors mRNA transcripts, thus resulting in excessive production of these factors. The elevation of growth-regulating factors in the plant leads to better tolerance against drought and salinity stress. Various research studies demonstrate how siRNA molecules adjust the transcription of

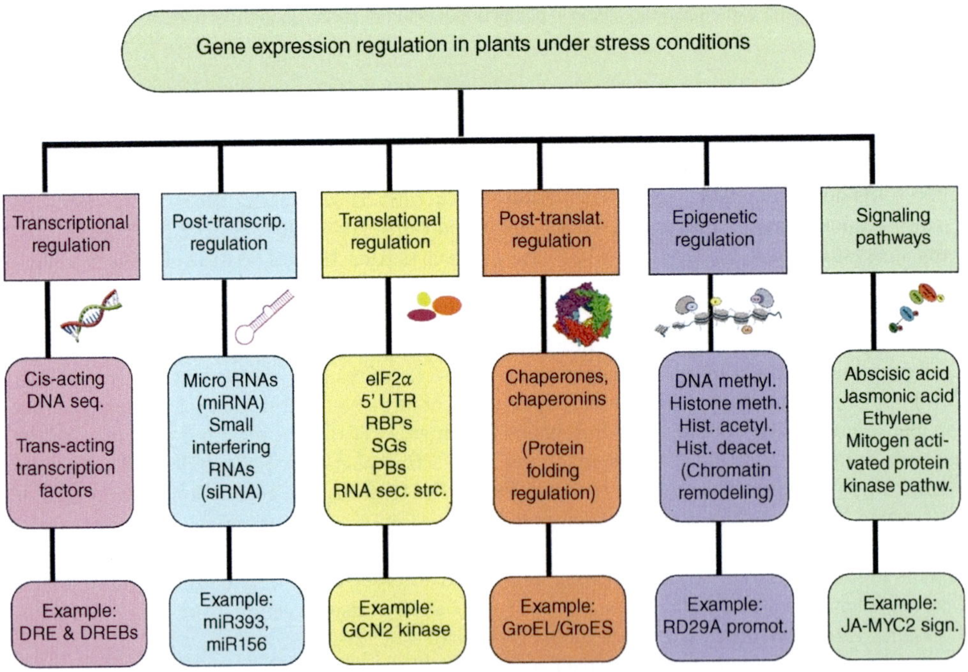

Fig. 3.2. Molecular pathways involved in gene expression regulation in plants under stress conditions. (Figure author's own.)

drought-resistance genes including ABA regulation factors (Jiang *et al.*, 2022).

These small RNAs direct chromatin-modifying complexes as they perform their role of chromatin remodeling. The targeted gene loci receive epigenetic modifications due to small RNAs recruiting DNA methylation enzymes and histone acetyltransferase enzymes, which results in gene expression regulation (Ali and Tang, 2025). *Arabidopsis* miR156 directs BRAHMA chromatin remodeling enzyme, which regulates the expression of stress-responsive genes (Jeyakumar *et al.*, 2020). RNA-binding proteins (RBPs) work alongside small non-coding RNAs in order to regulate gene expression regulation in plants, which experience stressful environmental conditions. The RBP RZ-1 connects with cold-resistant gene mRNA messages which enhances their stability thus boosting the translation and expression of cold-responsive genes (Ma *et al.*, 2021). Another RBP FCA supports the stability of drought-induced gene mRNA transcripts which promotes both their processing step and their export process. The binding of FCA activates two processes enhancing polyadenylation and transcript cleavage that results in more stable production and translation abilities for drought-induced gene transcripts. Plants under drought conditions use RBP FCA to regulate RD29A and RAB18 at the post-transcriptional level to optimize their adaptation mechanism (Fan *et al.*, 2024). The multiple regulatory levels of post-transcriptional gene expression regulation in plants modify stress responses, which allows plant survival in tough and stressful environmental conditions.

3.4 Translational Regulation of Gene Expression in Plants Under Stress

Plants remain immobile at their growth sites where they need to adjust their genetic expression quickly and precisely because environmental stress demands adaptation and survival. Under different stress conditions, plants use translational regulation as an additional layer of genetic expression management. Localized protein synthesis changes can be regulated rapidly under essential conditions because new transcripts are not always needed. The elements that make the translational regulation process function in plant cells are outlined below.

- eIF2 functions as the alpha subunit that forms eukaryotic translation initiation factor 2. This protein complex functions as a vital component to regulate protein synthesis under stress conditions. Enhanced expression through translation of stress-responsive genes becomes possible through eIF2α phosphorylation, together with decreased systemic protein synthesis transition (Son and Park, 2023). Plants activate this signaling option through administrative deficiency, which ultimately activates GCN2 kinase. The regulatory effects of eIF2α phosphorylation occur at a reduced level in plants than in mammals (Lokdarshi and Arnim, 2022).
- Scientific research has established that RNA secondary structures act as key elements in controlling plant gene expression translation during times of stress. The secondary RNA structures respond to temperature changes in a sensitive way to serve as regulatory cis-elements controlling mRNA transcript translation. The PIF7 mRNA translation is regulated by its 5' untranslated region, which produces a hairpin secondary structure that responds to environmental temperatures. The RNA secondary configuration serves as a fundamental element, which enables thermosensory processes and helps control gene expression of plants when they encounter environmental changes (Samtani *et al.*, 2022).
- RBPs serve as important elements that regulate translational gene expression stages in stressed plants to control gene expression patterns. RBPs selectively identify particular mRNA transcript regions with specific sequences and secondary structures, which they then enhance for transcription. Stress conditions lead to regulatory actions by specific RBPs on mRNAs of ET-mediated stress-responsive genes (Xu *et al.*, 2021). A second type of RBP functions in proliferating stem cells of plant organisms. WUS is a gene, coding a homeobox transcription factor. This transcription

factor is a master regulator of meristematic stem cells (Mayer *et al.*, 1998). The RBP La1 relocates from the nucleus into the cytoplasm under stress conditions to bind with WUS transcripts, which results in both increased transcription levels of WUS and increased proliferation of stem cells (Wu *et al.*, 2023).

- Under different stress situations plants utilize RNA modifications as an additional mechanism to control translation-based gene expression. Under environmental stress conditions RNA modifications, specifically N6-methyladenosine, enhance messenger (mRNA) stability and effectiveness along with determining its distribution to achieve precise plant responses (Hu *et al.*, 2022).
- Plants exposed to stress conditions show additional translation regulation of gene expression through physical structures including both processing bodies (PBs) and stress granules (SGs) in cytoplasmic areas. These bodies within the cytoplasm control the storage process and mRNA translation suppression alongside mRNA degradation functions. Challenging environmental conditions require the cytoplasmic structures to function as essential components in regulating gene expression through translation. (Kearly *et al.*, 2024). Plants use the combination of eIF2α with RNA secondary structures RNA modifications RBPs and PBs and SGs to perform complex and dynamic translational regulation of gene expression. Plants need translational gene expression regulation to survive during adverse conditions, as well as adapt to various stressful environments.

3.5 Epigenetic Regulation of Gene Expression in Plants Under Stress

Scientific research focused on epigenetic gene regulation in plants when exposed to environmental challenges has become significant. Plants regulate their gene expression patterns through two mechanisms, which include DNA methylation together with multiple histone modifications like histone acetylation phosphorylation methylation and ubiquitination. The complete description of these mechanisms exists in Chapter 5 "Epigenetic Regulation of Plant Stress Responses."

DNA methyltransferases (enzymes) use cytosine bases at promoter and coding regions of stress-related genes to apply methyl groups by DNA methylation procedures. During the DNA methylation process the structure of chromatin becomes denser because TFs experience decreased accessibility to DNA. DNA methylation patterns determine whether genes related to stress will become activated or silenced for gene expression. Under drought conditions the *Arabidopsis* plant experiences promoter activation of RD29A resulting in better drought tolerance through DNA methylation in its promoter region (Liu *et al.*, 2022b; Yadav *et al.*, 2024).

Histone modifications serve as vital regulators that control the gene expression of stress-responsive genes in translational processes within plant cells. The enzymatic activity of histone acetyltransferases produces open chromatin structure, which makes TFs more accessible as demonstrated through histone acetylation modification. Upregulation of stress-related genes expression occurs because of these changes. The activity of histone deacetylases brings about compact chromatin structure, which blocks stress-responsive genes from expression. The way genes controlling stress responses express themselves depends on the current state of environmental stress because these pressures modify histone acetylation patterns (Wang *et al.*, 2024).

DREB2A functions as the primary regulator for genes that respond to drought. The plant can achieve better drought tolerance under stress because drought elevates DREB2A gene expression through histone acetylation under drought conditions according to Li *et al.* (2021). Histone methylation functions as the second important molecule that controls the expression patterns of stress-responsive genes. Methylation standards found, along with particular histone components, determine whether genes become activated or silenced for transcription (Shi *et al.*, 2023). Drought-related transcription changes occur because H3K4me3 methylation on residue lysine 4 helps increase gene expression but H3K9me2/3 and H3K27me3 methylation causes gene repression (Park *et al.*, 2020; Dinkar *et al.*, 2024). Plants adapt better to environmental challenges when stress-altered histone modifications occur at stress-regulated regulatory elements.

3.6 Signalling Pathways and Gene Expression Regulation in Plants Under Stress

Plants, through evolution, have developed complex signal cascade systems that control genetic expression responses in different kinds of stress environments. The signaling pathways trigger different signaling cascades after detecting environmental stress signals that activate stress-responsive TFs. Through this activation of stress-responsive TFs the cell controls the expression of stress-responsive genes, which promotes improved stress tolerance in plants (Zhao *et al.*, 2021). Plants use ABA signaling as an example to regulate gene expression during stress situations. During high salinity combined with drought stress, plant hormone ABA levels increase as a response that activates special TFs called AREB/ABF family members. The TFs act upon stress-responsive genes through their promoter regions to improve RD29A and RAB18 expression rates (dehydrins-plant stress tolerance genes). Plants under drought stress conditions, along with saline environments, use the ABA-signaling pathway to turn on tolerance gene expression, which allows them to adapt to both water deficits and osmotic stress (Hussain *et al.*, 2021).

Plants under diverse stress conditions use the jasmonic acid (JA) signaling pathway as a pathway to control gene expression patterns. Under environmental stress situations including pathogen attacks and pest infections and damage to plants and exposure to cold temperatures and drought and salt conditions the plant accumulates the lipid-derived plant hormone JA (Hewedy *et al.*, 2023). Perception triggers biological hormone synthesis, which ends in hormone accumulation throughout the plant. Plant hormone accumulation triggers the activation of TFs referred to as JA-responsive factors where the MYC2 family of TFs serve as an example. The activated MYC2 TFs bond to stress-responsive gene promoter areas to initiate their expression. The dehydration-active protein MYC2 turns on as a result of JA accumulation during drought stress and then activates genes, including DRE-binding proteins, which offer osmotic stress tolerance (Ali and Baek, 2020).

The JA-MYC2 signaling pathway allows for cross-pathway communication with the ABA signaling pathway, which enhances plant responsiveness to biotic and abiotic stress factors (Khan *et al.*, 2023). At the post-transcriptional level, JA functions by inducing RBPs to control how stress-responsive mRNAs are located as well as their stability and translation (Sultana *et al.*, 2025). Plants use this system to control their gene expression programs quickly and adaptively in response to environmental conditions that vary and create urgent situations.

Plants use ethylene (ET) as a plant hormone to control gene expression through signal pathways in response to difficult environmental situations according to Huang *et al.* (2023). Under stress conditions originating from high salinity, drought and pathogen attacks, plants start the production of ET creating hormone accumulation within plant systems (Tang *et al.*, 2024).

The enhanced ET hormone activation triggers multiple signaling events, which begin with ET-binding to membrane receptors that initiate the succeeding transcription factor cascade. Plants utilize ET-responsive factors to handle stress responses that are members of the constituting transcription factor family. The DNA sequences in ET-responsive elements in stress-responsive gene promoters allow ET-responsive factors to bind them. The binding between ET-responsive TFs and their target areas either activates or suppresses selected stress-responsive genes depending on the specific type of ET-responsive factor active during the process. Among drought-response factors DREB2A stands as a master regulator that gets its activation signal through both ET pathway and ABA signaling pathway. The activation of DREB2A results in the elevated expression of RD29A gene and other drought tolerance-mediating genes (Ma *et al.*, 2024).

The ET signaling pathway functions as a regulator of gene expression control at the post-transcriptional level, although it previously demonstrated control at transcriptional levels. The process of ET signaling causes RBPs to emerge, which help manage the localization and stability and translation processes of mRNAs from stress-responsive genes (Duarte-Conde *et al.*, 2022). Apart from environmental control, plants gain better response management through this mechanism, which adds another level of regulatory function to stress responses. Under stressful situations plants use the mitogen-activated protein kinase (MAPK) signaling pathway to control gene expression patterns. Different stresses from both living and non-living environmental sources, including pathogen invasion, and oxidative and cold stress situations, activate the MAPK signaling cascade. Phosphor-activation of the MAPK signaling pathway enables the activation of downstream TFs. The *Arabidopsis* MAPK pathway MPK6 becomes activated through phosphor-activation and then recruits drought-responsive TFs to control the response of *Arabidopsis* to drought (Zhang and Zhang, 2022).

Apart from their involvement in transcriptional regulation these signaling pathways are also essential for post-transcriptional regulation of gene expression in plants under stress conditions. For example, the calcium-dependent protein kinase (CDPK) pathway is a signalling pathway that regulates the post-transcriptional gene expression of stress-responsive genes in the plants by influencing the stability and translation of these transcripts. CPDKs are activated by stress signal perception and subsequently phosphorylate and activate RBPs like cold shock proteins. Owing to these activated RBPs, the mRNA transcripts of cold shock proteins are bound as well as interacted with them, thus regulating their localization and processing, thus eliciting a fine-tuned response of plants to the cold stress. Post-transcriptional regulation of gene expression further enhances gene expression regulation in stress response to allow plants to modulate their gene expression rapidly and respond rapidly to environmental challenge. Bestowing plants with better survival and adaptation to a number of various environmental challenges (Ren *et al.*, 2023), allows plants to control the expression of genes related to stress via controlling localization and stability and translation of stress-related mRNAs.

Finally, the plants are able to regulate their gene expression programs and eventually their responses to complex environmental stress conditions by means of crosstalk among different stress regulating signaling pathways including the ABA, JA and ET signaling pathways. For instance, it has been reported that the ET and ABA signaling pathways have crosstalk and fine-tune the expression of plant genes codifying for stress response in high salinity and drought (Khan *et al.*, 2020). Plant signaling pathways are found to be important in regulating the plant stress response on the basis that they control the transcriptions and the post-transcriptional gene expression of the plant stress-related genes.

3.7 Post-translational Regulation of Gene Expression in Plants Under Stress

In addition to its role in proper folding of protein and regulation of gene expression, post-translational modifications (PTMs) would allow the plants to go into adaptation under stress conditions. In stress, the complex network expresses genes involved in genes responding and adapting to stress. For protein function, post-translational protein folding is critical and plays an important regulatory role at the gene expression regulation level of plant stress genes under stressful environmental conditions. In particular, plants sense stress conditions and adjust post-translational protein folding, a modulatory step in their stress response for adaptation to stressful environmental conditions. Proteins will not fold under different stress conditions or convert to their active form leading to accumulation of denatured and misfolded proteins (Howell, 2021).

There are different types of evolutionary strategies of plants to maintain protein homeostasis, for example, increasing the expression of chaperones, which include chaperonins and heat shock proteins (HSPs), which are involved in proper proteins folding. Micellar molecular chaperones are HSPs, which includes families of protein such as HSP90, HSP70 and small HSPs. Upon binding to partly folded or even unfolded proteins, they process these proteins to their three-dimensional structure (Mukhopadhyay *et al.*, 2024). GroEL or GroES complexes, as chaperonins, provide an enclosed environment where proteins can be folded into their correct structure without any interference from other components of the cell (Fatima *et al.*, 2021). Furthermore, there is a sensing response of misfolded proteins that leads to the activation of protein folding and protein degradation associated genes (e.g. genes for increased protein quality control such as Liu *et al.*, 2022a). This unfolded protein response (UPR) pathway is important for maintaining protein homeostasis in plants in the presence of stress.

Under stress conditions, like the pathogen attack, high temperature or drought, UPR is activated through the accumulation of denatured or misfolded proteins in the endoplasmic reticulum (ER) inside of plant cells. In plants, TFs are responding to activation of the UPR through coordinating signalling cascades and upregulating genes for protein quality control, folding and degradation (Read and Schröder, 2021). The bZIP family of TFs is one of the plant specific genes within the UPR pathway of *Arabidopsis*, represented by AtbZIP17, AtbZIP60 and AtbZIP28. These bZIP proteins are normally bound to an ER resident chaperone, for example BiP.

ER, under environmental stress conditions, accumulates misfolded proteins and triggers the release of bZIP TFs from BiP chaperones. Howell (2021) reports that the released bZIP TFs translocate into the nucleus. These TFs once get into the nucleus, they activate the expression of genes that code for molecular chaperones such as HSP90 and HSP70, as well as genes encoding for the components of the ER-associated or proteasomal degradation pathway involved in degradation of unfolded or misfolded proteins (Guo *et al.*, 2024). One example would be NAC transcription factor family associated with controlling unfolded protein in plants. NAC is a set of membrane-bound proteins, including NAM, ATAF1/2 and CUC2. Under stressing environmental conditions, they are released from ER and translocated into nuclear through cleavage. Inside the nucleus, they activate expression of the ER stress response and protein quality control genes including AtNAC089 and AtNAC062 membrane-bound NAC TFs (Yu *et al.*, 2021). As a result, this allows plants to respond to various environmental challenges through modulating their gene expression patterns and attaining cellular homeostasis by orchestrating transcriptional and post-transcriptional and protein folding processes. The genes are not solely determined by the presence or absence of other genes in a single dimension; rather, the regulation of genes in each major dimension supports responses to pressing environmental conditions that effectively and timely enhance plants' overall resilience and survival.

3.8 Transcription Factors' Modifications and Gene Expression Regulation in Plants Under Stress

PTMs of stress-responsive TFs are chemical modifications of these proteins performed upon their translation that enable these proteins to respond quickly to the environmental stress including cold, drought and high temperatures by modifying their localization, stability, interactions, activity, and therefore expression of, the stress-related genes and fine-tuning plant responses to these challenging environmental conditions. PTMs of stress-responsive TFs include phosphorylation, SUMOylation and ubiquitination (Han *et al.*, 2022).

Binding to environmental stresses leads to particular kinases reacting and, in response, phosphorylating stress-responsive TFs to activate stress signaling pathways. In fact, the phosphorylation of these TFs enhances their affinity to DNA for DNA binding, their nuclear localization increases, or their transcriptional activities increase, thus transcriptionally upregulating the stress-responsive genes. On the converse, dephosphorylation of stress-responsive TFs reverses their activation, and plants have a reversible and rapid response by stress conditions (Bao-xiang *et al.*, 2023).

A second type of PTMs of stress-responsive TFs under stressful environmental condition is ubiquitination, whereby the TFs are conjugated with ubiquitin moieties. Some selected E3ubiquitin ligases ubiquitinate stress-responsive TFs under stress conditions at the expense of proteasomal degradation of these targets. In order to prevent persistent activation of stress-responsive genes, such targeted degradation of stress-responsive TFs prevents such an activation (Doroodian and Hua, 2021). Conversely, deubiquitinating enzymes that de-ubiquitinate these TFs lead to stabilized stress-responsive TFs and prolonged stress response. This additional control layer in post-translational regulation of plants' stress response involves addition and removal of ubiquitin to stress-responsive TFs through ubiquitination and de-ubiquitination, respectively, to allow the plants to respond quickly and dynamically to changes in the intensity and duration of the stress stimulus (Skelly, 2022).

Attachment of the ubiquitin like small modifier proteins (SUMOylation) to the target proteins is critical in the regulation of the gene expression in the plants and modulating plant stress response in the stress conditions. SUMOylation of plants stress-responsive TFs under environmental stress occurs and results in changes of their stability, subcellular localization and activity (Singh *et al.*, 2022). For example, a plant transcription factor allowing the expression of drought-responsive genes is DREB2A. Under drought, DREB2A is SUMOylated to promote its transcriptional activity and stability to trigger upregulation of drought-responsive genes and acquiring more drought tolerance in plants (Zhang and Xia, 2023). In contrast, deSUMOylation reverses these effects so that stress-responsive genes and stress response in plants are reversible and dynamic (Ghimire *et al.*, 2020). Another level of regulation of gene expression in plant at post-translational level is through the SUMOylation/deSUMOylation balance, which allows the plants to tune the intensity and duration of its stress response.

3.9 Biotechnological Applications of Gene Expression Regulation of Plants Under Stress Conditions

In plants under stress conditions, gene expression regulation is complex and multi-layered in terms of the significance to crop improvement and biotechnology. Using these molecular signaling pathways in order to develop crop plants with enhanced stress tolerance, scientists can exploit insights of how environmental stress influences and controls transcriptional, post-transcriptional, translational and post-translational processes of gene expression programs. For example, one such example is the overexpression of some signaling proteins and TFs that control plant stress response. For example, the upregulation of bZIP TFs AtbZIP60, AtbZIP17 and AtbZIP17 by induced genes has been shown to enhance stress tolerance of transgenic plants to drought, heat and pathogen attack.

In an analogous fashion, NAC TFs, AtNAC089 and AtNAC062, upregulated expression, also upregulated expression of hundreds of ER stress-responsive genes, enhancing plant survival and performance at the extreme end of environmental conditions. In the same way that the mechanisms of regulation can be used to increase stress tolerance at post-transcriptional level of gene expression, such as mRNA stability and translational level, the mechanisms of regulation can also be used in plants. Stress-resilient plants may upregulate the expression RBPs, which stabilize RNA responsible for stress-responsive mRNA and modulate the expression of protein kinases responsible for processing and localizing mRNA. Moreover, engineering more resolute pathways of protein quality control may under stress contribute to better homeostasis in plants; for instance, the upregulation of chaperones such as HSP 90 and HSP 70.

Another approach for improving stress tolerance in plants and developing resilient crops is to enhance ER stress by disrupting UPR pathways. These in turn will enable researchers and biotechnologists to develop new approaches to the cultivation of new varieties of stress-tolerant crops in the future.

3.10 Conclusion

In this chapter we indicated the intricate and multi-level regulatory mechanisms by which gene expression is modified in the plant under stress conditions. Plants have evolved sophisticated routes of regulation of their stress responses, from transcriptional regulation to post-translational regulation, in the form of protein folding and modification. They are of interest from the standpoint of understanding these regulatory pathways for innovation in crop breeding and the development of stress-resistant crop varieties in biotechnology. By exploiting such pathways as upregulation of TFs of genes in response to stress, RBPs, and protein modifications and folding pathways, biotechnologists can genetically engineer crops that can endure stress and are more adaptive to a changing and challenging environment and climate.

As the global population continues to increase, food security and agricultural sustainability need to be more recognized, which also requires an understanding of gene expression regulation in plants under stress conditions, to play a role in the contribution of the engineering of stress-resistant crop varieties. Scientific knowledge built upon research has to be translated into inventions that will revolutionize agricultural practices, leading to rich harvests of crops, food security and sustainability of the environment for the generations to come.

References

Ahmad, Y., Haider, S., Iqbal, J., Abbasi, B.A., Yaseen, T. and Mahmood, T. (2022) The mechanisms of genome editing technologies in crop plants. In: *Principles and Practices of OMICS and Genome Editing for Crop Improvement.* Springer International Publishing, Cham, pp. 295–313.

Ahmad, Y., Haider, S., Iqbal, J., Naseer, S., Attia, K.A., Mohammed, A.A. and Mahmood, T (2024) In-silico analysis and transformation of OsMYB48 transcription factor driven by CaMV35S promoter in model plant – Nicotiana tabacum L. conferring abiotic stress tolerance. *GM Crops and Food* 15(1), 130–149.

Ali, Md. S. and Baek, K. (2020) Jasmonic acid signaling pathway in response to abiotic stresses in plants. *International Journal of Molecular Sciences* 21(2), 621. DOI: 10.3390/ijms21020621.

Ali, S. and Tang, Y. (2025) Non-coding RNA-mediated regulation of DNA methylation: Insights into plant epigenetic mechanisms. *Journal of Plant Growth Regulation* 44(2), 373–388. DOI: 10.1007/s00344-024-11462-0.

Ali, I., Salah, K.B.H., Sher, H., Ali, H., Ullah, Z., Ali, A. and Mahmood, T. (2022) Drought stress enhances the efficiency of floral dip method of Agrobacterium-mediated transformation in Arabidopsis thaliana. *Brazilian Journal of Biology* 84, e259326.

Asghar, A., Ullah, Z., Ali, T., Iqbal, J., Abbasi, B.A., Ijaz, S. and Mahmood, T. (2024) Introduction to nc-RNA and their crucial role in biological regulation under environmental stress. In: *ncRNAs: Mediated Regulation: The Unarguable Biological Function to Environmental Constrains.* Springer Nature Switzerland, Cham, pp. 1–46.

Bao-xiang, W., Sun, Z., Liú, Y., Xu, B., Li, J. *et al.* (2023) A pervasive phosphorylation cascade modulation of plant transcription factors in response to abiotic stress. *Planta* 258(4), 73. DOI: 10.1007/s00425-023-04232-x.

Bobadilla, R. and Berr, A. (2016) Histone Methylation-A Cornerstone for Plant Responses to Environmental Stresses? In *Abiotic and biotic stress in plants – Recent advances and future perspectives.* DOI: 10.5772/61733

Candela-Ferre, J., Diego-Martin, B., Pérez-Alemany, J. and Gallego-Bartolomé, J. (2024) Mind the gap: Epigenetic regulation of chromatin accessibility in plants. *Plant Physiology* 194(4), 1998. DOI: 10.1093/plphys/kiae024.

Dinkar, V., Pandey, S., Kumar, A., Shiv, A., Lal, D. *et al.* (2024) Epigenetic regulations under plant stress: A cereals perspective. *Environmental and Experimental Botany* 220, 105688. DOI: 10.1016/j.envexpbot.2024.105688.

Doroodian, P. and Hua, Z. (2021) The Ubiquitin switch in plant stress response. *Plants* 10(2), 246. DOI: 10.3390/plants10020246.

Duarte-Conde, J. A., Sans-Coll, G. and Merchante, C. (2022) RNA-binding proteins and their role in translational regulation in plants. *Essays in Biochemistry* 66(2), 87. DOI: 10.1042/ebc20210069.

Fan, S., Zhang, Y., Zhu, S. and Shen, L. (2024). Plant RNA-binding proteins: Phase separation dynamics and functional mechanisms underlying plant development and stress responses. *Molecular Plant* 17(4), 531. DOI: 10.1016/j.molp.2024.02.016.

Fatima, K., Naqvi, F. and Younas, H. (2021) A review: Molecular chaperone-mediated folding, unfolding and disaggregation of expressed recombinant proteins. *Cell Biochemistry and Biophysics* 79(2), 153. DOI: 10.1007/s12013-021-00970-5.

Ghimire, S., Tang, X., Zhang, N., Liu, W. and Si, H. (2020) SUMO and SUMOylation in plant abiotic stress. *Plant Growth Regulation* 91(3), 317. DOI: 10.1007/s10725-020-00624-1.

Guo, Z., Dzinyela, R., Yang, L. and Hwarari, D. (2024) bZIP Transcription factors: Structure, modification, abiotic stress responses and application in plant improvement. *Plants* 13(15), 2058. DOI: 10.3390/plants13152058.

Haider, S., Iqbal, J., Naseer, S., Yaseen, T., Shaukat, M., Bibi, H. and Mahmood, T. (2021) Molecular mechanisms of plant tolerance to heat stress: Current landscape and future perspectives. *Plant Cell Reports* 40(12) 2247–2271.

Haider, S., Raza, A., Iqbal, J., Shaukat, M. and Mahmood, T. (2022) Analyzing the regulatory role of heat shock transcription factors in plant heat stress tolerance: A brief appraisal. *Molecular Biology Reports* 49(6), 5771–5785.

Han, D., Yu, Z., Lai, J. and Yang, C. (2022) Post-translational modification: a strategic response to high temperature in plants. *aBIOTECH* 3(1), 49. DOI: 10.1007/s42994-021-00067-w.

Hernandez-Garcia, C.M. and Finer, J.J. (2013) Identification and validation of promoters and cis-acting regulatory elements. *Plant Science* 217, 109. DOI: 10.1016/j.plantsci.2013.12.007.

Hewedy, O.A., Elsheery, N.I., Karkour, A.M., Elhamouly, N.A., Arafa, R.A. *et al.* (2023) Jasmonic acid regulates plant development and orchestrates stress response during tough times. *Environmental and Experimental Botany* 208, 105260. DOI: 10.1016/j.envexpbot.2023.105260.

Howell, S.H. (2021) Evolution of the unfolded protein response in plants. *Plant Cell and Environment* 44(8), 2625. DOI: 10.1111/pce.14063.

Hu, J., Cai, J., Xu, T. and Kang, H. (2022) Epitranscriptomic mRNA modifications governing plant stress responses: underlying mechanism and potential application. *Plant Biotechnology Journal* 20(12), 2245. DOI: 10.1111/pbi.13913.

Huang, J., Zhao, X., Bürger, M., Chory, J. and Wang, X. (2023) The role of ethylene in plant temperature stress response. *Trends in Plant Science* 28(7), 808. DOI: 10.1016/j.tplants.2023.03.001.

Hussain, Q., Asim, M., Zhang, R., Khan, R., Farooq, S. and Wu, J. (2021) Transcription factors interact with ABA through gene expression and signaling pathways to mitigate drought and salinity stress. *Biomolecules* 11(8), 1159. DOI: 10.3390/biom11081159.

Ijaz, S., Iqbal, J., Abbasi, B. A., Yaseen, T., Ullah, Z., Alqahtani, T., *et al.* (2024). nc-RNA: A Key Player in Stress Response Pathways, a Molecular Perspective. In *ncRNAs: Mediated Regulation: The Unarguable Biological Function to Environmental Constrains*. Springer Nature, Cham, pp. 139–163.

Jeyakumar, J.M.J., Ali, A., Wang, W. and Thiruvengadam, M. (2020) Characterizing the role of the miR156-SPL network in plant development and stress response. *Plants* 9(9), 1206. DOI: 10.3390/plants9091206.

Jiang, J., Zhu, H., Li, N., Batley, J. and Wang, Y. (2022) The miR393-target module regulates plant development and responses to biotic and abiotic stresses. *International Journal of Molecular Sciences* 23(16), 9477. DOI: 10.3390/ijms23169477.

Joo, H., Baek, W., Lim, C.W. and Lee, S.C. (2020) Post-translational modifications of bZIP transcription factors in abscisic acid signaling and drought responses. *Current Genomics* 22(1), 4. DOI: 10.2174/1389202921999201130112116

Kearly, A., Nelson, A.D.L., Skirycz, A. and Chodasiewicz, M. (2024) Composition and function of stress granules and P-bodies in plants. *Seminars in Cell and Developmental Biology* 156, 167. DOI: 10.1016/j.semcdb.2022.11.008.

Khan, N., Bano, A., Ali, S. and Babar, M.A. (2020) Crosstalk amongst phytohormones from planta and PGPR under biotic and abiotic stresses. *Plant Growth Regulation* 90(2), 189. DOI: 10.1007/s10725-020-00571-x.

Khan, F.S., Goher, F., Paulsmeyer, M., Hu, C.-G. and Zhang, J. (2023) Calcium (Ca^{2+}) sensors and MYC2 are crucial players during jasmonates-mediated abiotic stress tolerance in plants. *Plant Biology* 25(7), 1025. DOI: 10.1111/plb.13560.

Lata, C., Yadav, A. and Pras, M. (2011) Role of Plant Transcription Factors in Abiotic Stress Tolerance. DOI: 10.5772/23172.

Li, S., He, X.J., Gao, Y., Zhou, C., Chiang, V.L. and Li, W. (2021) Histone acetylation changes in plant response to drought stress. *Genes* 12(9), 1409. DOI: 10.3390/genes12091409.

Liu, Y., Lv, Y., An, W., Guo, M., Li, Y. *et al.* (2022a) Unfolded protein response in balancing plant growth and stress tolerance. *Frontiers in Plant Science* 13, 1019414. DOI: 10.3389/fpls.2022.1019414.

Liu, Y., Wang, J., B, L. and Xu, Z. (2022b). Dynamic regulation of DNA methylation and histone modifications in response to abiotic stresses in plants. *Journal of Integrative Plant Biology* 64(12), 2252. DOI: 10.1111/jipb.13368.

Lokdarshi, A. and Arnim, A. G. von. (2022) Review: Emerging roles of the signaling network of the protein kinase GCN2 in the plant stress response. *Plant Science* 320, 111280. DOI: 10.1016/j.plantsci.2022.111280.

Ma, L., Cheng, K., Li, J., Deng, Z., Zhang, C. and Zhu, H. (2021) Roles of plant glycine-rich RNA-binding proteins in development and stress responses. *International Journal of Molecular Sciences* 22(11), 5849. DOI: 10.3390/ijms22115849

Ma, Z., Hu, L. and Jiang, W. (2024) Understanding AP2/ERF transcription factor responses and tolerance to various abiotic stresses in plants: A comprehensive review. *International Journal of Molecular Sciences* 25(2), 893. DOI: 10.3390/ijms25020893.

Marques, D.N., Reis, S.P. dos and Souza, C.R.B. de. (2017) Plant NAC transcription factors responsive to abiotic stresses. *Plant Gene* 11, 170. DOI: 10.1016/j.plgene.2017.06.003.

Mayer, K. F., Schoof, H., Haecker, A., Lenhard, M., Jürgens, G., & Laux, T. (1998) Role of WUSCHEL in regulating stem cell fate in the Arabidopsis shoot meristem. *Cell* 95(6), 805-815.

Mizoi, J., Shinozaki, K. and Yamaguchi-Shinozaki, K. (2011) AP2/ERF family transcription factors in plant abiotic stress responses [Review of AP2/ERF family transcription factors in plant abiotic stress responses]. *Biochimica et Biophysica Acta (BBA) – Gene Regulatory Mechanisms* 1819(2), 86. DOI: 10.1016/j.bbagrm.2011.08.004.

Mizoi, J., Shinozaki, K., & Yamaguchi-Shinozaki, K. (2012). AP2/ERF family transcription factors in plant abiotic stress responses. *Biochimica et Biophysica Acta (BBA) – Gene Regulatory Mechanisms*, 1819(2), 86–96.

Mukhopadhyay, R., Boro, P., Karmakar, K., Pradhan, P., Chowdhury, R.S. *et al.* (2024) Advances in the understanding of heat shock proteins and their functions in reducing abiotic stress in plants. *Journal of Plant Biochemistry and Biotechnology* 33(4), 474–491. DOI: 10.1007/s13562-024-00895-z.

Nakashima, K., Yamaguchi-Shinozaki, K. and Shinozaki, K. (2014) The transcriptional regulatory network in the drought response and its crosstalk in abiotic stress responses including drought, cold, and heat. *Frontiers in Plant Science*, 5, 170. DOI: 10.3389/fpls.2014.00170.

Park, S., Kim, G.W., Kwon, S.H. and Lee, J. (2020) Broad domains of histone H3 lysine 4 trimethylation in transcriptional regulation and disease. *FEBS Journal* 287(14), 2891. DOI: 10.1111/febs.15219.

Read, A. M. and Schröder, M. (2021) The unfolded protein response: An overview. *Biology* 10(5), 384. DOI: 10.3390/biology10050384.

Reddy, D.S., Mathur, P.B. and Sharma, K.K. (2013) Regulatory Role of Transcription Factors in Abiotic Stress Responses in Plants, p. 555. DOI: 10.1002/9783527675265.ch21.

Rehman, M. and Tanti, B. (2020) Understanding epigenetic modifications in response to abiotic stresses in plants. *Biocatalysis and Agricultural Biotechnology* 27, 101673. DOI: 10.1016/j.bcab.2020.101673.

Ren, H., Zhang, S., Zhong, M., Hussian, J., Tang, Y., Liu, S. and Qi, G. (2023) Calcium signaling-mediated transcriptional reprogramming during abiotic stress response in plants. *Theoretical and Applied Genetics* 136(10), 210. DOI: 10.1007/s00122-023-04455-2.

Saini, A., Li, Y.-F., Jagadeeswaran, G. and Sunkar, R. (2012) Role of microRNAs in Plant Adaptation to Environmental Stresses. In *MicroRNAs in plant development and stress responses*, pp. 219–232. Berlin, Heidelberg: Springer Berlin Heidelberg.

Samtani, H., Sharma, A., Khurana, J.P. and Khurana, P. (2022) Thermosensing in plants: Deciphering the mechanisms involved in heat sensing and their role in thermoresponse and thermotolerance. *Environmental and Experimental Botany* 203, 105041. DOI: 10.1016/j.envexpbot.2022.105041.

Shaw, B.P. (2018) Salt stress tolerance in plants: The role of miRNAs. *Advances in Plants and Agriculture Research* 8(6), 411–415. DOI: 10.15406/apar.2018.08.00360.

Shi, L., Cui, X. and Shen, Y. (2023) The roles of histone methylation in the regulation of abiotic stress responses in plants. *Plant Stress* 11, 100303. DOI: 10.1016/j.stress.2023.100303.

Singh, M., Singh, A., Yadav, N.R. and Yadav, D.K. (2022) Current perspectives of ubiquitination and SUMOylation in abiotic stress tolerance in plants. *Frontiers in Plant Science* 13. DOI: 10.3389/fpls.2022.993194.

Skelly, M. (2022) The emerging roles of deubiquitinases in plant proteostasis. *Essays in Biochemistry* 66(2), 147. DOI: 10.1042/ebc20210060.

Son, S. and Park, S.R. (2023) Plant translational reprogramming for stress resilience [Review of Plant translational reprogramming for stress resilience]. *Frontiers in Plant Science* 14, 1151587. DOI: 10.3389/fpls.2023.1151587.

Sultana, R., Imam, Z., Kumar, R.R., Banu, V.S., Nahakpam, S. *et al.* (2025) Signaling and defence mechanism of jasmonic and salicylic acid response in pulse crops: Role of WRKY transcription factors in stress response. *Journal of Plant Growth Regulation* 44(1) 5–21. DOI: 10.1007/s00344-023-11203-9.

Tan, H., Li, B. and Guo, H. (2020) The diversity of post-transcriptional gene silencing mediated by small silencing RNAs in plants. *Essays in Biochemistry* 64(6), 919. DOI: 10.1042/ebc20200006.

Tang, Q., Wei, S., Zheng, X., Tu, P. and Tao, F. (2024) APETALA2/ethylene-responsive factors in higher plant and their roles in regulation of plant stress response. *Critical Reviews in Biotechnology* 44(8), 1533. DOI: 10.1080/07388551.2023.2299769.

Todaka, D., Takahashi, F., Yamaguchi-Shinozaki, K., and Shinozaki, K. (2019). ABA-responsive gene expression in response to drought stress: cellular regulation and long-distance signaling. In: *Advances in Botanical Research* 92, pp. 83–113. Academic Press: London, UK.

Wang, L., Li, C.-H., Liu, Y., He, L., Li, P., Guo, J., Zhang, S., Zhao, B. and Guo, Y. (2024). Plant responses to abiotic stress regulated by histone acetylation. *Frontiers in Plant Science* 15, 1404977. DOI: 10.3389/fpls.2024.1404977.

Wu, H.-Y. L., Jen, J.J. and Hsu, P.Y. (2023) What, where, and how: Regulation of translation and the translational landscape in plants. *The Plant Cell* 36(5), 1540. DOI: 10.1093/plcell/koad197.

Xu, K., Wu, N., Yao, W., Li, X., Zhou, Y. and Li, H. (2021) The biological function and roles in phytohormone signaling of the F-box protein in plants. *Agronomy* 11(11), 2360. DOI: 10.3390/agronomy11112360.

Yadav, S., Meena, S., Kalwan, G. and Jain, P.K. (2024) DNA methylation: An emerging paradigm of gene regulation under drought stress in plants. *Molecular Biology Reports* 51(1), 311. DOI: 10.1007/s11033-024-09243-9.

Yamaguchi-Shinozaki, K. and Shinozaki, K. (2005) Organization of cis-acting regulatory elements in osmotic- and cold-stress-responsive promoters. *Trends in Plant Science* 10(2), 88. DOI: 10.1016/j.tplants.2004.12.012.

Yu, C., Cho, Y., Sharma, O. and Kanehara, K. (2021) What's unique? The unfolded protein response in plants. *Journal of Experimental Botany* 73(5), 1268. DOI: 10.1093/jxb/erab513.

Zhang, M. and Zhang, S. (2022) Mitogen-activated protein kinase cascades in plant signaling. *Journal of Integrative Plant Biology* 64(2), 301. DOI: 10.1111/jipb.13215.

Zhang, Y. and Xia, P. (2023) The DREB transcription factor, a biomacromolecule, responds to abiotic stress by regulating the expression of stress-related genes. *International Journal of Biological Macromolecules* 243, 125231. DOI: 10.1016/j.ijbiomac.2023.125231.

Zhao, B., Liu, Q., Wang, B. and Yuan, F. (2021) Roles of phytohormones and their signaling pathways in leaf development and stress responses. *Journal of Agricultural and Food Chemistry* 69(12), 3566. DOI: 10.1021/acs.jafc.0c07908.

Zhuang, J., Zhang, J., Hou, X., Wang, F. and Xiong, A. (2014) Transcriptomic, proteomic, metabolomic and functional genomic approaches for the study of abiotic stress in vegetable 3.crops. *Critical Reviews in Plant Sciences* 33(2), 225. DOI: 10.1080/07352689.2014.870420.

4 Metabolic Adjustment and Stress Tolerance in Plants

Muhammad Awais Ibrahim[1], Muhammad Muzamil Sattar[1], Rifat Bashir[1], Sadia Riaz[1], Javed Iqbal[2*], Banzeer Ahsan Abbasi[3], Farishta Zarshan[2] and Madiha Imtiaz[2]

[1]Department of Plant Sciences, Faculty of Biological Sciences, Quaid-i-Azam University, Islamabad 45320, Pakistan; [2]Department of Botany, Bacha Khan University, Charsadda 24420, Khyber Pakhtunkhwa, Pakistan; [3]Department of Botany, Rawalpindi Women University, Rawalpindi 46300, Pakistan

Abstract

The intricate systems of metabolic adaptation and stress resiliency in plants, animals, and microorganisms are studied in this chapter. It shows how organisms adapt to abiotic circumstances like drought, salinity, pH, temperature extremes, and oxidative damage, combined with biotic pressures such interspecies competition, herbivory, and disease attacks. Essential mechanisms include hormone modulation, metabolic reprogramming, and epigenetic alterations favoring survival in hazardous situations. This chapter discusses stress-resistance, the necessity of secondary metabolites, stress-related proteins, and energy management. It looks at usage in biotechnology and the agriculture sector, stressing resilient-to-stress crops and genetically altered animals to boost production. Future views address the concerns of climate change, the application of systems biology, ethical ramifications, and the requirement of multidisciplinary study. This detailed research stresses how stress response systems may be employed to assure industrial progress, environmental sustainability, and food security.

Keywords: plant stress responses, abiotic stress, biotic stress, metabolic adjustment, reactive oxygen species

4.1 Introduction

A key biological process that helps the adaptation of metabolic pathways in response to environmental cues, metabolic adjustments activate life and optimal performance. These adaptations comprise a variety of metabolic pathways, regulatory networks, and cellular responses that allow animals to adjust and have homeostasis in the face of changing environments (Fraire-Velazquez and Emmanuel, 2013). Major metabolic adjustments include changes in the synthesis of primary and secondary metabolites, energy metabolism, and the biosynthesis of protective compounds alongside osmolytes (Koza *et al.*, 2022). Stress tolerance, on the other hand, is the organic or learned level of a living thing to resist and adapt to stressful conditions like high-temperature, lack of water, malnutrition and exposure to toxins (Egamberdieva *et al.*, 2017). This tolerance

* Email: javed89qau@gmail.com

 Molecular Biology of Plant Stress Responses (Muhammad Waseem and Pingwu Liu eds)
DOI: 10.1079/9781800628823.0004

is managed through various mechanisms, including genetic, physiological, and biochemical responses that together lead to an organism's resistance (Hirt, 2012).

Recent developments in metabolic reprogramming and stress tolerance at the level of the topic attract particular attention from various scientific disciplines. In higher plants it comprise of alterations in photosynthetic productivity and ion homeostasis, together with hormonal signaling pathways (Suprasanna *et al.*, 2015). In lower vertebrates, metabolic modification facilitates a shift in adaptive responses owing to the control of hormones and the modulation of energy expenditure (Sakamoto and Murata, 2000). Aggressive environments necessitate fast metabolic reprogramming in microbial systems and are accomplished by changing membrane lipid composition, food intake mechanisms, and the activity of stress-responsive enzymes (Etesami and Beattie, 2017).

4.2 Importance in Biological Systems

Biological organisms undergo specific metabolic changes and develop certain stress tolerances that ensure their survival and functionality. In plants, the closing of these systems is essential for dealing with environmental stresses including drought, salt, and temperature changes. Plants accomplish this through metabolic reprogramming processes that involve production of protective osmolytes like proline and glycine betaine, along with activation of antioxidant defense mechanisms (Iqbal *et al.*, 2015). Moreover, plants use interactions with root bacteria to gain stress tolerance, which facilitates nutrient uptake and subsequent plant growth in unfavorable conditions (Kumar and Verma, 2017).

Even animals primarily rely on metabolic changes to control internal operations in response to external stresses like extreme temperatures, oxygen deprivation, and oxidative stress. Coordination of metabolic changes that enable energy mobilization and a degree of physiological flexibility is done by hormonal systems, specifically those involving cortisol and adrenaline (Tiwari and Lata, 2018). Resilience to long-lasting stress is supported by the metabolic flexibility, especially by the alternating dependency on glucose and lipid metabolism (Holmberg, 1996).

4.3 Role of Metabolic Adjustment in Plants

To survive and remain productive, plants have to undergo rapid alterations in their metabolism due to losing out to environmental challenges. Metabolic shifts in plants include spending resources on producing secondary metabolites, stress-responsive proteins, and antioxidant compounds to counter oxidative damage and maintain cellular homeostasis (Shiade *et al.*, 2024). Osmotic stress is normally countered through the accumulation of osmolytes, such as proline and glycine betaine (Iqbal *et al.*, 2015). Other than that, plants form beneficial relationships with microorganisms such as rhizobacteria and mycorrhizal fungi who aid in nutrient acquisition and relieve stress (Kumar and Verma, 2017).

4.4 Role of Metabolic Adjustment in Animals

Wild animals use a flexible metabolism to cope with a myriad of extreme conditions, including harsh temperatures, food shortage, and oxidative damage. In the case of increased environmental stress, the expression of certain protective proteins including heat shock proteins (HSPs) and antioxidant enzymes undergo change (Tiwari and Lata, 2018). Stress also affects behavior, which leads to regulated metabolic changes in which cortisol and adrenaline act as potent moderators (Holmberg, 1996). In addition, energy expenditure increases in order to deal with the stress, and active metabolism shifts from anabolic to catabolic processes (Sakamoto and Murata, 2000).

4.5 Role of Metabolic Adjustment in Microbes

Microbial metabolic adjustments are crucial for their survival in extreme environments such as high salinity, temperature changes, and food limitation.

Bacteria and fungi demonstrate incredible metabolic flexibility, enabling them to modify gene expression, enzyme activity, and cellular functions in response to environmental inputs (Lata *et al.*, 2018).

Microbial biofilms and spore formation are significant techniques used to tolerate protracted stress conditions (Corradi and Bonfante, 2012).

Additionally, metabolic associations with the host microbes contribute to their stress tolerance and ecological adaptation (Hou *et al.*, 2021).

4.6 Overview of Key Metabolic Pathways Involved

Metabolic adjustment encompasses numerous important mechanisms that enable organisms to adapt efficiently to stress. The key procedures include amino acid biosynthesis, lipid metabolism, glucose metabolism, and secondary metabolite generation. The carbohydrates metabolism, via glycolysis and the tricarboxylic acid (TCA) cycle, guarantees energy supply and maintains cellular redox homeostasis (Suprasanna *et al.*, 2015).

Lipid metabolism plays a vital role in membrane integrity and signaling under stress circumstances, with lipid peroxidation and fatty acid desaturation being significant processes (Iqbal *et al.*, 2015).

Amino acid production supports osmoprotection and cellular repair mechanisms, with proline and glutamine being particularly crucial stress-associated molecules (Holmberg, 1996). Secondary metabolic harvests such as flavonoids and terpenoids function as antioxidants and transmission molecules, which support improve oxidative damage and influence cellular operations (Kumar and Verma, 2017).

4.7 Relevance to Environmental and Physiological Stressors

Living things need to change their metabolic pathways to survive tough conditions. This ability is crucial when they face challenges like drought, high salt levels, extreme temperatures, or toxic heavy metals. To overcome these challenges, organisms must adapt their metabolism. This adaptation helps them stay alive and continue to grow or produce (Tiwari and Lata, 2018). For example, plants strengthen antioxidant defenses and osmoprotectant accumulation to cope with water stress, whereas bacteria adapt their membrane composition to exist in saline conditions (Hou *et al.*, 2021).

Physiological stresses, such as pathogen assaults, aging, and metabolic diseases, also demand metabolic adaptation to sustain cellular integrity and function (Corradi and Bonfante, 2012). Animals, for instance, rely on stress hormones to manage energy mobilization and immunological responses under physiological stress (Sakamoto and Murata, 2000).

Understanding the intricate link between metabolic pathways and stress responses is vital for devising measuresto promote stress tolerance in agriculture, medicine, and biotechnology (Lata *et al.*, 2018).

Enhanced stress tolerance can lead to greater crop yields, better human health management, and more efficient microbial applications in industrial processes (Koza *et al.*, 2022).

4.8 Types of Stress and Their Impact on Metabolism

4.8.1 Abiotic stress

Water scarcity hampers plant activity. A lower water intake reduces cell growth, limits photosynthesis, and leads to reactive oxygen species (ROS) formation. Plants switch on genes for drought response in reaction to dry circumstances. These genes increase the production of osmolytes, create antioxidant enzymes, and start stress hormone signals (Suprasanna *et al.*, 2015). Through the accumulation of osmoprotectants like proline and glycine betaine, and sugars, the cells retain stability and hydration (Iqbal *et al.*, 2015). The stress response systems of plants receive allocated resources, while growth-related processes decrease in order to conserve metabolic energy (Hou *et al.*, 2021). The regulation of stomatal closure through abscisic acid (ABA) production functions as a main mechanism for water loss control while improving water utilization (Kumar and Verma, 2017). Plants develop long-term stress tolerance through drought stress-induced changes in both membrane cell integrity and lipid peroxidation and metabolic processes (Tiwari and Lata, 2018).

Plants develop three main difficulties from salt stress, which are ion toxicity, osmotic stress and nutritional imbalances. The life processes of plants undergo modifications due to these problems during photosynthesis, together with protein synthesis and enzymatic operations. The plant survival mechanisms assist them in coping with salt-related stress. Plants generate types of protective substances known as osmoprotectants such as proline, polyamines and sugars. The maintenance of osmotic regulation alongside cellular health relies on osmoprotectants for plants (Koza *et al.*, 2022).

Ion transport paths, such as the salt-overly-sensitive (SOS) pathway, play a significant role in maintaining ion homeostasis by emission of excess sodium and keeping potassium inside cells (Egamberdieva *et al.*, 2017). Salinity encouraged oxidative stress leads to an overproduction of ROS, enhancing the activation of antioxidant defense mechanisms containing enzymes such as superoxide dismutase (SOD), catalase (CAT), and peroxidase (POD) (Fraire-Velázquez and Balderas-Hernández, 2013).

In addition to metabolic modifications, the remodeling of cell wall composition is required to adjust the structural rigidity and prevent ion loss (Hirt, 2012). Hormonal signaling pathways, notably those mediated by ABA and ethylene (ET), are also implicated in stress responses and strategies to cope with stress (Holmberg, 1996).

Extreme temperatures, whether high or low, stress metabolic processes notably by means of changes in general cellular activity, membrane flexibility, and enzyme activity. More ROS formation brought about by heat stress leads in time to oxidative damage on DNA, lipids, and proteins (Dutta *et al.*, 2018). Metabolic reactions to heat stress involve activation of HSPs, synthesis of relevant solutes including proline, and increased antioxidant protection (Zebral *et al.*, 2020). On the other hand, cold stress leads to membrane rigidification and interruption of metabolism, thereby requiring higher production of unsaturated fatty acids and cryoprotective molecules like sugars and polyols (Eutrópio *et al.*, 2015).

Imbalance between the antioxidant defense mechanism and ROS generation leads to oxidative stress. In stress scenarios, excess ROS production destroys cellular components and changes metabolic pathways (Frýzová *et al.*, 2017). Metabolic reactions in response to oxidative stress include initiation of enzymatic antioxidants SOD, CAT, and POD, as well as the non-enzymatic antioxidants ABA and glutathione (GSH) (Seneviratne *et al.*, 2017).

Plants and bacteria have a way to cope oxidative stress by increased secondary metabolites, as such phenolics and flavonoids. These compounds help them to live in harsh conditions (Dutta *et al.*, 2018).

When the heavy metals enter the plant system, they disturb the biochemical processes and, as a result, disturb the metabolic balance. Besides this, it can also cause changes in their nutritional state (Dubey *et al.*, 2018).

Activation of metal chelating compounds, including phytochelatins and metallothioneins, is an important major metabolic reaction in response to heavy metal pressure that binds the metal ions and detoxifies them (Sun *et al.*, 2014). Plants also enhance their antioxidant defenses in order to deal with ROS generated by heavy metal exposure (Ghori *et al.*, 2019). Additionally, changes in nutritional absorption and transport paths are also quite important to reduce metal toxicity (Hoque *et al.*, 2021). All these abiotic stress responses are shown in Fig. 4.1.

4.8.2 Biotic stress

Biotic stress caused by pathogens is an outbreak by fungi, viruses, bacteria, or other microbial agents that attack the plants. The key features of such infections are deprivation of nutrients, death of cellular structures, or otherwise control being taken of metabolic pathways.

Though it is not obvious, plants have extremely sophisticated defense systems, such as pathogen detection by pattern recognition receptors (PRRs) and effector-triggered immunity (ETI). These in turn result in the formation of ROS, phytoalexins, and pathogenesis-related proteins, which together constrain pathogen transmission (Rizvi *et al.*, 2019).

We would say that a plants way of stopping infections is through hypersensitive response (HR), which causes some cells to die in a certain area (Dangl and Jones, 2001). In particular,

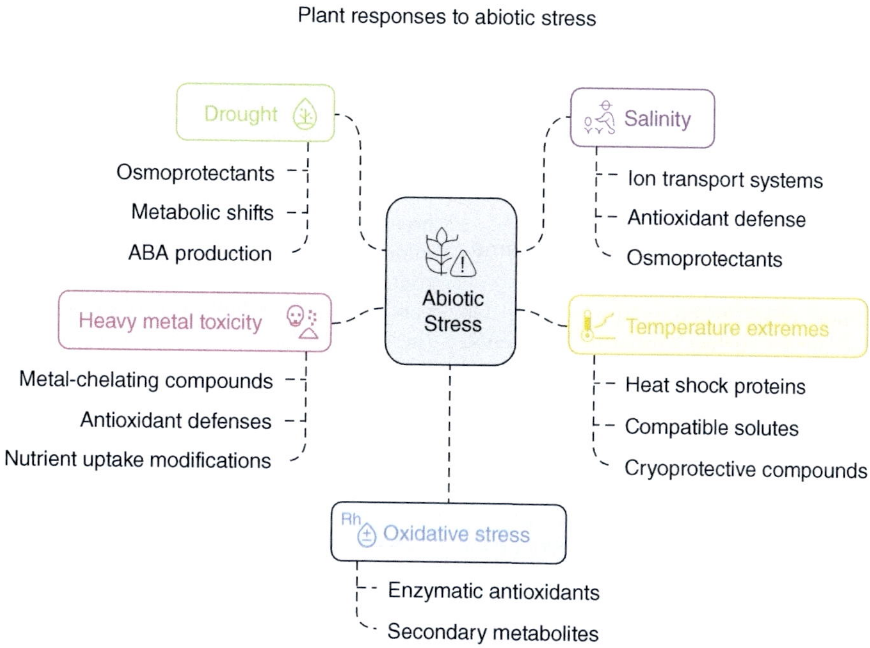

Fig. 4.1. Plant responses to abiotic stress. (Figure author's own.)

Pseudomonas syringae can impair plant defenses. It does this by making chemicals known as effectors, which weaken a plant's defenses. A compromised or hyperactive immune system facilitates the rapid spread of infection within the plant. Studies have demonstrated that secondary metabolites, particularly flavonoids and phenylpropanoids, play a critical role in reinforcing plant cell walls. These compounds contribute to the structural integrity of the cell wall, thereby limiting pathogen invasion and restricting the spread of infection. Additionally, they exhibit antimicrobial properties that inhibit the entry of microbes into plant cells (López-Gresa *et al.*, 2009). The systemic-acquired resistance (SAR) signaling molecules also provide protection to the unfit tissues by salicylic acid (SA) (Kushalap and Gunnaiah, 2013).

In many cases the propensity to infection can be determined by balance between host metabolism and pathogen virus factors (Azim, 2019). The sophisticated dynamics of plant to pathogen interaction has been advanced by progress in transcriptomics and metabolomics, offering attractive chances of genetic engineering of pathogen resistant crops (Pérez-Buno *et al.*, 2019).

The stress caused by feeding insects and vertebrates is known as herbivory-induced stress and ends in physical harm, followed by cascade of physiological disorders. There is an arsenal of constitutive and inducible defenses against herbivory for plants to use. Structural barriers such as lignified cell wall, trichomes, and cuticles constitute constitutive defenses that resist herbivores mechanically (Glombitza *et al.*, 2004).

Inducible defenses, which are triggered upon receipt of information identifying the presence of an herbivore, rely on the synthesis of secondary metabolites, including alkaloids, terpenoids, and phenolics providing both the modes of action of herbivore defense that induce physiological tolerance (reflexion and anti-digestive) and toxicity. These defenses include jasmonic acid (JA) signaling pathways, which are important for influencing these responses (Zhou *et al.*, 2023). This also stimulates the formation of volatile organic compounds (VOCs), which then attract natural predators of herbivores and thus form an indirect protective mechanism (Azim, 2019). In addition, herbivore saliva or oral secretions containing effectors may decrease plants defenses (Pérez-Bueno *et al.*, 2019). For example, certain enzymes secreted by caterpillars can block JA-mediated pathways, as discussed by Kushalappa and Gunnaiah (2013).

These interactions underscore the evolutionary arms race between plants and herbivores (Fig. 4.2).

4.8.3 Competition with other organisms

Competition arises when plants struggle for limited resources, such as light, water, nutrients, and space, with nearby plants or bacteria. This stress has a profound impact on plant growth and fitness. Aboveground, competition for light causes shadow avoidance responses, defined by elongation growth and changed leaf shape, mediated by phytochrome signaling (López-Gresa *et al.*, 2009). Belowground, root competition for nutrients such as nitrogen and phosphorus produces alterations in root architecture, including greater lateral root development (Glombitza *et al.*, 2004).

Allelopathy, whereby plants emit bioactive chemicals to hamper the development of rivals, further displays competitive strategies (Pérez-Bueno *et al.*, 2019). For example, phenolic chemicals released by some plants inhibit nearby root systems (Azim, 2019).

In microbial competition, beneficial rhizobacteria may outcompete pathogens by creating antimicrobial compounds and improving nutrition availability (Kushalappa and Gunnaiah, 2013). However, resource scarcity typically exacerbates trade-offs between competitive skill and stress tolerance (Zhou *et al.*, 2023). Recent breakthroughs in ecological genomics offer information on the genetic basis of competitive traits, allowing the design of agricultural cultivars optimized for high-density sowing without sacrificing yield.

4.9 Mechanisms of Metabolic Adjustment Under Stress

4.9.1 Primary metabolism alterations

One of the key stress routes is the breakdown of carbohydrates as a key moderator in energy balance and signaling. Signalling by sugars is also dramatically modified under stress conditions such as dry stress or pathogen infection. For example, trehalose-6-phosphate acts as a signalling compound which control sugar absorption and plant growth in response to external factors (Rizvi *et al.*, 2019). Plants resist energy-limited

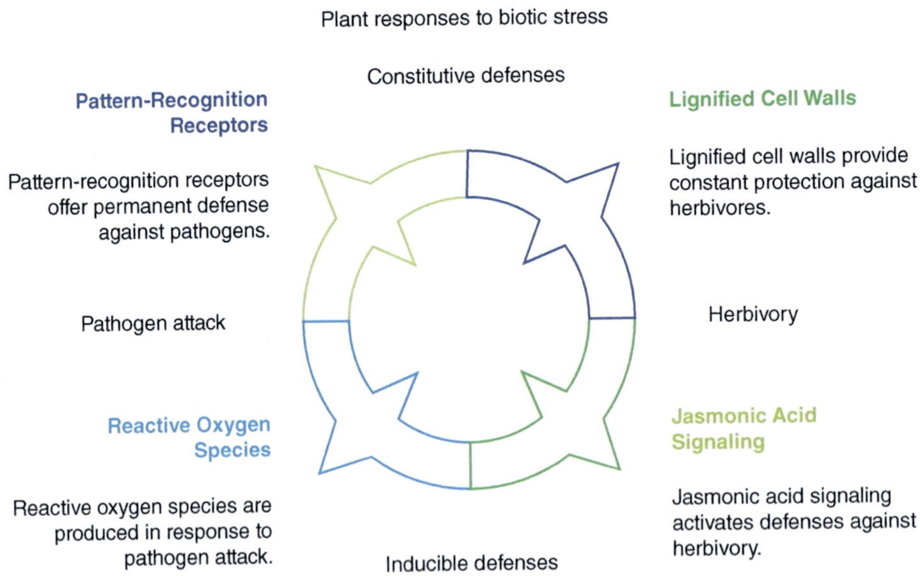

Fig. 4.2. Plant responses to biotic stress. (Figure author's own.)

stress due to the breakdown of starch to increase supply of glucose for glycolysis and osmolyte synthesis. This process is performed by enzymes such as amylase and glucan phosphorylase (Kushalappa and Gunnaiah, 2013). Furthermore, sucrose is also a signaling molecule that leads to the regulation of stress-respondent genes through the pathway (Zhou *et al.*, 2023). In general, stress-inspired carbohydrate redistribution usually remains enough to guarantee cellular energy balance (Pérez-Bueno *et al.*, 2019). In addition, Chinese transmission interacts with hormonal channels including ABA to increase stomatal closure and osmotic adjustment (Azim, 2019). A combination of these networks reveals the complexity of glucose metabolism under stress.

The stress conditions a large extent of protein metabolism, contributing to a quick rate of protein turnover and the stimulation of protein quality control systems. Proteolysis, through the ubiquitin–proteasome system (UPS), facilitates the elimination of damaged or misfolded proteins, hence maintaining cellular homeostasis (Dangle and Jones, 2001). E3 ubiquitin ligases facilitate selective ubiquitylation in proteostasis, marking damaged proteins for destruction. Molecular chaperones, especially HSPs, facilitate proper folding and inhibit aggregation during thermal and oxidative stress. Stress-induced proteolysis reclaims amino acids for energy generation and the synthesis of secondary metabolites. Proline increases during stress, functioning as an osmoprotectant and ROS scavenger. The unfolded protein response (UPR) in the endoplasmic reticulum (ER) is activated during stress to reestablish proteostasis. Collectively, these pathways are crucial in stress-resilience and recuperation.

The lipid plays an important role in signaling to preserve the integrity of the metabolic membrane and under stress conditions. Stressful modifications in lipid structure, such as an increase in unsaturated fatty acids, promoting membrane liquidity and stability (Kushalappa and Gunnaiah, 2013). Phospholipase enzyme membranes accelerate the rupture of the lipids, releasing signaling molecules such as phosphatidic acid (PA), which replaces stress-posting routes (Zhou *et al.*, 2023). Additionally, JA derived from linolenic acid is an important lipid-based hormone that is implicated in vegetarian and pathogen protection (Rizvi *et al.*, 2019).

Under oxidative stress, lipid peroxidation leads to the formation of reactive aldehydes, which serve as secondary messengers in stress signaling (Pérez-Bueno *et al.*, 2019). In addition, glycolipids and sterols contribute to membrane integrity under cold or dehydration stress (Azim, 2019). The dynamic nature of lipid metabolism indicates the stress reactions of the plant and its importance in the adaptation mechanism (Fig. 4.3).

4.9.2 Secondary metabolism adjustments

The secondary metabolic process plays an important role in reducing stress-related damage through the formation of antioxidants such as flavonoids, phenolics, and carotenoids. Flavonoids, especially quercetin and kaempferol, serve as ROS scavengers and protect cells and tissues from oxidative stress-inspired damage (Balan *et al.*, 2018). Phenolic chemicals generated through the phenylpropanoid passage, strengthen cell walls and act as antimicrobial agents, reducing pathogen growth (Hasham *et al.*, 2019). Carotenoids, such as beta-carotene, extinguish not only singlets, but also act as a pioneer for biological notes such as ABA, which affects stress reactions (Lehmann *et al.*, 2009). The activation of these antioxidant routes displays their significant participation in maintaining redox homeostasis under stress conditions (Jozefczuk *et al.*, 2010).

Stress is associated with metabolites such as proline and glycine betaine, which are important for osmotic improvement and protection against stress. Proline works as an osmoprotectant, which supports proteins and membranes neutralizing ROS (Kannojia *et al.*, 2017). Its synthesis is controlled by enzyme pyoline-5-carboxylate synthetase (P5Cs), which is extended during drought and salt stress (Iqbal *et al.*, 2015). Glycine betaine, another suitable soluble, strengthens the clever structures of enzymes and protects photo systems under environmental stress (Fraira-Welázquez and Balderas-Herndez, 2013). These metabolites also improve the activity of antioxidant enzymes, which has a sympathetic effect on stress reduction (Borges *et al.*, 2014).

Hormonal control is important for stress reactions, with ABA, ET and SA. The ABA accumulates during the stress of dry conditions and high salinity, increases the stomatal closure and

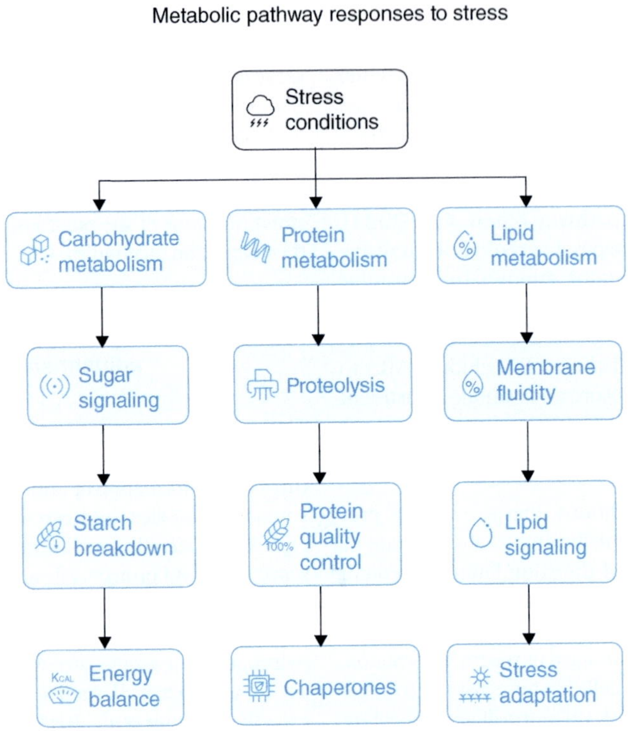

Fig. 4.3. Primary metabolic pathway response to stress. (Figure author's own.)

triggers stress-exquisite genes (Bhat *et al.*, 2010). ET controls senses and stress-inspired regulated cell death, while also interacts with JA in vegetarian protection (Pereira *et al.*, 2018). SA, on the other hand, is important for SAR for infection, increases the expression of pathogenesis-related proteins (Detrich *et al.*, 2011). These hormones often display cross stocks, fine-tuning stress reactions, based on stress (Borges *et al.*, 2014). Understanding hormonal signaling routes gives a platform to design stress-tolerant crops.

4.9.3 Energy management strategies

Under stress, managing energy rotates when balanced at ATP synthesis and demands to maintain cellular activity by preserving resources. During stress, plants typically prevent energy-intensive processes, such as cell division and growth, in favor of activities related to survival (Lehmann *et al.*, 2009). Glycolysis and the oxidative pentose phosphate pathway (PPP) generate ATP and reducing equivalents such as NADPH, which are important for stress-related biosynthesis and detoxification (Jozefczuk *et al.*, 2010). Under low oxygen conditions, fermentation activities, such as ethanol synthesis, can be started to create anaerobic (Pereira *et al.*, 2018).

Sucrose-non-fermenting-1-related protein kinase-1 (SnRK1) plays an important role in regulating metabolic flow by blocking the ATP-ubhog processes and activating catabolic reactions (Dietrich *et al.*, 2011). Additionally, the ATP generation in mitochondria and chloroplasts can be properly tuned to match under low photosynthetic rates.

Mitochondria play an important role in stress adaptation by changing energy production and redox balance. Under stress, there is often a regrouping of mitochondrial electron transport chain (ETC) activity to limit the formation of ROS, while improving ATP synthesis (Bhat *et al.*, 2010).

Alternative oxidases (AOXs) and uncoupling proteins (UCPs) are activated to reduce ETC over-reduction, decreasing excessive ROS build-up (Fraire-Velázquez and Balderas-Hernández, 2013). Stress also stimulates the expression of genes associated with the stress encoding components of the mitochondria and the stress encoding of the ETC and the TCA cycle (Hashem *et al.*, 2019).

Moreover, mitochondria link closely with other organelles, like chloroplasts and peroxisomes, to control energy metabolism and ROS detoxification (Borges *et al.*, 2014). Mitochondrial membrane ability and dynamic ups and downs in respiratory rate reflect the plant's ability to effectively react to environmental challenges (Iqbal *et al.*, 2015). These adjustments underline the adaptive flexibility of mitochondrial metabolism in maintaining energy and redox balance during stress.

4.10 Molecular and Genetic Regulation of Stress Tolerance

Transcription factors (TFs) perform an important function in regulating gene expression under stress conditions, allowing plants to adjust environmental restrictions. Dehydration-ex-element binding (DREB) is important for family abiotic stress reactions. DREB proteins regulate stress-inducible genes and enhance tolerance to drought, salinity and cold stress (Bhat *et al.*, 2010).

Similarly, WRKY transcription controls both biotic and abiotic stress reactions by regulating pathogen-related genes and ROS-scavenging routes (Table 4.1). For example, WRKY70 combines SA and JA signals to balance the plants defense mechanisms (Borges *et al.*, 2014).

MYB TF is important in secondary metabolic regulation and ROS signaling, assisting in the synthesis of anthocyanin and flavonoids under stress (Fraire-Velázquez and Balderas-Hernández, 2013). These TF families often operate co -interactive, forming complex regulatory systems that increase stress reactions and maintain life. In addition, TF cross-regulations are identified in the interaction of NAC, DREB, and bZIP families, which act on shared target genes to promote drought and heat tolerance (Jozefczuk *et al.*, 2010). The TF further increases the TF mediation stress reactions by maintaining epigenetic amendments, such as histone acetylation and DNA methylation, and transcriptional memory (Dietrich *et al.*, 2011).

Signal transduction paths serve as important condensers for stress detection and reaction. The mitogen-active protein is a protected mechanism from which mitogen-activated protein kinase (MAPK) cascade transfers external stress signals to intracellular reactions. MAPK modules regulate many stress-existed genes involved in development, apoptosis and defense. For example, the activation of MAPK3 and MAPK6 is related to the expression of the pathogen-related proteins after ROS production and pathogen infection (Lehmann *et al.*, 2009). MAPKs also modulate hormonal pathways—for example, ET and ABA signalling—depending on the stress context (Jian-peng *et al.*, 2019).

Calcium signaling also plays an important component in stress optimization. Calcium ions (Ca^{2+}) act as secondary messengers, activating calcium-binding proteins such as calmodulins and the CBL-CIPK complex, which modify ion transport and gene expression (Kannojia *et al.*, 2017). Calcium-dependent protein kinases (CDPKs) mediate calcium signals in changing metabolic routes and defenses. Interestingly, calcium waves are rapidly spread throughout the plant tissues increasing systemic communication under abiotic stress reactions like drought and salt (Glombitza *et al.*, 2004). Cross stock further improves the specialty and sensibility of cross stock stress response between MAPK and calcium courses, permitting plants to guarantee that a compound variety of abiotic or biotic stresses must be replied to (Hashem *et al.*, 2019). New components of these routes, including recent MAPK phosphate and calcium binding proteins regarding the ability to manufacture stress flexible crops, are advancing in molecular biology (Table 4.1) (Zhou *et al.*, 2023).

4.11 Epigenetic Modifications and Stress-responsive Gene Expression

4.11.1 Epigenetic modifications

Epigenetic modifications play an important role in changing plant reactions to environmental stress by affecting gene expression without

Table 4.1. Key transcription factors in stress tolerance. (Table author's own.)

Transcription factor family	Stress type targeted	Mechanism	Example organisms
DREB	Drought, salinity, cold	Enhances stress-inducible genes	*Arabidopsis*
WRKY	Biotic, abiotic stress	Modulates pathogen-related genes	Rice, *Arabidopsis*
MYB	Secondary metabolism	Regulates flavonoid synthesis	Maize, *Arabidopsis*
NAC	Drought, salinity	Mediates osmoprotectant accumulation	Wheat

changing the underlying DNA sequence. DNA methylation, a methyl group, in addition to the locations of cytosine, generally reduces gene expression and is dynamically modified in stress conditions. For example, dry stress generates changes in the methylenation pattern, affecting the gene involved in the efficiency of water use (Balan *et al.*, 2018). Histone changes, which include acetylation, methylation, and phosphorylation, contribute to stress tolerance by modifying chromatin structure and access. Stress-inspired histone acetylation, intermediate by histone acetyltransferase (HATS) promotes the expression of stress-respondent genes (Lehmann *et al.*, 2009). Conversely, histone deacetylation, mediated by histone deacetylases (HDAC) suppresses additional gene activity, preserves energy under extended stress (Pereira *et al.*, 2018). DNA in the management of emerging research epigenetic memory underscores the link between methylation and histone changes, which store the "stress memory" and react more strongly to the recurrent tension (Borges *et al.*, 2014). Identifying such passages increases the ability to produce crops with better stress tolerance using methods of epigenetic changes.

4.11.2 Stress-responsive gene expression networks

Stress-respondent gene expression allows network plants to identify and react by activating some transcriptional routes. Major TFs, such as NAC, WRKY, and MYB, coordinate these networks by binding promoter areas of genes (Dietrich *et al.*, 2011). For example, the NAC TF family is important in modifying drought and salinity reactions, while WRKY TFS plays an important role in the pathogen defense. ABA signaling system is an important regulator of reactions to stress, which combines inputs from inputs such as dehydration and cold to activate the downstream gene (Jozefczuk *et al.*, 2010).

The CBF/DREB transcriptional cascade is preferably triggered under cold stress, which promotes the level of cold-regulated (COR) gene expression, offering cold tolerance (Fraire-Velázquez and Balderas-Hernández, 2013). Additionally, the stress-respondent expression of the gene is enhanced by modifying access to epigenetic regulators, such as chromatin remodelers (Iqbal *et al.*, 2015). Advance in transcriptomics unveil the complex mobility of these networks for coordinating the systemic and local stress reaction of the plants (Hashem *et al.*, 2019). However, these studies also provide insight into how to increase tolerance to excessive stress in plants through technological innovations.

4.12 Adaptive Strategies and Trade-offs

The immediate adaptation involves the biochemical and physical changes to enhance the homeostasis of the cell and short-term metabolic reactions to ensure instant survival. These responses encompass turning on antioxidant defense mechanisms, osmotic shifts, as well as the expression of different stress-related genes (Koza *et al.*, 2022).

Most short -term reactions in plants are the accumulation of osmoprotectants, such as proline and glycine betaine, to deal with cellular dehydration (Hou *et al.*, 2021). Conversely, long-term metabolic reactions include structural and functional changes adding flexibility to the periodic stress (Etesami and Beattie, 2017). To adapt to these changes, there can be a change in the architecture of the root structure or a redistribution of energy and the production of stress-related proteins (Korenblum and Aharoni, 2019).

Short-term reactions immediately prevent the death of an organism, while long-term adaptations are necessary for the continued development and production of organisms repeated through a stress condition (Lata *et al.*, 2018). It has been concluded in studies that animals who have a balance between short- and long-term reflexes have a great potential of being rich in uncertain conditions (Tiwari and Lata, 2018). One of the most basic obstacles is the business of organisms between growth and stress tolerance. However, there are many cases in which under stress settings and resources, including both energy and nutrients, are transferred from the development processes to protect processes, thus contributing to low biomass accumulation and reproductive production (Lata *et al.*, 2018). In the example plants under drought or salt often have defense chemicals and stress activities (Koza *et al.*, 2022). This change gives the plant a better existence but detracts from the production and productivity (Etesami and Beattie, 2017).

Animals also experience an analogical trade-closure, as metabolic energy is redirected from growth and reproduction such as immune function and reactions to stress such as thermoregulation (Tiwari and Lata, 2018).

Microorganisms display trade-bands through changed metabolic flux that favor their survival under adverse conditions, often at the cost of decreasing proliferation and development of the colony (Korenblum and Aharoni, 2019). An optimal balance between growth and stress tolerance is important for long -term stability in both natural and managed ecosystems (Hou *et al.*, 2021).

4.13 Biotechnological and Agronomic Implications

Genetic engineering plays an important role in increasing stress tolerance through the introduction of stress-tension between animals. Advance in biotechnology has facilitated the manufacture of transgenic plants with better drought, salinity, and temperature tolerance by use of osmoprotectants, antioxidant enzymes, and regulatory proteins (Thomson *et al.*, 2018). Gene-editing tools, such as CRISPR-CAS9, have further enhanced accuracy in changing stress-respondent routes to achieve desirable phenotypes (Nowicka *et al.*, 2018).

Genetic changes focus on water consumption efficiency, nutritional absorption, and stress-enhancing signaling routes, providing more flexibility for environmental challenges (Bárzana and Carvajal, 2020).

4.13.1 Metabolic engineering approach

Metabolic engineering gives another way to stress tolerance by strengthening metabolic routes to promote stress-resistance. It changes metabolic flow to promote the production of protective chemicals such as polyamine, flavonoids, and suitable solutions (Sun *et al.*, 2015). Development in biotechnology has facilitated the design of specific metabolic routes that can promote the reaction to stress efficiency without development and losing production (Nielsen and Keasling, 2016). These strategies are particularly beneficial in biotechnology for agriculture, where stress-resistant crops can be engineered to guarantee food supply under the changing climatic conditions (Deparis *et al.*, 2017).

4.14 Case Studies and Practical Applications

Arabidopsis thaliana acts as an important model to study the stress-resistance mechanisms due to its well analyzed genome and ease of genetic manipulation (Fig. 4.4).

Research has demonstrated that genes such as AtSAP13 promote resistance to many innovative stresses, including salt and drought through the activation of genes related to stress (Dixit *et al.*, 2017). Additionally, the IAN gene cluster (IAN2 to IAN6) is displayed to negatively affect heat tolerance in *Arabidopsis*, by replacing the protein reaction routes and cell death regulation (Lu *et al.*, 2020).

The relevance of HSP17.6B, a HSP, for example, has been worked on in *Arabidopsis* with an emphasis on maintenance of chlorophyll and further increase in temperature tolerance through reduction of electrolyte loss under high

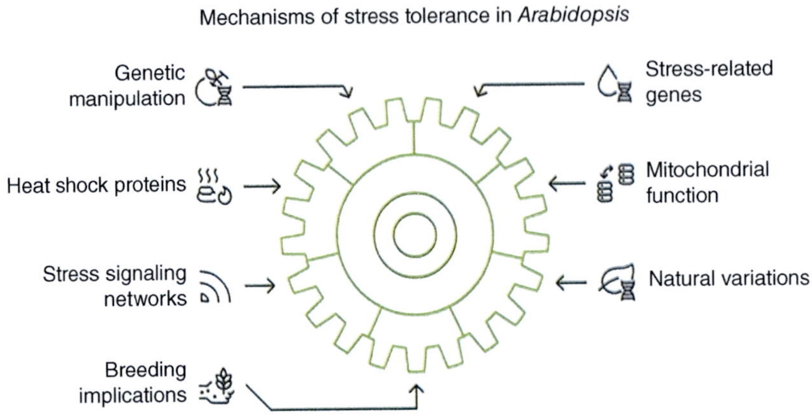

Fig. 4.4. Mechanism of stress tolerance in *Arabidopsis*. (Figure author's own.)

temperature (Tawab *et al.*, 2019). Finally, mitochondrial fragmentation is critical for long-term heat tolerance through maintenance of the complex genes such as ELM1, mitochondria functionality, and stopping cell death (Tsukimoto *et al.*, 2021).

The discovery of STCH4/REIL2 genes that increase cold tolerance by affecting ribosomal RNA metabolism and CBF protein translation in *Arabidopsis* models has also been helpful (Yu *et al.*, 2020).

Cai *et al.* (2021) recently provided a diagram illustrating that TF OXS2 is important to salt stress tolerance, and its overexpression promotes the expression of salt answer genes. Another noteworthy finding includes the *Arabidopsis thaliana* C24 accession, which shows higher drought and salt tolerance and is characterized with its unique genetics and adequate water use efficiency (Bechtold *et al.*, 2018). The essential variety among *Arabidopsis* accessions provides impressive information for producing climate resistant crops. Moreover, it testifies that overexpression of genes such as AVP1 and PP2A-C5 increase *Arabidopsis* tolerance to multiple abiotic stimuli by improving water retention and ion homeostasis (Fig 4.4) (Sun *et al.*, 2018).

4.15 Application in Agriculture and Industry

The applications of the stress tolerance system are immediate to farming and manufacturing, which enables the production of hard crops and enhances the industrial processes. Genetic engineering and reproductive strategy in agriculture are used to improve the plant's stress-related symptoms and improve their drought resistance, salt tolerance and insect resistance to enhance the agricultural production and stability. For instance, CRISPR and CAS9 were applied to modify genes such as the OsdreB1 gene in rice for improved drought tolerance (Zhang *et al.*, 2023). Further, marker-assisted selection (MAS) has promoted growth of climate robust crops, such as a maize hybrid with improved ability to utilize water and heat tolerances (Glombitza *et al.*, 2004). Stress-tolerant strains are used in the industry for bio-fuel production, bioremediation, and even drug synthesis.

One example is ethanol production from genetically converted *Saccharomyces cerevisiae* strains with increased oxidative stress tolerance (Pereira *et al.*, 2018). Similarly, the revised algae are exploited in biodiesel manufacturing to bear high heat and salt (Kushalappa and Gunnaiah, 2013). These applications underline the promise of implementing stress tolerance systems for global food, agricultural stability, and the supply of industrial advances (Table 4.2).

Much of the stress-to-height farming has been engineered to address the difficulties given by environmental conditions. Notable examples include varieties of salt-tolerant rice such as Pokkali and CSR30, which thrive in saline soil and are made using drought-tolerant corn hybrids, such as DTMA lines, and marker-assisted breeding (Balan *et al.*, 2018). In addition,

Table 4.2. Applications in agriculture and industry. (Table author's own.)

Application area	Example organisms/crops	Biotechnology approach	Outcome
Agriculture	Drought-resistant rice, maize	CRISPR-Cas9 editing, transgenics	Enhanced yield, resilience
Biofuel production	Stress-tolerant algae	Metabolic engineering	Improved biodiesel output
Bioremediation	Metal-tolerant microbes	Genetic modification	Efficient pollutant cleanup

heat-tolerant wheat cultivation, especially HD2967, displays adequate yields under rising temperatures (Lehmann *et al.*, 2009). In addition, barley cultivation with better drought and salinity tolerance in a golden environment is created for the desert environment (Rani *et al.*, 2016). Other examples include transgenic tomato plants, which express HSP70 for better thermal tolerance and are resistant to vegetarianism through JA pathway changes for soybean cultivation (Azim, 2019). Recently, progress in molecular reproduction has developed quinoa cultivation with severe salinity and high-altitude UV exposure with better flexibility (Glombitza *et al.*, 2004). This farming portrays the effectiveness of adding stress-sensitivity properties in current agricultural functions, allowing crop production in marginal areas and challenging climatic conditions.

4.16 Future Perspectives and Challenges

The effect of climate change is increasing the frequency and severity of abiotic and organic stresses, causing great damage to global food production. Rising temperatures, irregular rainfall patterns, and increasing salt in the soil require the design of crops capable of understanding multiple difficulties simultaneously (Glombitza *et al.*, 2004). Including climate models with crop breeding programs is important to remove and manage future stress conditions. Strategies such as breeding heat-resistant wheat and flood-tolerant rice are emerging as important tools for maintaining food security (Lehmann *et al.*, 2009).

- **Integration of systems biology approaches:** The system presents a complete view of stress reactions by merging biology genome, transcriptomics, proteomics and metabolomics data. This method allows the discovery of significant regulatory networks and stress-tension-vigorous routes, which allow customized crop development (Zhang *et al.*, 2023). Computer technologies such as machine learning are employed to quickly predict gene functions and interactions, which further enhances the generation of flexible agricultural farming (Balan *et al.*, 2018).
- **Ethical and ecological considerations:** The increasing acceptance of genetically-modified organisms (GMOs) presents moral and ecological problems. Potential concerns, such as wild relatives require the results on gene flow and non-target species, rigorous considerations (Kushalappa and Gunnaiah, 2013). The public acceptance of GMO's depends on honest facts about their benefits and risks, as well as the appropriate regulatory framework to ensure biosafety (Rani *et al.*, 2016).
- **Need for interdisciplinary research:** Addressing complex stress challenges requires cooperation in subjects including plant biology, environmental science, engineering, and policy making. Interdisciplinary research supports progress, such as the construction of stress-to-height bio-engineered crops and permanent agricultural technologies (Zhang *et al.*, 2023). By combining diverse abilities, the global farming system can effectively adapt to future challenges.

4.17 Conclusion

This chapter reflects the important role of metabolic changes in allowing organisms to avoid biotic and abiotic stresses. Major findings suggest that organisms exploit complex metabolic systems, such as glucose metabolism, protein folding, and lipid signaling, to buffer against stress.

Additionally, secondary metabolites and hormonal modulation antioxidants promote resistance through defense and systemic signaling.

The value of metabolic adjustment rests in its normal application in biological systems. In plants, it increases the existence between photosynthesis, ion homeostasis, and osmoprotectant production between dry conditions, high salinity, and temperature extremes.

Animals depend on hormonal systems to control energy metabolism, but microorganisms display incredible flexibility through metabolic plasticity. These systems not only maintain life but also protect ecological balance and productivity in various scenarios. Despite this progress, many areas are more and more researchable. A combination of system biology techniques provides a possible passage to decompose the refined regulatory network running metabolic processes. Additionally, moral and ecological ideas, especially GMOs, require open information and stringent biosafety laws. Interdisciplinary research is important to solve the difficulties caused by climate change, which require imaginative techniques to generate stress-tolerant organisms and permanent activities. Overall, understanding and using metabolic variations provides a framework for resolving international concerns in agriculture, industry, and environmental protection, underlining its value in preserving life in adverse conditions.

References

Azim, K. (2019) Organic amendments to alleviate plant biotic stress. In: Ansari, R., Mahmood, I. (eds) *Plant Health Under Biotic Stress*. Springer, Singapore, pp. 147–165 https://doi.org/10.1007/978-981-13-6043-5_8

Balan, B., Marra, F.P., Caruso, T. and Martinelli, F. (2018) Transcriptomic responses to biotic stresses in Malus x domestica: A meta-analysis study. *Scientific Reports* 8(1), 1970. DOI: 10.1038/s41598-018-19348-4.

Bárzana, G. and Carvajal, M. (2020) Genetic regulation of water and nutrient transport in water stress tolerance in roots. *Journal of Biotechnology* 324, pp.134–142.

Bechtold, U., Ferguson, J.N. and Mullineaux, P.M. (2018) To defend or to grow: Lessons from Arabidopsis C24. *Journal of Experimental Botany* 69(11), 2809–2821. DOI: 10.1093/jxb/ery106.

Bhat, A.G., Vashisht, R. and Chandra, N. (2010) Modeling metabolic adjustment in Mycobacterium tuberculosis upon treatment with isoniazid. *Systems and Synthetic Biology* 4(4), 299–309. DOI: 10.1007/s11693-011-9075-6.

Borges, A.A., Jiménez-Arias, D., Expósito-Rodríguez, M., Sandalio, L.M. and Pérez, J.A. (2014) Priming crops against biotic and abiotic stresses: MSB as a tool for studying mechanisms. *Frontiers in Plant Science* 5, 642. DOI: 10.3389/fpls.2014.00642.

Cai, J., Liu, T., Li, Y. and Ow, D.W. (2021) A C-terminal fragment of Arabidopsis OXIDATIVE STRESS 2 can play a positive role in salt tolerance. *Biochemical and Biophysical Research Communications* 556, 23–30. DOI: 10.1016/j.bbrc.2021.03.127.

Corradi, N. and Bonfante, P. (2012) The Arbuscular Mycorrhizal Symbiosis: Origin and Evolution of a Beneficial Plant Infection. *PLoS Pathog* 8(4): e1002600. https://doi.org/10.1371/journal.ppat.1002600

Dangl, J.L. and Jones, J.D.G. (2001) Plant pathogens and integrated defence responses to infection. *Nature* 411(6839), 826–833. DOI: 10.1038/35081161.

Deparis, Q., Claes, A., Foulquié-Moreno, M.R. and Thevelein, J.M. (2017) Engineering tolerance to industrially relevant stress factors in yeast cell factories. *FEMS Yeast Research* 17(4). DOI: 10.1093/femsyr/fox036.

Dietrich, K., Weltmeier, F., Ehlert, A., Weiste, C., Stahl, M., Harter, K. and Dröge-Laser, W. (2011) Heterodimers of the Arabidopsis Transcription Factors bZIP1 and bZIP53 Reprogram Amino Acid Metabolism during Low Energy Stress. *The Plant Cell* 23(1), 381–395. DOI: 10.1105/tpc.110.075390.

Dixit, A., Tomar, P., Vaine, E., Abdullah, H., Hazen, S. and Dhankher, O.P. (2017) A stress-associated protein, AtSAP13, from Arabidopsis thaliana provides tolerance to multiple abiotic stresses. *Plant Cell and Environment* 41(5), 1171–1185. DOI: 10.1111/pce.13103.

Dubey, S., Shri, M., Gupta, A., Rani, V. and Chakrabarty, D. (2018) Toxicity and detoxification of heavy metals during plant growth and metabolism. *Environmental Chemistry Letters* 16(4), 1169–1192. DOI: 10.1007/s10311-018-0741-8.

Dutta, S., Mitra, M., Agarwal, P., Mahapatra, K., De, S., Sett, U. and Roy, S. (2018) Oxidative and genotoxic damages in plants in response to heavy metal stress and maintenance of genome stability. *Plant Signaling and Behavior* 1–49. DOI: 10.1080/15592324.2018.1460048.

Egamberdieva, D., Wirth, S.J., Alqarawi, A.A., Abd_Allah, E.F. and Hashem, A. (2017) Phytohormones and beneficial microbes: essential components for plants to balance stress and fitness. *Frontiers in Microbiology* 8, 2104. DOI: 10.3389/fmicb.2017.02104.

Etesami, H. and Beattie, G.A. (2017) Plant-Microbe interactions in adaptation of agricultural crops to abiotic stress conditions. In: Kumar, V., Kumar, M., Sharma, S., Prasad, R. (eds) *Probiotics and Plant Health*, pp. 163–200. Springer eBooks, Singapore. DOI: 10.1007/978-981-10-3473-2_7.

Eutrópio, F.J., Ramos, A.C., Da Silva Folli-Pereira, M., De Aquino Portela, N., Santos, J.B.D. (2015) Heavy metal stress and molecular approaches in plants. In: Ahmad, P. (ed.), *Plant Metal Interaction Emerging Remediation Techniques*, pp. 531–543. Elsevier Science eBooks, Netherlands. DOI: 10.1016/b978-0-12-803158-2.00022-9.

Fraire-Velázquez, S. and Balderas-Hernández, V.E. (2013). Abiotic stress in plants and metabolic responses. In: Vahdati, K. and C. Leslie (eds) *Abiotic Stress*. IntechOpen. https://doi.org/10.5772/54859

Ghori, N., Ghori, T., Hayat, M.Q., Imadi, S.R., Gul, A., Altay, V. and Ozturk, M. (2019) Heavy metal stress and responses in plants. *International Journal of Environmental Science and Technology* 16(3), 1807–1828. DOI: 10.1007/s13762-019-02215-8.

Glombitza, S., Dubuis, P., Thulke, O., Welzl, G., Bovet, L. *et al.* (2004) Crosstalk and differential response to abiotic and biotic stressors reflected at the transcriptional level of effector genes from secondary metabolism. *Plant Molecular Biology* 54(6), 817–835. DOI: 10.1007/s11103-004-0274-3.

Hashem, A., Tabassum, B. and Abd_Allah, E.F. (2019) Bacillus subtilis: A plant-growth promoting rhizobacterium that also impacts biotic stress. *Saudi Journal of Biological Sciences* 26(6), 1291–1297. DOI: 10.1016/j.sjbs.2019.05.004.

Hirt, H. (2012) Rôle des micro-organismes bénéfiques pour aider les plantes à acquérir une tolérance aux stresses environnementaux. *Biologie Aujourd Hui* 206(4), 285–290. DOI: 10.1051/jbio/2012028.

Holmberg, N. (1998) Metabolic Engineering: Approaches Towards Improved Stress Tolerance in Microorganisms and Plants. (Thesis) Available at: https://www.semanticscholar.org/paper/Metabolic-engineering%3A-Approaches-Towards-Improved-Holmberg/0bb0110a61fa979b94c331013b3a17d59f1bf710?utm_source=consensus (date accessed 03/09/2025).

Hoque, M.N., Tahjib-Ul-Arif, M., Hannan, A., Sultana, N., Akhter, S. *et al.* (2021) Melatonin modulates plant tolerance to heavy metal stress: morphological responses to molecular mechanisms. *International Journal of Molecular Sciences* 22(21), 11445. DOI: 10.3390/ijms222111445.

Hou, S., Wolinska, K.W. and Hacquard, S. (2021) Microbiota-root-shoot-environment axis and stress tolerance in plants. *Current Opinion in Plant Biology* 62, 102028. DOI: 10.1016/j.pbi.2021.102028.

Iqbal, N., Nazar, R. and Khan, N.A. (2015) Osmolytes and plants Acclimation to changing environment: Emerging Omics technologies. In: Iqbal, N., Nazar, R. and Khan, N.A. (eds) *Osmolytes and Plants Acclimation to Changing Environment: Emerging Omics Technologies*. Springer, India, pp. 129–153. DOI: 10.1007/978-81-322-2616-1.

Jozefczuk, S., Klie, S., Catchpole, G., Szymanski, J., Cuadros-Inostroza, A. *et al.* (2010) Metabolomic and transcriptomic stress response of Escherichia coli. *Molecular Systems Biology* 6(1), 364. DOI: 10.1038/msb.2010.18.

Kannojia, P., Sharma, P.K., Kashyap, A.K., Manzar, N., Singh, U.B. *et al.* (2017) Microbe-Mediated Biotic stress management in plants. In: D.P. Singh, H.B. Singh, and R. Prabha (eds) *Plant-Microbe Interactions in Agro-Ecological Perspectives: Volume 2: Microbial Interactions and Agro-Ecological Impacts*. Springer, Singapore, pp. 627–648. DOI: 10.1007/978-981-10-6593-4_26.

Korenblum, E. and Aharoni, A. (2019) Phytobiome metabolism: beneficial soil microbes steer crop plants' secondary metabolism. *Pest Management Science* 75(9), 2378–2384. DOI: 10.1002/ps.5440.

Koza, N., Adedayo, A., Babalola, O. and Kappo, A. (2022) Microorganisms in plant growth and development: roles in abiotic stress tolerance and secondary metabolites secretion. *Microorganisms* 10(8), 1528. DOI: 10.3390/microorganisms10081528.

Kumar, A. and Verma, J.P. (2017) Does plant-microbe interaction confer stress tolerance in plants: A review? *Microbiological Research* 207, 41–52. DOI: 10.1016/j.micres.2017.11.004.

Kushalappa, A. C. and Gunnaiah, R. (2013) Metabolo-proteomics to discover plant biotic stress resistance genes. *Trends in Plant Science* 18(9), 522–531. DOI: 10.1016/j.tplants.2013.05.002.

Lata, R., Chowdhury, S., Gond, S. and White, J. (2018) Induction of abiotic stress tolerance in plants by endophytic microbes. *Letters in Applied Microbiology* 66(4), 268–276. DOI: 10.1111/lam.12855.

Lehmann, M., Schwarzländer, M., Obata, T., Sirikantaramas, S., Burow, M. *et al.* (2009) The metabolic response of arabidopsis roots to oxidative stress is distinct from that of heterotrophic cells in culture and highlights a complex relationship between the levels of transcripts, metabolites, and flux. *Molecular Plant* 2(3), 390–406. DOI: 10.1093/mp/ssn080.

López-Gresa, M.P., Maltese, F., Bellés, J.M., Conejero, V., Kim, H.K., Choi, Y.H. and Verpoorte, R. (2009) Metabolic response of tomato leaves upon different plant–pathogen interactions. *Phytochemical Analysis* 21(1), 89–94. DOI: 10.1002/pca.1179.

Lu, S., Zhu, T., Wang, Z., Luo, L., Wang, S. *et al.* (2020) Arabidopsis immune-associated nucleotide-binding genes repress heat tolerance at the reproductive stage by inhibiting the unfolded protein response and promoting cell death. *Molecular Plant* 14(2), 267–284. DOI: 10.1016/j.molp.2020.11.010.

Nielsen, J. and Keasling, J.D. (2016) Engineering cellular metabolism. *Cell* 164(6), 1185–1197. DOI: 10.1016/j.cell.2016.02.004.

Nowicka, B., Ciura, J., Szymańska, R. and Kruk, J. (2018) Improving photosynthesis, plant productivity and abiotic stress tolerance – current trends and future perspectives. *Journal of Plant Physiology* 231, 415–433. DOI: 10.1016/j.jplph.2018.10.022.

Pereira, T., Vilaprinyo, E., Belli, G., Herrero, E., Salvado, B. *et al.* (2018) Quantitative operating principles of yeast metabolism during adaptation to heat stress. *Cell Reports* 22(9), 2421–2430. DOI: 10.1016/j.celrep.2018.02.020.

Pérez-Bueno, M.L., Pineda, M. and Barón, M. (2019) Phenotyping plant responses to biotic stress by chlorophyll fluorescence imaging. *Frontiers in Plant Science* 10, 1135. DOI: 10.3389/fpls.2019.01135.

Rani, V., Deep, G., Singh, R.K., Palle, K. and Yadav, U.C. (2016) Oxidative stress and metabolic disorders: Pathogenesis and therapeutic strategies. *Life Sciences* 148, 183–193. DOI: 10.1016/j.lfs.2016.02.002.

Rizvi, A., Shankar, A., Chatterjee, A., More, T.H., Bose, T. *et al.* (2019) Rewiring of metabolic network in mycobacterium tuberculosis during adaptation to different stresses. *Frontiers in Microbiology* 10, 2417. DOI: 10.3389/fmicb.2019.02417.

Sakamoto, A. and Murata, N. (2000) Genetic engineering of glycinebetaine synthesis in plants: current status and implications for enhancement of stress tolerance. *Journal of Experimental Botany* 51(342), 81–88. DOI: 10.1093/jexbot/51.342.81.

Seneviratne, M., Rajakaruna, N., Rizwan, M., Madawala, H.M.S.P., Ok, Y.S. and Vithanage, M. (2017) Heavy metal-induced oxidative stress on seed germination and seedling development: a critical review. *Environmental Geochemistry and Health* 41(4), 1813–1831. DOI: 10.1007/s10653-017-0005-8.

Shiade, S.R.G., Zand-Silakhoor, A., Fathi, A., Rahimi, R., Minkina, T., Rajput, V.D., Zulfiqar, U. and Chaudhary, T. (2024) Plant metabolites and signaling pathways in response to biotic and abiotic stresses: Exploring bio stimulant applications. *Plant Stress* 12, 100454.

Sun, X., Yu, G., Li, J., Jia, P., Zhang, J. *et al.* (2014) A Heavy Metal-Associated Protein (AcHMA1) from the Halophyte, Atriplex canescens (Pursh) Nutt., confers tolerance to iron and other abiotic stresses when expressed in Saccharomyces cerevisiae. *International Journal of Molecular Sciences* 15(8), 14891–14906. DOI: 10.3390/ijms150814891.

Sun, L., Pehlivan, N., Esmaeili, N., Jiang, W., Yang, X. *et al.* (2018) Co-overexpression of AVP1 and PP2A-C5 in Arabidopsis makes plants tolerant to multiple abiotic stresses. *Plant Science* 274, 271–283. DOI: 10.1016/j.plantsci.2018.05.026.

Sun, X., Shen, X., Jain, R., Lin, Y., Wang, J. *et al.* (2015) Synthesis of chemicals by metabolic engineering of microbes. *Chemical Society Reviews* 44(11), 3760–3785. DOI: 10.1039/c5cs00159e.

Suprasanna, P., Nikalje, G.C. and Rai, A.N. (2015) Osmolyte accumulation and implications in plant abiotic stress tolerance. In: N. Iqbal, R. Nazar, and N.A. Khan (eds) *Osmolytes and Plants Acclimation to Changing Environment: Emerging Omics Technologies*. Springer, India, pp. 1–12. DOI: 10.1007/978-81-322-2616-1_1.

Tawab, F., Munir, I., Ullah, F. and Iqbal, A. (2019) Over-Expressed HSP 17.6B, Encoding HSP20-Like Chaperones Superfamily Protein, Confers Heat Stress Tolerance in Arabidopsis Thaliana. *Pakistan Journal of Botany*, 51, 3–36.

Thomson, K., Babar, M.A., Erickson, J.E., Mulvaney, M., Beecher, C. and MacDonald, G. (2018) Comparative physiological and metabolomics analysis of wheat (*Triticum aestivum* L.) following post-anthesis heat stress. *PLoS one* 13(6), e0197919. DOI: 10.1371/journal.pone.0197919.

Tiwari, S. and Lata, C. (2018) Heavy metal stress, signaling, and tolerance due to plant-associated microbes: An overview. *Frontiers in Plant Science* 9, 452. DOI: 10.3389/fpls.2018.00452.

Tsukimoto, R., Isono, K., Kajino, T., Iuchi, S., Shinozawa, A. *et al.* (2021) Mitochondrial fission complex is required for long-term heat tolerance of Arabidopsis. *Plant and Cell Physiology* 63(3), 296–304. DOI: 10.1093/pcp/pcab171.

Yu, H., Kong, X., Huang, H., Wu, W., Park, J. *et al.* (2020) STCH4/REIL2 confers cold stress tolerance in Arabidopsis by promoting RRNA processing and CBF protein translation. *Cell Reports* 30(1), d229–242.e5. DOI: 10.1016/j.celrep.2019.12.012.

Zebral, Y.D., Da Silva Fonseca, J., Roza, M., Costa, P.G., Robaldo, R.B. and Bianchini, A. (2020) Combining elevated temperature with waterborne copper: Impacts on the energy metabolism of the killifish Poecilia vivipara. *Chemosphere* 253, 126631. DOI: 10.1016/j.chemosphere.2020.126631.

Zhang, Y. and Xia, P. (2023) The DREB transcription factor, a biomacromolecule, responds to abiotic stress by regulating the expression of stress-related genes. *International Journal of Biological Macromolecules* 243, 125231. https://doi.org/10.1016/j.ijbiomac.2023.125231

Zhou, Z., Windhorst, A., Schenke, D. and Cai, D. (2023) RNAseq-based working model for transcriptional regulation of crosstalk between simultaneous abiotic UV-B and biotic stresses in plants. *Genes* 14(2), 240. DOI: 10.3390/genes14020240.

5 Epigenetic Regulation of Stress Responses in Plants

Hajira Bibi[1], Muhammad Abubakr[2], Javed Iqbal[3], Banzeer Ahsan Abbasi[4], Muhammad Anas[5]* and Syeda Anber Zahra[6]

[1]*Department of Biochemistry, Faculty of Biological Sciences, Quaid-i-Azam University, Islamabad 45320, Pakistan;* [2]*Department of Botany, Division of Science and Technology, University of Education, Lahore, Pakistan;* [3]*Department of Botany, Bacha Khan University, Charsadda 24420, Khyber Pakhtunkhwa, Pakistan;* [4]*Department of Botany, Rawalpindi Women University, Rawalpindi 46300, Pakistan;* [5]*Department of Soil Science and Plant Nutrition, University of Selcuk Campus, Konya 42079, Turkiye;* [6]*Institute of Biomaterials and Biomedical, School of Food and Pharmacy, Shanghai Zhongqiao Vocational and Technical University, Shanghai, China*

Abstract

This chapter is about epigenetic regulation in plants responding to environmental stresses. Fluctuating environmental conditions initiate epigenetic changes, which include DNA methylation, small RNA-mediated regulation, and histone modification, allowing plants to dynamically change gene expression. The mechanisms are outlined, as well as their effect on stress-responsive genes. The roles of epigenetic regulation of plant stress responses are played by histone modifications such as acetylation, deacetylation, and methylation. Stress-responsive gene expression is regulated by histone acetyltransferases and deacetylases, which alter chromatin structure and accessibility. Stress-related gene regulation also comes from histone methylation patterns. The examples of conserved nature are presented based on model plants, such as *Arabidopsis*, and crops, such as rice and apple trees. In addition, the chapter also deals with interactions of different epigenetic mechanisms, like the involvement of small RNA (sRNA) in stress response gene regulation, among others. This dynamic epigenetic control allows plants to quickly vary their physiology for a wide variety of abiotic stresses, a condition that is essential for plant stress physiology. The chapter makes essential points regarding the intricacies in the molecular mechanisms of stress resilience in plants.

Keywords: epigenetic regulation, environmental stresses, DNA methylation, plant stress responses

5.1 Introduction to Epigenetic Regulation

Plants are sessile beings and, as a result, plenty of environmental stresses happen to them, including drought and high salinity, as well as extremely high fluctuations in temperature. Plants have built up a very complicated mechanism for them to use to identify and face such challenges (Sun *et al.*, 2021, Ahmad *et al.*, 2022, Ahmad *et al.*, 2024).

* Email: muhammed.anas@selcuk.edu.tr

 Molecular Biology of Plant Stress Responses (Muhammad Waseem and Pingwu Liu eds)
DOI: 10.1079/9781800628823.0005

An important role in orchestrating plant stress responses has been established for epigenetic regulation, which refers to an alteration in chromatin structure or the DNA not involving the underlying genetic sequence (Sun *et al.*, 2021). Plants utilize epigenetic mechanisms for rapid and dynamic regulation of the programs of gene expression in response to environmental conditions. As such, this epigenetic mechanism allows plants to rapidly adapt to different abiotic stresses by modulating stress-responsive genes (Sun *et al.*, 2021, Ali *et al.*, 2022). Therefore, DNA methylation can block stress response genes, and histone modifications can be activators or repressors of stress-responsive gene transcription. In addition, small RNAs can silence genes that negatively control the stress responses and enhance the stress resilience of the plant as the plant is still able to protect itself from stress (Lujan and Dinkova, 2021, Haider *et al.*, 2021, Haider *et al.*, 2022).

Epigenetic regulation of plant stress responses has been summarized in Table 5.1 and illustrated in Fig. 5.1.

5.2 Histone Modifications and Stress Responses

Histone modifications play a role in regulating epigenetic regulation of plant stress response. Groups of core protein components of nucleosomes called histones can be subject to a number of post-translational modifications (PTMs) such as acetylation, methylation, phosphorylation, ubiquitination, and SUMOylation where SUMO stands for "small ubiquitin-like modifier". Chromatin structure is modified depending on the nature of histone modification and makes the transcription of stress-responsive genes accessible.

5.2.1 Histone acetylation

Histone acetylation is generally considered to occur with histone acetyltransferases and transcriptional activation. Elevated histone acetylation of the promoters of stress-responsive genes may occur under stress conditions. This increases the expression of stress-responsive genes and allows the plant to respond effectively to stress (Li *et al.*, 2021). For example, in *Arabidopsis*, induction of drought stress results in the induction of GCN5 histone acetyltransferase, essentially increasing acetylation of histones H3 and H4 (Gan *et al.*, 2021), thus leading to the expression of stress-responsive genes such as RD29A and COR15A. Therefore, RD29A and COR15A encode proteins involved in central regulation of plant response to abiotic stresses such as drought and cold to enable the plants to survive and thrive under these drought conditions (Asghar *et al.*, 2024; Ijaz *et al.*, 2024; Shilpa *et al.*, 2024).

Similarly, crops and fruit trees that economically matter have similar histone acetylation patterns. In rice, the histone acetyltransferase OsHAC701 is upregulated by drought stress, which consequently results in increased acetylation of histone H3 at lysine 9 (H3K9ac) and the upregulation of drought-responsive genes such as OsDREB2A and OsLEA3 (Wang *et al.*, 2024).

5.2.2 Histone deacetylation

Conversely, histone deacetylation catalysed by histone deacetylases is usually coupled with transcriptional repression. When the plant is under stress conditions, HDAC-mediated deacetylation of histones can silence the expression of genes related to growth and development while the plant transitions to surviving the stress (Cui *et al.*, 2023). For example, the *Arabidopsis* HDAC HDA6 deacetylates histones at the promoters of stress-responsive genes such as RD29A and COR15A and represses their expression in non-stress conditions (Chung *et al.*, 2021).

In rice it is the same story, drought stress incites overexpression of histone deacetylase OsHDAC1, which induces deacetylation of histones H3 and H4 on promoters of drought-responsive genes, OsDREB2A and OsLEA3, repression of such genes and adaptation to stress (Todaka *et al.*, 2012; Kumar *et al.*, 2013).

In apple trees, the MdHDAC3 is activated in response to cold stress (Wang *et al.*, 2023). Thus, it removes acetyl groups of histones H3 and H4 present at the promoters of MdCBF1 and MdCOR15A cold-responsive genes, silencing their transcription and in turn allowing plant machinery to be used for a cold-stress response (Białoskórska *et al.*, 2024).

Table 5.1. Different mechanisms of epigenetic regulation of stress responses in plants. (Table author's own.)

Epigenetic modification	Stress	Plant	Enzyme	Example	Effect
Histone acetylation	Drought	*Arabidopsis*	Histone acetyltransferase	Increased acetylation of histones H3 and H4	Expression of stress-responsive genes (*RD29A, COR15A*) (Shilpa *et al.*, 2024)
	Drought	Rice	Histone acetyltransferase	Increased acetylation of histone H3 at lysine 9	Upregulation of drought-responsive genes (*OsDREB2A, OsLEA3*) (Wang *et al.*, 2024)
Histone deacetylation	Non-stress conditions	*Arabidopsis*	Histone deacetylase	Deacetylate histones at the promoters of stress-responsive genes	Repressing the expression of stress-responsive genes (*RD29A and COR15A*) (Chung *et al.*, 2021)
	Drought	Rice	Histone deacetylase	Deacetylation of histones H3 and H4 at promoters of drought-responsive genes (*OsDREB2A, OsLEA3*)	Repression of growth genes and adaptation to the stress (Kumar *et al.*, 2013; Todaka *et al.*, 2012)
	Cold	Apple	Histone deacetylase	Deacetylation of histones H3 and H4 located at the promoters of the cold responsive genes (*MdCBF1, MdCOR15A*)	Silencing growth genes and adapting to stress (Białoskórska *et al.*, 2024)
Histone methylation	Drought	Rice	Histone methyl transferase	Trimethylation of histone H3 at lysine 4 at the promoters of drought-responsive genes *OsDREB2A, OsLEA3*)	Upregulation of drought-responsive genes and repression of growth genes (Singh *et al.*, 2022b)
	Non-stress conditions	Apple	Histone methyl transferase	Trimethylation of histone H3 at lysine 27 of cold-responsive genes (*MdCBF1, MdCOR15A*)	Suppression of cold-responsive genes (Fennell, 2014)
Histone phosphorylation	Pathogen attack	*Arabidopsis*	Mitogen-activated protein kinase (MAPK)	MPK4 phosphorylates histone H3 at serine 28 of defense-related genes	Onset of plant immune responses (Kumar *et al.*, 2020)
	Pathogen attack	Rice	MAPK	MAPK *OsMPK3* phosphorylates histone H3 at serine 10	Upregulation of pathogen defense genes and increased disease resistance (Zhou *et al.*, 2024)

	Pathogen attack	Apple	MAPK	MdMPK1 phosphorylates histone H3 at serine 28	Upregulating the genes related to the defense of the plant against fungal pathogens (Sun *et al.*, 2007)
	Heat	*Arabidopsis*	CIPK25 kinase (calcineurin β-like interacting protein kinase 25)	Phosphorylation of histone H2B at serine 14	Upregulates heat-responsive genes, adapting plant to high temperatures (Khadka *et al.*, 2020)
	Heat	Rice	homolog of *OsCIPK25*	Phosphorylation of histone H2B at serine 14	Plant heat stress tolerance (Khadka *et al.*, 2020)
Histone ubiquitination	Drought	*Arabidopsis*	E3 ubiquitin ligase HUB1	Monoubiquitination of histone H2B at lysine 143	Activation of drought-responsive genes (*RD29A, COR15A*) (Li *et al.*, 2024b); Sharma *et al.*, 2023)
	Salt stress	*Arabidopsis*	E3 ligase PUB26	Ubiquitination of histone H2B	Upregulation of stress-responsive genes and improved salt tolerance (Yung *et al.*, 2021)
	Drought stress	Rice	E3 ubiquitin ligase OsPUB56	Ubiquitination of histone H2B at lysine 143	Activation of drought-responsive genes (*OsDREB2A, OsLEA3*), enhancing the plant's drought tolerance (Li *et al.*, 2024b)
	Cold stress	Apple	E3 ubiquitin ligase MdPUB15	Monoubiquitination of histone H2B at lysine 143	Upregulation of cold-responsive genes such as *MdCBF1* and *MdCOR15A* (Kurpisz and Pawłowski, 2022)
	Cold	*Arabidopsis*	E3 ubiquitin ligase HOS1	Polyubiquitination of histone H2B at lysine 143	Silencing of genes that have a role in cell proliferation and development (Li *et al.*, 2024b)
Histone SUMOylation	Drought	*Arabidopsis*	SUMO E3 ligase SIZ1	SUMOylation of histone H4	Enhanced expression of drought-responsive genes (Singh *et al.*, 2023a)
	Salinity	Rice	SUMO E3 ligase OsSIZ1	SUMOylation of histones	Salt-induced activation of genes that confer the ability of the plant to tolerate higher salinity levels (Sharma *et al.*, 2023)
	Cold	Apple	SUMO E3 ligase MdSIZ1	SUMOylation of histones	Activation of cold-responsive genes (Li *et al.*, 2023)
	Heat	*Arabidopsis*	SUMO ligase	PolySUMOylation on histone H2B	Upregulation of heat-responsive genes (Vertegaal, 2022)

Continued

Table 5.1. Continued.

Epigenetic modification	Stress	Plant	Enzyme	Example	Effect
	Pathogen attack	*Arabidopsis*	SUMO ligase SIZ1	MonoSUMOylation of histone H3	Activation of defense genes against pathogenic microbial infection, enhancing immunity to such infections (Vertegaal, 2022)
DNA methylation	Drought	*Arabidopsis*	DNA methyltransferase	Increased DNA methylation in the promoters of stress-responsive genes, such as *RD29A* and *DREB2A*	Downregulation of growth genes, which helps the plant conserve resources and prioritize stress response mechanisms (Chung *et al*., 2021)
	Heat	Apple	DNA methyltransferase	High DNA methylation in the promoters of heat-responsive genes, such as *HSP70* and *HSP90*	Upregulation of these genes and the ability of the plant to adapt to heat stress (Lang *et al*., 2024)
	Pathogen infection	*Arabidopsis*	DNA methyltransferase	Elevation in DNA methylation on the promoters of defense-associated genes	Enhanced immunity (Huang and Jin, 2022)
	Salinity	Rice	DNA methyltransferase	Increased DNA methylation in the promoter regions of salt-responsive genes, like *NHX1* and *SOS1*	Adaptation of the plant to high-salinity environments (Saddhe *et al*., 2021; Ullah *et al*., 2022)
	Pathogen attack	*Arabidopsis*, rice, maize	DNA methyltransferase	Decreased DNA methylation in the promoters of defense-related genes	Enhancement of the plant's immune response (Yin *et al*., 2024)
Small RNA-mediated histone modification	Oxidative stress	*Arabidopsis*	miRNA miR398 downregulated	Upregulation of its target genes, *CSD1* and *CSD2*, which encode copper/zinc superoxide dismutase (SOD) involved in the plant's antioxidant defense system (AOS)	Increased immune response of the plant (Li *et al*., 2022)
	Drought	Rice	miR169 is reduced	Its target gene, NUCLEAR FACTOR Y A expression is increased	Drought tolerance is increased (Singh *et al*., 2022b)

Chromatin remodeling	Drought	*Arabidopsis*	Chromatin remodeling complex SWI/SNF	SWI/SNF is recruited to the promoters of drought-responsive genes (*RD29A* and *DREB2A*)	Activation of drought-responsive genes (Nguyen *et al*., 2022)
	Heat	Rice	Chromatin remodeling complex *CHR729*	*CHR729* targets promoters of heat-responsive genes, like *HSP70* and *HSP90*, facilitating nucleosome remodeling and the upregulation of these genes under high-temperature conditions	Increased heat-stress tolerance (Song *et al*., 2021)
	Pathogen attack	Apple	Chromatin remodeling complex SWI/SNF	Targets the promoters of defense-related genes such as *PR1* and *PR2*, which activate the genes and enhances the immune response of the plant	Increased pathogen resistance (Duan *et al*., 2018)
Small peptides	Pathogen attack or wounding		5–50 amino acids in length such as “elicitor peptides” and “phytosulfokines”	Induce defense responses in the plant, such as the production of reactive oxygen species (ROS) and the activation of defense-related genes	Increased pathogen resistance (Wang *et al*., 2022)
Signaling pathways	Biotic and abiotic stresses		Phytohormones such as abscisic acid (ABA) signaling pathway, jasmonic acid (JA) signaling pathway, ethylene (ET), salicylic acid (SA) signaling pathway	Recruits DNA methyltransferase and histone-modifying enzymes	Improves stress tolerance (Raza *et al*., 2020; Kaya *et al*., 2024; Mishra *et al*., 2024)

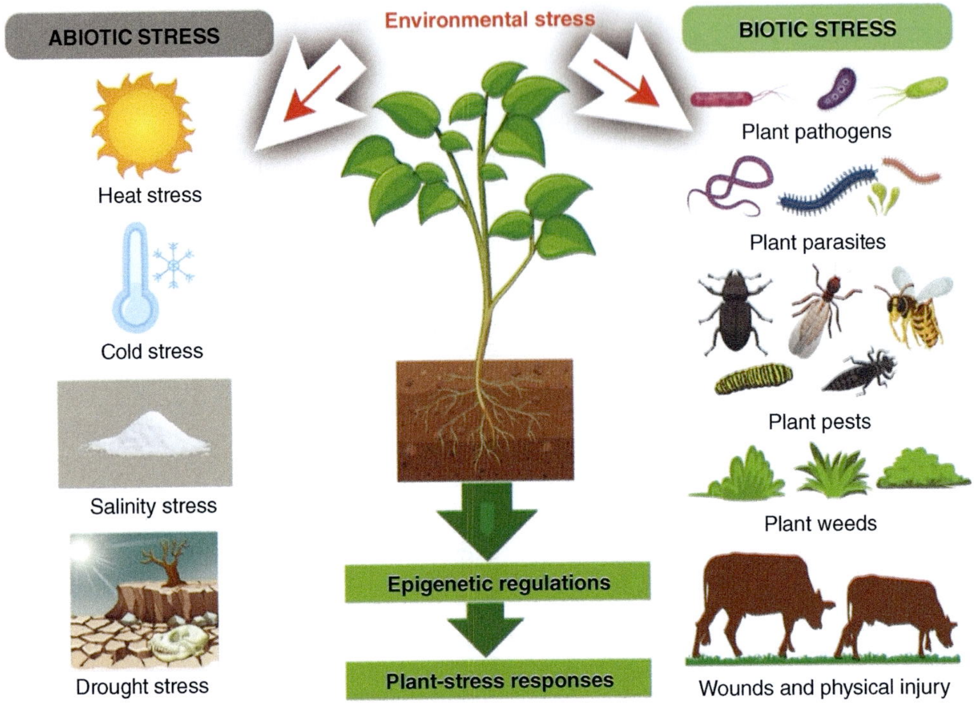

Fig. 5.1. Different mechanisms of epigenetic regulation of stress responses in plants. (Figure author's own.)

5.2.3 Histone methylation

Histone methylation is another important aspect involved in the plant's responses to stresses besides histone acetylation. In the case of histone methylation, gene expression can be modified in a variable way depending on which lysine or arginine is modified and to what degree (Zhou *et al.*, 2020). For example, histone H3 monomethylation, dimethylation, and trimethylation at lysine 4 are usually associated with active transcription, whereas histone H3 monomethylation, dimethylation, and trimethylation at lysine 27 are normally associated with transcriptional repression (Borkiewicz, 2021).

The dynamic regulation of these histone methylation patterns in response to environmental stresses is shown to play an important role in the epigenetic control of stress-responsive gene expression. Under drought stress, trimethylation of histone H3 at lysine 4 is increased in promoters of the drought-responsive genes, OsDREB2A and OsLEA3, that are upregulated in rice (Singh *et al.*, 2022a). In a similar manner, cold-responsive genes such as MdCBF1 and MdCOR15A in apple trees are repressed when histone H3 is trimethylated at lysine 27 under non-stress conditions (Fennell, 2014).

5.2.4 Histone phosphorylation

Other plant species may be involved in other stress responses by having histone phosphorylation (Nunez-Vazquez *et al.*, 2022). In response to pathogen attack, the MPK4 mitogen-activated protein kinase (MAPK) of *Arabidopsis* phosphorylates the H3 histone at serine 28. This phosphorylation is indicated with the activation of defense-related genes and the commencement of the plant immune response (Kumar *et al.*, 2020). In rice, the MAPK OsMPK3 also phosphorylates histone H3 at serine 10, which results in the upregulation of pathogen defense genes and increased disease resistance (Zhou *et al.*, 2024).

MdMPK1 phosphorylates histone H3 at serine 28 in apple trees, leading to the expression of genes that are involved in defense of the plant against fungal pathogens (Sun *et al.*, 2007). Additionally, it has been demonstrated that histone phosphorylation is involved in heat stress response. CIPK25 kinase, phosphorylates histone H2B on serine 14 in *Arabidopsis*, which upregulates heat-responsive genes and allows plants to adapt to high temperature (Khadka *et al.*, 2020). The OsCIPK25 homolog also phosphorylates histone H2B at serine 14 in rice and its phosphorylation contributes to the plant's heat stress tolerance.

These examples demonstrate that histone phosphorylation conserves a function in stress-responsive gene expression across plant species, including the sites of phosphorylation, and associated kinases are essential for epigenetic control of stress adaptation (Khadka *et al.*, 2020).

5.2.5 Histone ubiquitination

The epigenetic modification of histone ubiquitination is also very important for the plant stress response. It is a covalent conjugation of ubiquitin molecules to the lysine residues on histone proteins. The type and extent of ubiquitination on a given target might have various effects on the gene expression associated with this modification (Chang *et al.*, 2019). However it has been demonstrated that histone ubiquitination can respond to stress-responsive genes in plants. HUB1 is an E3 ubiquitin ligase that participates with the monoubiquitination of histone H2B at lysine 143 during drought stress in *Arabidopsis* (Li *et al.*, 2024b). The type of ubiquitination mark it makes is connected with the activation of the drought-responsive genes, RD29A and COR15A, allowing the plant to react effectively to water deficit (Sharma *et al.*, 2023).

Like PUB26, the E3 ligase, the histone H2B ubiquitination in response to salt stress is increased in response of stress-responsive genes and increases salt tolerance (Yung *et al.*, 2021). Under drought stress, the E3 ubiquitin ligase OsPUB56 ubiquitinates H2B at lysine 143 in rice, allowing for the activation of drought-responsive genes like OsDREB2A and OsLEA3 to increase the plants drought tolerance (Li *et al.*, 2024b). In apple trees, the E3 ligase MdPUB15 is required for the monoubiquitination of histone H2B at lysine 143 following cold stress. It regulates the activity of cold-responsive genes, such as MdCBF1 and MdCOR15A, mediates the upregulation of cold-responsive genes and enhances the capacity of the plant to survive low-temperature conditions (Kurpisz and Pawłowski, 2022).

Moreover, histone ubiquitination can also lead to the transcriptional repression of growth genes and thus the plant can allocate resources to stress adaptation. In *Arabidopsis*, for example, the E3 ligase HOS1 targets the lysyl residue (and tandem repeats) of histone H2B in lysine 143 to facilitate the ubiquitination of histone H2B and gene silencing in response to cold stress, which consists of cell proliferation and development. The plant is left free to direct growth repression on its stress response mechanisms (Khadka *et al.*, 2020). The regulator of rice (OsPUB56) acts as an E3 ubiquitin ligase to ubiquitinate histone H2B at lysine 143 and downregulate growth development genes upon drought stress so as to shift the plant's focus toward drought stress response.

In apple trees, the E3 ligase also participates in another mechanism, wherein H2B in lysine 143 is mono-ubiquitin at cold stress to participate in reducing the expression of a number of growth-related genes and allowing the plant to survive better under cold temperature stress conditions (Melo *et al.*, 2021).

These examples show that histone ubiquitination in epigenetic control of stress responses in plants has various functional roles in modulating gene expression patterns to enhance tolerance and adaptation to stress. These epigenetic mechanisms are briefly described in Fig. 5.2.

5.2.6 Histone SUMOylation

One of the important post-translational histone modifications involved in plant stress responses is histone SUMOylation. For instance, small ubiquitin-like modifier proteins can be covalently attached to specified lysine residues of histone proteins so as to change their interactions with DNA as well as other chromatin-linked elements (Augustine and Vierstra, 2018). This histone SUMOylation is also reported in many plants, including *Arabidopsis*, rice, and apple trees, and plays a role in regulating stress responses to drought, heat, pathogens, and salinity.

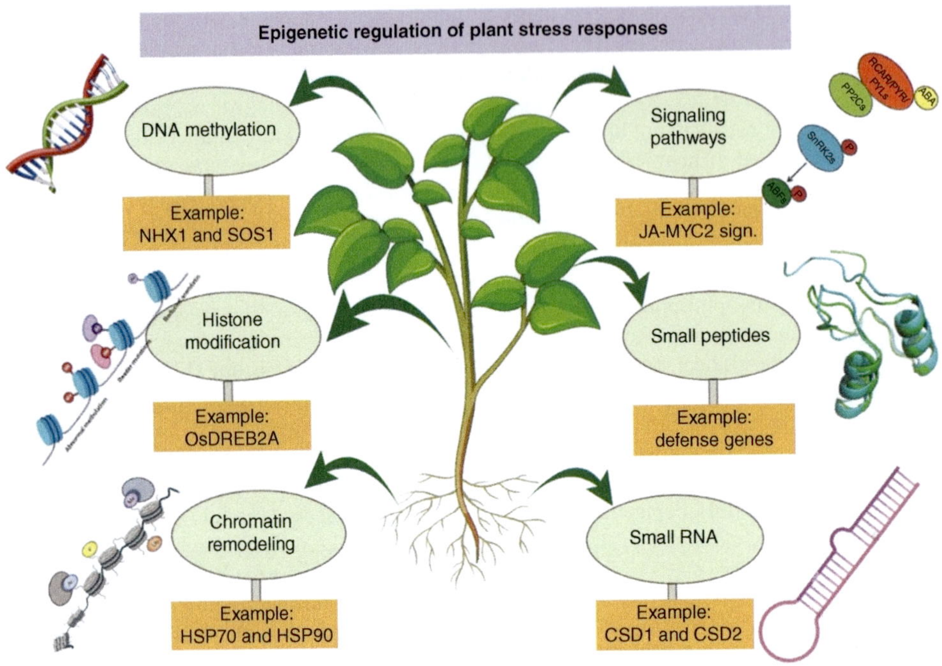

Fig. 5.2. Different mechanisms for epigenetic regulation of stress responses in plants with examples of genes. (Figure author's own.)

Histone SUMOylation can have opposing effects on gene expression, from gene activation or repression of stress-responsive genes in plants and thus fine tuning of the stress adaptation mechanisms in plants (Singh *et al.*, 2022; Li *et al.*, 2023; Sharma *et al.*, 2023). Histone SUMOylation has recently been implicated to play a role in the regulation of stress-responsive genes in plants.

SUMO moieties may, under stress conditions, attach to histones and cause transcription, activation, or repression depending on the targeted genes and stress response. This is demonstrated by the SUMOylation of histone H4 by the SUMO E3 ligase SIZ1 enhancing the expression of drought-responsive genes and therefore making a plant better adapted to drought conditions (Singh *et al.*, 2023a). Similarly, in rice, OsSIZ1, an E3 ligase for SUMOylation of histones, is shown to trigger the SUMOylation of histones in response to salt and that this correlates with salt-induced activation of genes encoding the traits to tolerate greater salinity levels. On the other hand, the SIZ1-mediated SUMOylation of histone H2B leads to repression of growth promoting genes as a developmental process as secondary to the stress response mechanism (Sharma *et al.*, 2023).

SUMOylation of histones by the SUMO E3 ligase MdSIZ1 is required for the activation of cold-responsive genes in the apple fruit tree to enhance cold tolerance (Li *et al.*, 2023). In addition, the stress regulatory mechanism involves the participation of the degree of histone SUMOylation. In *Arabidopsis*, the upregulation of heat-responsive genes was correlated with polySUMOylation on histone H2B. Together with polySUMOylation, it aided *Arabidopsis* survival against heat shock. However, as previously mentioned, the monoSUMOylation of H3 histone by SIZ1 is also involved in the activation of defense genes upon pathogenic microbial infection and immunity to such a pathogenic infection (Vertegaal, 2022).

These examples show that histone SUMOylation is an important process, which empowers the plant to control its gene expression in a very fine way and places the plant at an advantageous position in coping with various stressful conditions.

Finally, it can be concluded that plant stress responses are dependent on epigenetic states induced by the addition of acetylation, methylation, phosphorylation, and ubiquitination, as well as SUMOylation, to diverse histone modifications. Therefore, these histone modifications are dynamic regulators of stress-responsive genes and allow the plant to rapidly respond to and acclimate to a variety of environmental stresses. Thus it is necessary to histone modifications allow for the epigenetic regulation of the fine tuning of the stress response pathways through the modulation of gene access to the stress genes, as well as through recruitment and repelling of the chromatin-associated factors.

To gain an understanding of how these histone modifications function together and how they influence stress tolerance, it is necessary to develop strategies to enhance plant stress resilience through improved sustainable agricultural productivity.

5.3 DNA Methylation and Stress Responses

Another major epigenetic path for stress response in plants is DNA methylation. DNA methylation is the process of attaching a methyl group to the cytosine residues in the DNA sequence by DNA methyltransferase. It can change the pattern of expression of the gene in a cell. These DNA methyltransferase enzymes become activated in response to many stress conditions such as drought, heat, pathogen attack, and salinity, causing an increase in DNA methylation and gene expression response to stress to assist the plants adaptation to stresses (Liu and He, 2020).

It has been demonstrated that DNA methylation has an essential role in plants' responses to different kinds of stresses, including drought, heat, pathogen assault, and salt stress. For example, in the crop plant *Arabidopsis* drought stress induces increased DNA methylation in the promoter regions of stress-responsive genes such as RD29A and DREB2A which downregulates the expression of these genes. The downregulation of these genes result in resources saving to the plant which prioritizes the stress response mechanism (Chung *et al.*, 2021). As in the major fruit tree crop apple, high DNA methylation in the promoters of heat-responsive genes, such as HSP70 and HSP90, was increased by high temperature stress in the plant state. These changes contributed to the upregulation of these genes and the plant's ability to adapt to heat stress (Lang *et al.*, 2024).

In *Arabidopsis*, researchers found an elevation in DNA methylation on the promoters of defense-associated genes to be a mechanism by which the genes are induced during pathogen infection and by which the immune component of such plants is elevated (Huang and Jin, 2022).

In rice, it has also been reported that increased DNA methylation in salt-responsive gene promoter regions, such as NHX1 and SOS1, upregulates these genes and enables rice to adapt to a high salt environment (Saddhe *et al.*, 2021; Ullah *et al.*, 2022).

DNA methylation has also been shown to regulate plant response to biotic stress, for example during pathogen infection. DNA methylation in the promoters of defense-related genes has decreased in the model plant *Arabidopsis*, as well as in crop plants like rice and maize, when pathogens are present and the genes involved get activated for the enhancement of the plants immune to pathogens.

This stress-induced DNA demethylation can be passed along to the offspring of the plant increasing its pathogen resistance in *Arabidopsis*, rice, and maize. For example, these various mechanisms of DNA methylation represent the ways in which the latter can modulate gene expression profiles to best cope with conditions of different stresses (e.g. drought, heat, pathogen invasion, etc.). Therefore, the center of research for creating a strategy to produce crops that are more stress-resistant with a sustainable agriculture is the intricacy of interaction between DNA methylation and stress-inducible regulation of genes.

5.4 Small RNA-mediated Histone Modification

MicroRNAs (miRNAs) and small interfering RNAs (siRNAs) play an important role in directing the epigenetic regulation of the plant stress response through small RNAs. SiRNA molecules can affect gene expression patterns by degrading target mRNAs and by translational repression, as well as through regulating DNA and histone modifications (Prasad *et al.*, 2019).

The expression of some miRNAs and siRNAs may be changed under stressful conditions that are responsible for the post-transcriptional regulation of stress-responsive genes (Dzinyela *et al.*,

2023). For instance, the miR398 in *Arabidopsis* is downregulated under oxidatively stress for the possibility of upregulation of its target genes, CSD1 and CSD2, which encode copper/zinc superoxide dismutase (SOD) involved in the plant antioxidant defense system (AOS) (Li *et al.*, 2022). The drought stress reduces miR169 and increases its target gene, NUCLEAR FACTOR Y A, which is a transcription factor (TF) that increases drought tolerance in rice (Singh *et al.*, 2023a).

Small RNAs can also drive further DNA and histone modification such as DNA methylation and histone deacetylation in order to control expression of stress-responsive genes. SiRNAs specifically regulate DNA methylation by the RNA-directed DNA methylation pathway. In this case the siRNAs hit the genomic loci and covalently bind to DNA methylases for binding and transferring the methyl group to cytosine residues. To adapt to the water deficit conditions, the plant uses the siRNA-directed DNA methylation pathway to silence the promoter of the drought-responsive gene to prevent transcription (Kanno and Habu, 2011).

In rice, there is another related mechanism where siRNA-mediated histone deacetylation pathways were found to target salt-responsive gene promoters that got repressed via transcription and allowed the plants to adapt to saline conditions (Liu *et al.*, 2014).

Further, transgenerational changes in the expression of small RNAs due to stress are common and the plant's offspring may therefore enjoy improved tolerance of stress. Moreover, transgenerational changes in the expression of small RNAs due to stress are common, and the plant's offspring may therefore enjoy improved tolerance of stress. For example, drought stress induces changes in the levels of miR156, and miR159 in *Arabidopsis*. These altered levels of miR156, and miR159 in *Arabidopsis* are inherited to the progeny of the plant, resulting in increased drought/chill tolerance in the offspring (Zhang *et al.*, 2022).

All the above examples show the key role of small RNAs in epigenetic control of plant stress responses for the fine tuning of gene expression patterns to survive better in different kinds of stresses. To improve the stress resilience of important crop plants, it is important to understand how small RNAs mediate stress responses in plants.

5.5 Chromatin Remodeling

The integrity of chromatin remodeling is an integral process of the epigenetic regulation of plant stress responses. The basis to this dynamic mechanism is to remodel the nucleosome, a basic unit in chromatin, to strengthen gene expression changes in response to certain environmental stresses (Chinnusamy and Zhu, 2009). Chromatin remodeling in response to a wide variety of abiotic and biotic stresses has been shown to also take place in plants under drought, heat, salinity, nutrient deficiency, and pathogen attack. Upon stress these specific chromatin remodeling complexes are recruited to stress-responsive genes' promoters and change nucleosome positioning and histone modifications to activate or repress gene-expression (Luo *et al.*, 2011).

For instance, the SWI/SNF chromatin remodeling complex is shown to be recruited to the promoters of drought-responsive genes RD29A and DREB2A, exerting nucleosome eviction and subsequently leading to these genes being activated in response to water deficit stress (Nguyen *et al.*, 2022). Likewise, in rice, it was observed that the chromatin remodeling complex CHR729 targets the promoters of heat-responsive genes, for example HSP70 and HSP90, allowing for the nucleosome remodeling and u regulation of these genes under high-temperature conditions (Song *et al.*, 2021).

Chromatin remodeling has been reported in defense responses to biotic and abiotic stresses in the important fruit tree crop apple. There is evidence for the chromatin remodeling complex SWI/SNF targeting the promoters for defense-related genes, like PR1 and PR2, so that the gene can be activated and the immune response of the plant increased under pathogen attack. Additionally, DNA methylation and histone deacetylation at the promoters of salt-responsive genes such as NHX1 and SOS1 have been showing elevated levels in apple under salinity conditions. This helps to repress the transcription of these genes and the plant's ability to acclimate to high salinity environment (Duan *et al.*, 2018). Overall, these examples demonstrate how chromatin remodeling is an important aspect of the epigenetic regulation of plant stress response in different crop plants and fruit trees. Modulating the accessibility and structure of the chromatin allows chromatin remodeling complexes to activate or repress stress-responsive genes to give plants ability to adapt to many of the environmental challenges. This dynamic process of chromatin remodeling is a critical way by which plants can adjust their transcriptional programs, making them more stress resilient and more sustainable for agriculture production.

5.6 Small Peptides and Stress Responses

In most cases, short sequences of amino acids that are usually 5–50 amino acids in length, play an important role in plant stress responses. The precursor proteins contain small peptides. These small peptides are involved in stress-related processes, namely signaling, defense, and adaptation (Takahashi *et al.*, 2019). The plant elicitor peptides (PEPs) (Bartels *et al.*, 2013) are one of the subclasses of small peptides that are involved in plant stress responses and are derived from larger precursor proteins, the so-called precursor peptides of plant elicitor peptides (PROPEPs).

When pathogens attack or wound the plant, peps are also released leading to the induction of defense responses in the plant, including the production of reactive oxygen species (ROS) and defense-related genes (Bartels and Boller, 2015; Wang *et al.*, 2022).

Another set of plant stress-responsive small peptides, termed as phytosulfokines, are known to be derived from larger precursor proteins and have been shown to be involved in plant responses to abiotic stress such as drought and salinity (Tsintzou and Madesis, 2024). Usually, these peptides are formed by different proteases namely proteolytic cleavage of larger precursor proteins. Specific genes may encode the precursor protein, or the precursor protein may be produced by processing of other proteins, such as those belonging to primary metabolism or cell wall modifications. These molecules are therefore of great significance in the understanding of how epigenetic regulation of plant adaptation to environmental challenges involves diverse small peptides with respect to their presence and their role in the stress responses (Ren *et al.*, 2024). Their effects are mediated by small signaling peptides such as plant elicitor peptides, like their phytosulfokines, and these use specific receptor-ligand interactions.

Plant elicitor peptide receptors (PEPRs) are cell surface receptors that belong to the leucine rich repeat receptor kinase family of receptors, and which are the receptor for plant elicitor peptides (PEPs). Upon the binding of this receptor to ligand, this receptor-ligand binding induces MAPKs, production of ROS, and upregulation of defense-related genes (Ranatnuge and Hasara, 2024).

Like phytosulfokine peptides, the phytosulfokine receptor binds and signals downstream of the phytosulfokine receptor, a leucine-rich repeat receptor kinase that modulates plant growth and stress responses. Therefore, these small signaling peptides act as molecular messengers and they convey information about stress conditions, and the triggering of appropriate physiological and transcriptional responses in the plant to enhance its adaptation to environmental challenges (Li *et al.*, 2024a).

In fact, small peptides were seen to control transcriptional programs in receptor ligand complexes and in downstream signaling cascades and are therefore indispensable to the plants' ability to dynamically recognize chromatin during sensing and responding to various biotic and abiotic stress types.

5.7 Stress Signaling Pathways and Epigenetic Regulation

Thus, plants have evolved a series of complex signaling pathway to convert the environmental signals into the epigenetic response. The pathways between these external stimuli and a change in the plant's epigenome and transcriptional profile, as well as its adaptive responses, are referred to as stress signaling pathways (Gupta *et al.*, 2024). The abscisic acid (ABA) signaling pathway is one of the most studied signaling pathways in the process of stress-induced epigenetic regulation. ABA is a central phytohormone that responds to abiotic stresses, including drought stress, salinity, or high and low temperatures. Upon being perceived, these stresses caused increases in levels of ABA which initiates a signaling cascade and activation of ABA receptors, protein kinases, and TFs. Moreover, this signaling network has been shown to affect the activity of epimutations including DNA methyltransferases and histone-modifying enzymes to change the epigenome of the plant in response to existing stress conditions (Kaya *et al.*, 2024).

The jasmonic acid (JA) signaling pathway is also another important signaling pathway related to the plant's defense against biotic stressors such as pathogen attack and herbivory (Yang *et al.*, 2019). JA formation and signalling is associated with histone modifications, chromatin remodelling, and the generation of siRNAs that result in the transcriptional changes in defense related genes (Raza *et al.*, 2020).

Ethylene (ET) is another important signal transduction pathway that is connected to a plant's response to stress involving its biotic and abiotic stresses. Regulation of histone deacetylases, as well as DNA methyltransferases by ET biosynthesis and signaling, contributes to changes in the chromatin structure and expression pattern and adaptation as well as survival of the plant in such stress situations (Kaya *et al.*, 2024). Additionally, in the plant's defense against biotic stresses such as pathogen attack, the salicylic acid (SA) signaling pathway is important (Chen *et al.*, 2020). Upregulation of histone acetyltransferases in SA biosynthesis and signaling has been correlated with the SA-induced upregulation in defense related genes and positive regulation of plant immune response (Mishra *et al.*, 2024).

5.8 Epigenetic Memory and Inheritance of Stress Events

One peculiar plant characteristic is the ability to somehow "remember" exposure to stress and respond more quickly, more strongly to similar exposures later on. The epigenetics manner to generate this epigenetic memory of stress events could be affected by various mechanisms of epigenetic modifications, including DNA methylations, histone modification, and the siRNAs activity (Kinoshita and Seki, 2014).

The transgenerational inheritance of DNA methylation patterns is one of the best studied examples of epigenetic stress memory in plants. It is known that DNA methylation profiles are undergoing changes with abiotic stresses such as drought or high salinity and that they are eventually maintained and replicated in each cell division. Even after the stress is gone, such modified methylation patterns are transmitted to subsequent generations (Quadrana and Colot, 2016). Stable epigenetic inheritance of DNA methylation marks from stressed parent plants to their offspring to prepare the offspring to be more efficiently stressed by the same stress experienced by the parent. These specific mechanisms lie in the inheritance of DNA methylation patterns in the key stress-responsive genes that can be rapidly activated, along with the stress response pathways in the plants in the future, if they experience the same type of environmental challenge (Cao and Chen, 2024).

Histone modifications have also been implicated in the epigenetic memory of stress event, in addition to DNA methylation. Histone acetylation, methylation, and phosphorylation have been observed to change under such stress-induced conditions in plants, and such changes can be kept stable by holding this histone variant deposition with histone acetyltransferases, histone methyltransferases, histone kinases and the inheritance of the cells divisions. Instead, these histone marks become stably transferred and are then read on the cells of the subsequent generations, thereby contributing to the plants' greater stress response in the long term.

Promoter-associated modifications, such as increased H3K4 methylation or decreased H3K9 acetylation, can be selectively enriched or depleted at the promoters of stress-responsive genes. Then, these epigenetic marks prime the chromatin landscape, making the chromatin landscape ready to respond quickly in response to similar environmental challenges and thereby "precipitate" or enable more rapid activation or repression of these genes if and when the plant encounters similar stress (Oberkofler *et al.*, 2021).

In the establishment of epigenetic stress memory in plants, siRNAs are crucial. Upon exposure to a drought, salinity or pathogen attack stress-responsive genes are induced. This, in turn, activates stress-responsive genes, and the resultant stress-responsive siRNAs may target and silence expression of these stress-related genes. These stress-responsive siRNAs then pass on the silencing patterns to the following generations by means of epigenetic inheritance and get the plant's genome ready for a faster and more efficient response the following time that same stress was encountered in the future (Ali *et al.*, 2020).

It establishes transcriptional and post-transcriptional gene silencing gene by using siRNAs to make gene memory. Various stresses, such as drought, salinity, and pathogen attack, inspire the expression of stress-responsive genes in plants (Tan *et al.*, 2020). In this stress-induced activation, the stress-responsive siRNAs are formed, which can also target and silence the expression of genes relevant to that stress. They recruit the epigenetic silencing machinery, dominant miRNA off targets, namely DNA methyltransferases and histone modifying enzymes, to the promoters of stress-responsive genes via siRNAs. Such action sets up repressive epigenetic marks on the DNA at the point of transcription (DNA methylation), and the histone at the point of packaging and histone deacetylation that represses the transcription of those genes (Dinkar *et al.*, 2024). Epigenetic modifications of the siRNAs will then be stably inherited across cell divisions and even transferred to the plant offspring, thereby conditioning the genome to react more quickly and more violently if similar stress is subsequently experienced during their future. Such a memory of epigenetic stress mediated by siRNAs allows the plant to spare energy and resources in the absence of stress, and react quickly when stress is reintroduced, therefore improving its fitness and resilience to harsh environmental conditions (Junaid *et al.*, 2024).

5.9 Agricultural Applications of Epigenetic Regulation

Epigenetic regulation is a path through which a plant can harness its power to develop new strategies to increase the crop's stress resilience and promote sustainable agriculture (Samantara *et al.*, 2021).

The application to epigenetic engineering of plants for increasing plant stress tolerance is one of the key agricultural applications (Abdulraheem *et al.*, 2024). Direct control over epigenetic marks that affect stress-responsive genes, including DNA methylation and histone modifications, has allowed researchers to now modulate expression of stress-responsive genes. It serves to prime the plant's transcriptional machinery for a better adaptation to environmental stresses (Singh *et al.*, 2023b). Transgenic or genome editing technologies, such as CRISPR-Cas9 (Kumar *et al.*, 2023), may be used for the targeted modulation of epigenetic regulators, like those that include the *bona fide* DNA methyltransferases and the *bona fide* histone-modifying enzymes. Epigenetic engineering strategies have been explored in model plant species and crop plants to improve the drought stress and salt stress tolerance and pathogen resistance, respectively.

Overexpression of stress-responsive genes (Gan *et al.*, 2021) has been used to improve the drought resistance of rice by increasing the overexpression of the GCN5, a histone acetyltransferase. As in several crop plants, DNA methylation pattern manipulation has been connected with the enhanced abiotic and biotic stress tolerance. Besides direct epigenetic engineering, epigenetic biomarkers linked to responses to stress can be also used to design non-transgenic strategies to improve crop stress tolerance. This represents a useful tool for epigenetic information, without the transfer of foreign genetic material, in the field of sustainable agriculture. Natural variations in the DNA methylation patterns and histone modification signatures are revealed through epigenomic profiling and are indicative of superior stress tolerance in crop plants. These epigenetic variations can be leveraged by the marker-assisted selection (MAS) or as an epigenetic breeding to develop stress-resistant cultivars without using the genetic transformation (Kakoulidou *et al.*, 2021). Additionally, non-transgenic approaches could consist of treatment applications of epigenetic modulators, like DNA methylation inhibitors or histone deacetylase inhibitors, to induce a favorable epigenetic pattern and better stress adaptability of the plant. In these approaches epigenetic variation is used, and the manipulation of the epigenetic methods is used so as to improve agronomic traits in a more sustainable way, relying on the plant's own epigenetic machinery. Epigenomic data can be further integrated with other "–omic" technologies such as transcriptomics and metabolomics to better understand non-transgenic strategies to inform of the complex regulatory networks underlying stress-induced epigenetic changes in plants (Chao *et al.*, 2023).

5.10 Conclusion and Future Outlook

The study of epigenetic plant stress has rapidly advanced in the understanding of the intricate control of plant gateway to environment stress during adaptation and plant resilience to environmental challenges. The performance of cutting edge epigenomic profiling techniques, in integration with other high throughput –omics approaches, has enabled unprecedented picture of dynamic interaction of epigenetic modifications with transcriptional, proteomic, and metabolic changes in diverse stress conditions. By filling in the blanks produced by these systems level analyses, some of the key epigenetic modulators, such as DNA methyltransferases, histone modifying enzymes, and chromatin remodeling factors, have been identified as playing a central role in orchestrating the plant's stress responses.

By systematically characterizing these epigenetic regulators and their associations with other regulatory components, these epigenetic regulators have been characterized in the basis for a more complete understanding of stress-induced epigenetic changes and their functional consequences in plants.

To look towards the future, it is research on epigenetic stress that may have most potential in enhancing our knowledge of how plants adapt and how stable they are. However there is much more exciting research going on these days into the integrative multiple omissions approaches, epigenetic engineering for stress tolerance, epigenetic memory, transgenerational inheritance, epigenomic single cell applications, the interplay between epigenetics and plant epigenetics, and the microbiome. Together, it will help make agriculture more resilient and sustainable in the face of global climate change, as well as other environmental stressors.

While this field continues to refine the intricate epigenetic regulatory networks that control plant stress responses, a greater understanding of plant biology will not only enhance our fundamental understanding of which biological processes are important for a better understanding of the plants we depend on for our food, but will also create a wealth of knowledge that can be used to ensure crop productivity and the safeguarding of food security, and support for sustainable agriculture.

References

Abdulraheem, M.I., Xiong, Y., Moshood, A.Y., Cadenas- Pliego, G., Zhang, H. and Hu, J. (2024) Mechanisms of plant epigenetic regulation in response to plant stress: recent discoveries and implications. *Plants* 13(2), 163. DOI: 10.3390/plants13020163.

Ahmad, Y., Haider, S., Iqbal, J., Abbasi, B.A., Yaseen, T. and Mahmood, T. (2022) The mechanisms of genome editing technologies in crop plants. In: *Principles and Practices of OMICS and Genome Editing for Crop Improvement*. Springer International Publishing, Cham, pp. 295–313.

Ahmad, Y., Haider, S., Iqbal, J., Naseer, S., Attia, K.A., Mohammed, A.A. and Mahmood, T. (2024) In-silico analysis and transformation of OsMYB48 transcription factor driven by CaMV35S promoter in model plant–Nicotiana tabacum L. conferring abiotic stress tolerance. *GM Crops and Food* 15(1), 130–149.

Ali, M., Javaid, A., Naqvi, S.H., Batcho, A., Kayani, W.K. *et al.* (2020) Biotic stress triggered small RNA and RNAi defense response in plants. *Molecular Biology Reports* 47(7), 5511. DOI: 10.1007/s11033-020-05583-4.

Ali, I., Salah, K.B.H., Sher, H., Ali, H., Ullah, Z., Ali, A. and Mahmood, T. (2022) Drought stress enhances the efficiency of floral dip method of Agrobacterium-mediated transformation in Arabidopsis thaliana. *Brazilian Journal of Biology* 84, e259326.

Asghar, A., Ullah, Z., Ali, T., Iqbal, J., Abbasi, B.A., Ijaz, S. and Mahmood, T. (2024) Introduction to nc-RNA and their crucial role in biological regulation under environmental stress. In: *ncRNAs: Mediated Regulation: The Unarguable Biological Function to Environmental Constrains*. Springer Nature Switzerland, Cham, pp. 1–46.

Augustine, R.C. and Vierstra, R.D. (2018) SUMOylation: re-wiring the plant nucleus during stress and development. *Current Opinion in Plant Biology* 45, 143. DOI: 10.1016/j.pbi.2018.06.006.

Bartels, S. and Boller, T. (2015) Quo vadis, Pep? Plant elicitor peptides at the crossroads of immunity, stress, and development. *Journal of Experimental Botany* 66(17), 5183. DOI: 10.1093/jxb/erv180.

Bartels, S., Lori, M., Mbengué, M., Verk, M. van, Klauser, D. *et al.* (2013) The family of Peps and their precursors in Arabidopsis: differential expression and localization but similar induction of pattern-triggered immune responses. *Journal of Experimental Botany* 64(17), 5309. DOI: 10.1093/jxb/ert330.

Białoskórska, M., Rucińska, A. and Boczkowska, M. (2024) Molecular mechanisms underlying freezing tolerance in plants: Implications for cryopreservation. *International Journal of Molecular Sciences* 25(18), 10110. DOI: 10.3390/ijms251810110.

Borkiewicz, L. (2021) Histone 3 Lysine 27 trimethylation signature in breast cancer. *International Journal of Molecular Sciences* 22(23), 12853. DOI: 10.3390/ijms222312853.

Cao, S. and Chen, Z. J. (2024) Transgenerational epigenetic inheritance during plant evolution and breeding. *Trends in Plant Science* 29(11), 1203–1223. DOI: 10.1016/j.tplants.2024.04.007.

Chang, Y., Zhu, C., Jiang, J., Zhang, H., Zhu, J. and Duan, C. (2019) Epigenetic regulation in plant abiotic stress responses. *Journal of Integrative Plant Biology* 62(5), 563. DOI: 10.1111/jipb.12901.

Chao, H., Zhang, S., Hu, Y., Ni, Q., Xin, S. *et al.* (2023) Integrating omics databases for enhanced crop breeding. *Berichte aus der medizinischen Informatik und Bioinformatik/Journal of Integrative Bioinformatics* 20(4), 20230012. DOI: 10.1515/jib-2023-0012.

Chen, J., Clinton, M., Qi, G., Wang, D., Liu, F. and Fu, Z.Q. (2020) Reprogramming and remodeling: transcriptional and epigenetic regulation of salicylic acid-mediated plant defense. *Journal of Experimental Botany* 71(17), 5256. DOI: 10.1093/jxb/eraa072.

Chinnusamy, V. and Zhu, J. (2009). Epigenetic regulation of stress responses in plants. *Current Opinion in Plant Biology* 12(2), 133. DOI: 10.1016/j.pbi.2008.12.006.

Chung, S., Kwon, C. and Lee, J. (2021) Epigenetic control of abiotic stress signaling in plants. *Genes and Genomics* 44(3), 267. DOI: 10.1007/s13258-021-01163-3.

Cui, X., Dard, A., Reichheld, J. and Zhou, D. (2023) Multifaceted functions of histone deacetylases in stress response. *Trends in Plant Science* 28(11), 1245. DOI: 10.1016/j.tplants.2023.06.006.

Dinkar, V., Pandey, S., Kumar, A., Shiv, A., Lal, D. *et al.* (2024) Epigenetic regulations under plant stress: A cereals perspective. *Environmental and Experimental Botany* 220, 105688. DOI: 10.1016/j.envexpbot.2024.105688.

Duan, C., Zhu, J. and Cao, X. (2018) Retrospective and perspective of plant epigenetics in China. *Journal of Genetics and Genomics* 45(11), 621. DOI: 10.1016/j.jgg.2018.09.004.

Dzinyela, R., Alhassan, A.R., Kiani- Pouya, A., Rasouli, F., Yang, L. and Movahedi, A. (2023) Role of small RNAs in plant stress response and their potential to improve crops. *Crop and Pasture Science* 74(12), 1116. DOI: 10.1071/cp22385.

Fennell, A. (2014) Genomics and functional genomics of winter low temperature tolerance in temperate fruit crops. *Critical Reviews in Plant Sciences* 33(2), 125. DOI: 10.1080/07352689.2014.870410.

Gan, L., Wei, Z., Yang, Z., Li, F. and Wang, Z. (2021) Updated mechanisms of GCN5—the monkey king of the plant kingdom in plant development and resistance to abiotic stresses. *Cells* 10(5), 979. DOI: 10.3390/cells10050979.

Gupta, S., Kaur, R., Upadhyay, A., Chauhan, A. and Tripathi, V. (2024) Unveiling the secrets of abiotic stress tolerance in plants through molecular and hormonal insights. *3 Biotech* 14(10), 252. DOI: 10.1007/s13205-024-04083-7.

Haider, S., Iqbal, J., Naseer, S., Yaseen, T., Shaukat, M., Bibi, H. and Mahmood, T. (2021) Molecular mechanisms of plant tolerance to heat stress: current landscape and future perspectives. *Plant Cell Reports* 40(12), 2247–2271.

Haider, S., Raza, A., Iqbal, J., Shaukat, M. and Mahmood, T. (2022) Analyzing the regulatory role of heat shock transcription factors in plant heat stress tolerance: A brief appraisal. *Molecular Biology Reports* 49(6), 5771–5785.

Huang, C.-Y. and Jin, H. (2022) Coordinated epigenetic regulation in plants: A potent managerial tool to conquer biotic stress. *Frontiers in Plant Science* 12, 795274. DOI: 10.3389/fpls.2021.795274.

Ijaz, S., Iqbal, J., Abbasi, B.A., Yaseen, T., Ullah, Z., Alqahtani, T. and Mahmood, T. (2024) nc-RNA: A key player in stress response pathways, a molecular perspective. In: *ncRNAs: Mediated Regulation: The Unarguable Biological Function to Environmental Constrains*. Springer Nature Switzerland, Cham, pp. 139–163.

Junaid, M.D., Chaudhry, U.K., Şanlı, B.A., Gökçe, A.F. and Gökçe, Z.N.Ö. (2024) A review of the potential involvement of small RNAs in transgenerational abiotic stress memory in plants. *Functional and Integrative Genomics* 24(2), 74. DOI: 10.1007/s10142-024-01354-7.

Kakoulidou, I., Avramidou, E.V., Baránek, M., Brunel- Muguet, S., Farrona, S. *et al.* (2021) Epigenetics for crop improvement in times of global change. *Biology* 10(8), 766. DOI: 10.3390/biology10080766.

Kanno, T. and Habu, Y. (2011) siRNA-mediated chromatin maintenance and its function in Arabidopsis thaliana. *Biochimica et Biophysica Acta (BBA) – Gene Regulatory Mechanisms* 1809(8), 444. DOI: 10.1016/j.bbagrm.2011.05.002.

Kaya, C., Uğurlar, F. and Adamakis, I.S. (2024) Epigenetic modifications of hormonal signaling pathways in plant drought response and tolerance for sustainable food security. *International Journal of Molecular Sciences* 25(15), 8229. DOI: 10.3390/ijms25158229.

Khadka, J., Pesok, A. and Grafi, G. (2020) Plant histone HTB (H2B) variants in regulating chromatin structure and function. *Plants* 9(11), 1435. DOI: 10.3390/plants9111435.

Kinoshita, T. and Seki, M. (2014) Epigenetic memory for stress response and adaptation in plants. *Plant and Cell Physiology* 55(11), 1859. DOI: 10.1093/pcp/pcu125.

Kumar, K., Kumar, M., Kim, S., Ryu, H. and Cho, Y. (2013) Insights into genomics of salt stress response in rice. *Rice* 6(1), 27. DOI: 10.1186/1939-8433-6-27.

Kumar, K., Raina, S.K. and Sultan, S.M. (2020) Arabidopsis MAPK signaling pathways and their crosstalks in abiotic stress response. *Journal of Plant Biochemistry and Biotechnology* 29(4), 700. DOI: 10.1007/s13562-020-00596-3.

Kumar, M., Prusty, M.R., Pandey, M.K., Singh, P., Bohra, A., Guo, B. and Varshney, R.K. (2023) Application of CRISPR/Cas9-mediated gene editing for abiotic stress management in crop plants. *Frontiers in Plant Science* 14, 1157678. DOI: 10.3389/fpls.2023.1157678.

Kurpisz, B. and Pawłowski, T.A. (2022) Epigenetic mechanisms of tree responses to climatic changes. *International Journal of Molecular Sciences* 23(21), 13412. DOI: 10.3390/ijms232113412.

Lang, L., Wang, Z.W., Liu, B., Chang-qing, S.H.E.N., Jing-yi, T.U. *et al.* (2024) The effects and mechanisms of heat stress on mammalian oocyte and embryo development. *Journal of Thermal Biology* 124, 103927. DOI: 10.1016/j.jtherbio.2024.103927.

Li, S., He, X.J., Gao, Y., Zhou, C., Chiang, V.L. and Li, W. (2021) Histone acetylation changes in plant response to drought stress. *Genes* 12(9), 1409. DOI: 10.3390/genes12091409.

Li, J., Song, Q., Zuo, Z.-F. and Liu, L. (2022) MicroRNA398: A master regulator of plant development and stress responses. *International Journal of Molecular Sciences* 23(18), 10803. DOI: 10.3390/ijms231810803.

Li, X., Ma, Z., Song, Y., Shen, W., Yue, Q. *et al.* (2023) Insights into the molecular mechanisms underlying responses of apple trees to abiotic stresses. *Horticulture Research* 10(8), uhad144. DOI: 10.1093/hr/uhad144.

Li, Y., Di, Q., Luo, L. and Liang, Y. (2024a) Phytosulfokine peptides, their receptors, and functions. *Frontiers in Plant Science* 14, 1326964. DOI: 10.3389/fpls.2023.1326964.

Li, Y., Ma, T., Jiang, J., Hahn, M. and Yin, Y. (2024b) The dynamics and functional mechanisms of H2B mono-ubiquitination. *Crop Health* 2(1), 1. DOI: 10.1007/s44297-023-00022-9.

Liu, J. and He, Z. (2020) Small DNA methylation, big player in plant abiotic stress responses and memory. *Frontiers in Plant Science* 11, 595603. DOI: 10.3389/fpls.2020.595603.

Liu, X., Yang, S., Zhao, M., Luo, M., Yu, C. *et al.* (2014) Transcriptional repression by histone deacetylases in plants. *Molecular Plant* 7(5), 764. DOI: 10.1093/mp/ssu033.

Lujan, E. and Dinkova, T.D. (2021) Time to wake up: Epigenetic and small-RNA-mediated regulation during seed germination. *Plants* 10(2), 236. DOI: 10.3390/plants10020236.

Luo, M., Liu, X., Singh, P., Cui, Y., Zimmerli, L. and Wu, K. (2011) Chromatin modifications and remodeling in plant abiotic stress responses. *Biochimica et Biophysica Acta (BBA) – Gene Regulatory Mechanisms* 1819(2), 129. DOI: 10.1016/j.bbagrm.2011.06.008.

Melo, F.V., Oliveira, M.M., Saibo, N.J.M. and Lourenço, T. (2021) Modulation of abiotic stress responses in rice by E3-ubiquitin ligases: A promising way to develop stress-tolerant crops. *Frontiers in Plant Science* 12, 640193. DOI: 10.3389/fpls.2021.640193.

Mishra, S., Roychowdhury, R., Ray, S., Hada, A., Kumar, A. *et al.* (2024) Salicylic acid (SA)-mediated plant immunity against biotic stresses: An insight on molecular components and signaling mechanism. *Plant Stress* 11, 100427. DOI: 10.1016/j.stress.2024.100427.

Nguyen, N.H., Vu, N.T. and Cheong, J. (2022) Transcriptional stress memory and transgenerational inheritance of drought tolerance in plants. *International Journal of Molecular Sciences* 23(21), 12918. DOI: 10.3390/ijms232112918.

Nunez-Vazquez, R., Desvoyes, B. and Gutiérrez, C. (2022) Histone variants and modifications during abiotic stress response. *Frontiers in Plant Science* 13, 984702. DOI: 10.3389/fpls.2022.984702.

Oberkofler, V., Pratx, L. and Bäurle, I. (2021) Epigenetic regulation of abiotic stress memory: maintaining the good things while they last. *Current Opinion in Plant Biology* 61, 102007. DOI: 10.1016/j.pbi.2021.102007.

Prasad, A., Sharma, N., Muthamilarasan, M., Rana, S. and Prasad, M. (2019) Recent advances in small RNA mediated plant-virus interactions. *Critical Reviews in Biotechnology* 39(4), 587. DOI: 10.1080/07388551.2019.1597830.

Quadrana, L. and Colot, V. (2016) Plant Transgenerational Epigenetics. *Annual Review of Genetics* 50(1), 467. DOI: 10.1146/annurev-genet-120215-035254.

Ranatnuge, N.P. and Hasara, W.P.T. (2024) Plant Elicitor Peptides as Amplifiers of Immune Responses Against Biotic Stressors, p. 1. DOI: 10.1007/978-981-97-6374-0_1.

Raza, A., Charagh, S., Zahid, Z., Mubarik, M.S., Javed, R., Siddiqui, M.H. and Hasanuzzaman, M. (2020) Jasmonic acid: A key frontier in conferring abiotic stress tolerance in plants. *Plant Cell Reports* 40(8), 1513. DOI: 10.1007/s00299-020-02614-z.

Ren, G., Zhang, Y., Chen, Z., Xue, X. and Hai, F. (2024) Research progress of small plant peptides on the regulation of plant growth, development, and abiotic stress. *International Journal of Molecular Sciences* 25(7), 4114. DOI: 10.3390/ijms25074114.

Saddhe, A.A., Mishra, A.K. and Kumar, K. (2021) Molecular insights into the role of plant transporters in salt stress response. *Physiologia Plantarum* 173(4), 1481. DOI: 10.1111/ppl.13453.

Samantara, K., Shiv, A., Sousa, L.L. de, Sandhu, K.S., Priyadarshini, P. and Mohapatra, S.R. (2021) A comprehensive review on epigenetic mechanisms and application of epigenetic modifications for crop improvement. *Environmental and Experimental Botany* 188, 104479. DOI: 10.1016/j.envexpbot.2021.104479.

Sharma, M., Sidhu, A.K., Samota, M.K., Gupta, M., Koli, P. and Choudhary, M. (2023) Post-translational modifications in histones and their role in abiotic stress tolerance in plants. *Proteomes* 11(4), 38. DOI: 10.3390/proteomes11040038.

Shilpa, K., Thakur, R. and Prasad, P. (2024) Epigenetic regulation of abiotic stress responses in plants. *Biochimica et Biophysica Acta (BBA) – General Subjects* 1868(9), 130661. DOI: 10.1016/j.bbagen.2024.130661.

Singh, M., Singh, A., Yadav, N.R. and Yadav, D.K. (2022a) Current perspectives of ubiquitination and SUMOylation in abiotic stress tolerance in plants. *Frontiers in Plant Science* 13. DOI: 10.3389/fpls.2022.993194.

Singh, P. K., Indoliya, Y., Agrawal, L., Awasthi, S., Deeba, F. *et al.* (2022b) Genomic and proteomic responses to drought stress and biotechnological interventions for enhanced drought tolerance in plants. *Current Plant Biology* 29, 100239. DOI: 10.1016/j.cpb.2022.100239.

Singh, A., Jain, D., Pandey, J.L., Yadav, M., Bansal, K.C. and Singh, I.K. (2023a) Deciphering the role of miRNA in reprogramming plant responses to drought stress. *Critical Reviews in Biotechnology* 43(4), 613–627. DOI: 10.1080/07388551.2022.2047880.

Singh, R.R., Mahajan, M., Das, S. and Kumar, V. (2023b) Protein SUMOylation: Current updates and insights to elucidate potential roles of SUMO in plants. *South African Journal of Botany* 157, 398. DOI: 10.1016/j.sajb.2023.04.022.

Singh, V.K., Ahmed, S., Saini, D.K., Gahlaut, V., Chauhan, S. *et al.* (2023c) Manipulating epigenetic diversity in crop plants: Techniques, challenges and opportunities. *Biochimica et Biophysica Acta (BBA) - General Subjects* 1868(2), 130544. DOI: 10.1016/j.bbagen.2023.130544.

Song, Z., Liu, J. and Han, J. (2021) Chromatin remodeling factors regulate environmental stress responses in plants. *Journal of Integrative Plant Biology* 63(3), 438. DOI: 10.1111/jipb.13064.

Sun, J., Chen, H., Espino, P.S. and Davie, J. (2007) Phosphorylated serine 28 of histone H3 is associated with destabilized nucleosomes in transcribed chromatin. *Nucleic Acids Research* 35(19), 6640. DOI: 10.1093/nar/gkm737.

Sun, C., Ali, K., Yan, K., Fiaz, S., Dormatey, R., Bi, Z. and Bai, J. (2021) Exploration of epigenetics for improvement of drought and other stress resistance in crops: A review [Review of exploration of epigenetics for improvement of drought and other stress resistance in crops: A review]. *Plants* 10(6), 1226. DOI: 10.3390/plants10061226.

Takahashi, F., Hanada, K., Kondo, T. and Shinozaki, K. (2019) Hormone-like peptides and small coding genes in plant stress signaling and development. *Current Opinion in Plant Biology* 51, 88. DOI: 10.1016/j.pbi.2019.05.011.

Tan, H., Li, B. and Guo, H. (2020) The diversity of post-transcriptional gene silencing mediated by small silencing RNAs in plants. *Essays in Biochemistry* 64(6), 919. DOI: 10.1042/ebc20200006.

Todaka, D., Nakashima, K., Shinozaki, K. and Yamaguchi-Shinozaki, K. (2012) Toward understanding transcriptional regulatory networks in abiotic stress responses and tolerance in rice. *Rice* 5(1), 6. DOI: 10.1186/1939-8433-5-6.

Tsintzou, G.P. and Madesis, P. (2024) Peptide Elicitors for Defense Against Abiotic Stress, p. 19. DOI: 10.1007/978-981-97-6374-0_2.

Ullah, M.A., Abdullah-Zawawi, M., Zainal-Abidin, R., Sukiran, N.L., Uddin, Md. I. and Zainal, Z. (2022) A Review of Integrative Omic Approaches for Understanding Rice Salt Response Mechanisms. *Plants* 11(11), 1430. DOI: 10.3390/plants11111430.

Vertegaal, A.C.O. (2022) Signalling mechanisms and cellular functions of SUMO. *Nature Reviews Molecular Cell Biology* 23(11), 715. DOI: 10.1038/s41580-022-00500-y.

Wang, A., Guo, J., Wang, S., Zhang, Y., Lu, F. *et al.* (2022) BoPEP4, a C-terminally encoded plant elicitor peptide from broccoli, plays a role in salinity stress tolerance. *International Journal of Molecular Sciences* 23(6), 3090. DOI: 10.3390/ijms23063090.

Wang, C., Yan, J., Hu, B., Wang, R., Miao, B. *et al.* (2023) Genome-wide analysis of histone deacetylases in Apple and functional analysis of MdHDA6 in drought and salt stress responses. *Scientia Horticulturae* 320, 112196. DOI: 10.1016/j.scienta.2023.112196.

Wang, L., Li, C.-H., Liu, Y., He, L., Li, P. *et al.* (2024) Plant responses to abiotic stress regulated by histone acetylation. *Frontiers in Plant Science* 15. DOI: 10.3389/fpls.2024.1404977.

Yang, J., Duan, G., Li, C., Liu, L., Han, G., Zhang, Y. and Wang, C. (2019) The crosstalks between jasmonic acid and other plant hormone signaling highlight the involvement of jasmonic acid as a core component in plant response to biotic and abiotic stresses. *Frontiers in Plant Science* 10. DOI: 10.3389/fpls.2019.01349.7

Yin, M., Wang, S., Wang, Y., Wei, R., Liang, Y., Zuo, L. *et al.* (2024). Impact of abiotic stress on rice and the role of DNA methylation in stress response mechanisms. *Plants* 13(19), 2700.

Yung, W., Li, M., Sze, C., Wang, Q. and Lam, H. (2021) Histone modifications and chromatin remodelling in plants in response to salt stress. *Physiologia Plantarum* 173(4), 1495. DOI: 10.1111/ppl.13467.

Zhang, Y., Zhou, Y., Zhu, W., Liu, J. and Fang, C. (2022) Non-coding RNAs fine-tune the balance between plant growth and abiotic stress tolerance. *Frontiers in Plant Science* 13. DOI: 10.3389/fpls.2022.965745.

Zhou, H., Liu, Y., Liang, Y., Zhou, D., Li, S. *et al.* (2020) The function of histone lysine methylation related SET domain group proteins in plants. *Protein Science* 29(5), 1120. DOI: 10.1002/pro.3849.

Zhou, X., Lei, Z. and An, P. (2024) Post-translational modification of WRKY transcription factors. *Plants* 13(15), 2040. DOI: 10.3390/plants13152040.

6 Biotic Stress Responses: Pathogen and Pest Resistance

Salma Noureen[1], Javed Iqbal[1*], Banzeer Ahsan Abbasi[2], Tabassum Yaseen[1], Ghulam Murtaza[3], Rashid Iqbal[4,5] and Kamran Akbar[1]

[1]Department of Botany, Bacha Khan University, Charsadda 24420, Pakistan; [2]Department of Botany, Rawalpindi Women University, Rawalpindi 46300, Pakistan; [3]School of Agriculture, Yunnan University, Kunming, Yunnan 650504, China; [4]Department of Agronomy, Faculty of Agriculture and Environment, The Islamia University of Bahawalpur 63100, Pakistan; [5]Department of Life Sciences, Western Caspian University, Baku, Azerbaijan

Abstract

Owing to their sessile lifestyle, plants are continuously exposed to a broad range of environmental stresses. In addition to abiotic stresses, plants face the threat of infection by pathogens (including bacteria, fungi, viruses and nematodes) and attack by herbivore pests. Biotic stress is one of the major environmental factors affecting plants. Viruses, fungi, bacteria, insects and other pathogens represent a major constraint to agricultural productivity and a serious threat to vegetable and cereal crops. At first interface plants use trichomes, hairs and wax as physical defenses against pathogen attacks. This results in direct metabolic reactions and triggers the synthesis of stress proteins, as well as the activation of gene expression. Plants have developed a variety of defense mechanisms including danger-associated molecular patterns (DAMPs) and pathogen-associated molecular patterns (PAMPs). Small RNAs (sRNAs) are non-coding RNA molecules of about 20–30 nucleotides (nt) length and they control gene expression in eukaryotes by a process known as RNA silencing. Plants also have the ability to develop resistant gene-mediated immunity, known as ETI (effector-triggered immunity), to combat pathogen effectors. Histone modification, accumulation of sRNA, DNA methylation and chromatin remodeling are the signaling pathways through which transgenerational resistance can be brought on by pathogen attack. These epigenetic practices organize the plant stress responses and fine-tune the balance of plant growth and defense in plants. However, the majority of mechanisms underlying plant tolerance to combined stresses are still unknown and further studies are required to address them. In addition, studies are also required to uncover the complex mode of signaling pathways that underline the response of plants to various stress.

Keywords: biotic stress, chromatin remodeling, defensive mechanisms, DNA methylation, histone modification, host plant, molecular responses, pathogen effectors

6.1 Introduction

Stress is "any ecological factor possibly hostile to living creatures," (Jena *et al.*, 2020). Any stimulus that, when applied excessively, kills the organ or tissue in question (Linke *et al.*, 2024). The stress concept has also been used to describe environmental and adverse restrictions in plants (Koyro

* Email: javed89qau@gmail.com

DOI: 10.1079/9781800628823.0006

and Breckle, 2024). However, the notion of stress in humans and animals differs significantly from that of stress in plants (Liu *et al.*, 2025). Books and scholarly articles about plant stress and stress detection, significantly increased during the last ten years (McKersie and Lesheim, 2013). Such as stress and stress managing in cultivated plants, plant adaptation to environmental stress, plant stress reviews, proceedings of symposia, plant biochemistry, plant physiology, ecophysiology and journals of botany etc. (Márquez *et al.*, 2025). Additionally, *Vegetation Stress*, edited by Lichtenthaler (1996), contains more than 90 innovative contributions on stress effects and stress detection in plants (Cudjoe *et al.*, 2024). A unified definition of plant stress is necessary because majority of authors used the term "plant stress" in a very broad sense. This process is still ongoing and could continue at an even faster pace in the future (Challoumis, 2024).

A variety of stressors with various modes of action frequently cause the same or at least very comparable general reactions in the plant, in addition to extremely particular consequences (Du *et al.*, 2025). However, not every slight alteration to a plant's growth response, metabolic pathway, or development pattern should be interpreted as a stress effect, stress coping strategy and stress reaction (Choudry *et al.*, 2024). In other words, "stress" would not be applicable to quick reformations of metabolic fluxes, rates of transpiration as brought by differences in the photon flux density, an uncertain change in temperature, an increase/decrease in air humidity or photosynthetic rates and so on (Lichtenthaler, 1996).

Plants require unique stress avoidance and adaptation strategies since they are confined to their habitat and are unable to escape from the numerous environmental stresses (Lamalakshmi *et al.*, 2017). In reaction to shifting ecological conditions, plants are accustomed to and typically react flexibly to such consistently recurring variations in cell metabolism and physiological processes (Hofmann and Todgham, 2010). This also holds true for the daily variations in growth patterns, metabolic activity, and the processes of cell division and cell differentiation, all of which manifest at the day and night transitions (Aiello *et al.*, 2020). These changes can only be understood as a reorientation of development and metabolic responses based on the preferred activity of specific metabolic processes during the day or at night (Shahid *et al.*, 2024). In addition to rapid acclimation, plants can also react to ecological fluctuations by developing long lasting alterations to low and high light growth conditions (Zhang *et al.*, 2022). These adaptations can alter a variety of parameters, including leaf size, stomatal density, chloroplast ultrastructure and function by raising the levels of stress metabolites and photo shielding enzymes (Zhang *et al.*, 2022). These changes could happen in a day or two, or no more than a week (Weigelt *et al.*, 2021). In addition to avoiding stress limitations, plants with such light adaption responses, optimize their development and metabolism to the constantly altering environment (Demmig-Adams *et al.*, 2014).

The adaptation responses to high and low light levels frequently manifest not just in reaction to varying irradiance but also in response to a number of additional stressors (Morales and Kaiser, 2020). Even though they may quickly adapt to certain metabolic fluxes and established definite stress-enduring mechanisms, plants are frequently subjected to abrupt long- and short-term stress events that ultimately result in minimal cell activity and plant growth (Shi *et al.*, 2022). If plants' stress repair mechanisms are overused, this can result in serious damage, which ultimately leads to the cell death (Prasad *et al.*, 2022).

There are a number of natural stressors that, dependent on their duration and severity, can weaken plants and harm them. Each of these stressors can also be categorized as either biotic or abiotic (Quan, 2025). Various authors have given different definition of stress for plants in order to differentiate among characteristic acclimation and adaptation responses, stressors, stress restrictions and stress effects (Borowitzka, 2018).

6.2 Plant Stress

The two phrases that best sum up the ideas of Hans Selye are "All agents can act as stressors, producing both stress and specific action" and "There exist stressor specific response and nonspecific general response" (Haag, 2013). Any factor that disrupts, limits, or dangerously speeds up a plant's regular metabolic processes is referred to as plant stress (Sharma *et al.*, 2024). According to Larcher (2003) plant stress is "a specific condition in which plants demands are increasing, leading to an early decline in functions, followed by regularization and enhanced resistance" (Kuderov *et al.*, 2024).

Lichtenthaler (1996) developed the idea of plant stress "by differentiating between distress and eustress, comprising the regeneration phase of plants, which happens when stressors are removed. Distress is a severe stress condition that damages and harms the plant, eustress is a stimulating and activating stress, best for plant development" (Ahmed *et al.*, 2024). Even over an extended period of time, a modest stressor has no negative consequences and can boost a plant's physiological activity and cell metabolism (Hasanuzzaman *et al.*, 2013). For the plant, this kind of mild stimulating stress is beneficial (Holmgren and Scheffer, 2010). In any event, it is important to keep in mind that "stress is a dose-dependent matter" (Haag, 2013).

Abiotic or environmental variables including drought, nutrient deficiencies, severe temperatures, radiation and pollution, as well as biotic agents for example viruses, bacteria, fungi and oomycetes can lead to plant stress (Choudhary *et al.*, 2021) (Fig. 6.1).

Plants have developed a simpler yet more robust immune system than animals (Li *et al.*, 2020). Plant resistances can be categorized into physical resistances, included pre-existing barriers like plant bark, waxy cuticles, trichomes, thorns, spines, thicker leaves and cell walls and induced resistances, included the release of volatile organic compounds that prevent pathogenic attack, the amassing of signaling hormones and toxic chemicals (Kumar *et al.*, 2021). Due to their various resistance reactions, controlled by complicated signaling pathways, plants can redirect their growth effectively (Mishra and Bae, 2019).

A stressor, such an herbicide, can increase plant growth and metabolism at relatively low concentrations (Velini *et al.*, 2010). Therefore, it is possible for diffident dosages of a xenobiotic to have the opposite impact from higher concentrations (Arini *et al.*, 2023). However, if the stressor

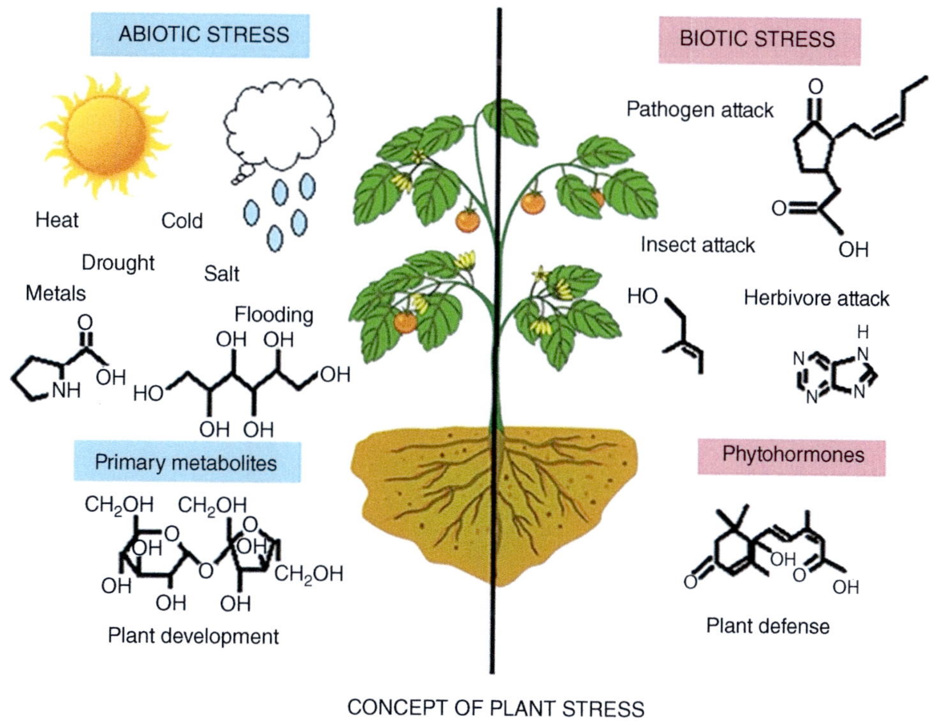

Fig. 6.1. Nature's trails, both seen and unseen where biotic and abiotic stresses shape a plant's resilience, crafting strength from adversity. (Figure author's own.)

is not eliminated, the same xenobiotics may harm the plant and cause in death (Shahid *et al.*, 2024). In the sense of a distress, harmful stressor and extraordinary doses of stress are damaging to plants' physiology. According to this theory, genuine stress occurs when a stressor reaches a threshold that the plant can no longer withstand (Issawi, 2018). More research is needed in this area since, although it appears to be evident, the applicability of the "the stressor dose and stress effects relationship" has not yet been established in any conditions (Kang *et al.*, 2024).

6.3 Phases Induced by Stress

By taking into consideration various concepts of plant stress, Selye (1936) proposed that there are four phases of the plant's stress responses (Gechev *et al.*, 2006). Prior to being exposed to stress, the plants are in a typical physiological state that is optimal within the constraints imposed by light, water, and mineral supply conditions (Bhattacharya and Bhattacharya, 2021).

When stressors or complicated stress events are removed, as long as the damage was not too great, the four phases of the stress response (Fig.6.2) will be followed (Bonanno *et al.*, 2010).

The four stress response phases are:

I. **Response phase**
This is the alarm reaction and occurs at the start of the stress. During this phase there is a decline of vitality, deviation of the functional norm and catabolic processes exceed anabolism.

II. **Restitution Phase**
This is the resistance phase and occurs under continuing stress. During this phase repair processes, hardening (reactivation) and adaptation processes occur.

III. **End Phase**
This is the stage of exhaustion and occurs due to long-term stress. During this phase high stress intensity, an increase in adaptation capacity and chronic disease/death occur.

IV. **Regeneration Phase**
During this stage, when the stressor is removed and as long as the damage is not too great, partial or full regeneration of the physiological function occurs.

When stress starts, plants react by reducing one or more biological functions, including photosynthesis, metabolite transport, and ion uptake (Lichtenthaler, 1996). The plants deviate from their systematic biological standard as a result of a decrease in metabolic activities and their

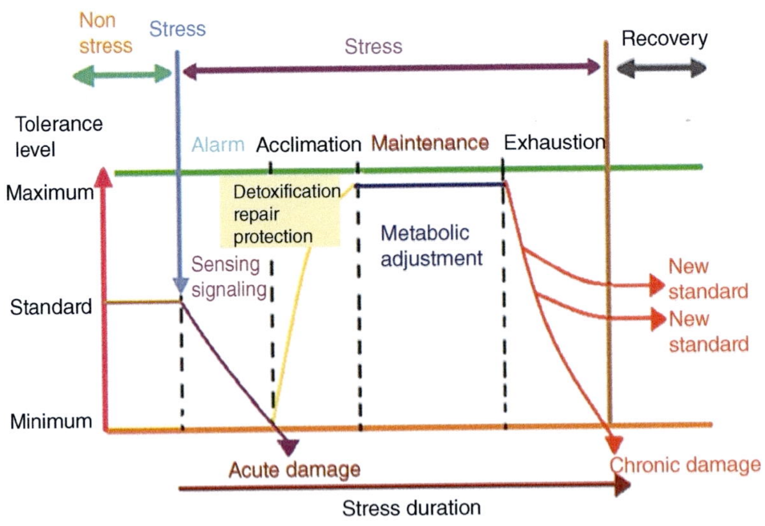

Fig. 6.2. From struggle to survival, stress unfolds in phases each challenge a stepping stone towards growth and resilience. (Figure author's own.)

vitality also lowers (Li *et al.*, 2024a). Plants with a low resistance or weak stress-tolerance mechanisms, experienced acute damage and senescence quickly (Wu *et al.*, 2018b). However, during this alarming phase, the majority of plants use their stress-preventing strategies, including long-term morphological adaptations, repair process activation, and metabolic flux acclimation (Kanaani, 2022). The General Alarm Syndrome (GAS) will cause a hardening of the plants by establishing a new optimal physiological state (Kumar *et al.*, 2023).

During the end phase, when plants experience prolonged stress, the physiology of plant gradually disappears (Baseer *et al.*, 2024). However, when stressors are eliminated at the appropriate moment before senescence processes take over, plants can regenerate through the "regeneration phase" (Kesineni *et al.*, 2023).

Both internal and external factors determined that for which duration the plant will remain in a new biological condition (Weyers and Paterson, 2001). This is definitely not too prolonged for field plants. Endogenous modifications to a plant's development program have always been linked to modifications to its physiological activities (Silveira and Sousa, 2024). It is important to remember that because there are numerous stressors, plants might be exposed to stress on a regular basis. As a result, a plant experiences stress and strain on a regular basis (Galviz *et al.*, 2022). However, persistent stress and strain may not always harm a plant (Kranner *et al.*, 2010). If the stress duration and intensity are not excessively high and prolonged, the plants can adjust themselves within a defined range (minimum and maximum) by resistance (Lichtenthaler, 1996). In these situations, signs of damage are not noticeable. External signals are represented by all biotic and abiotic stimuli, both natural and man-made (Zhang *et al.*, 2024a).

The plant parts (leaves, root, stem, and flowers) have a variety of ways to perceive signals (Das, 2024). On one hand, this results in direct metabolic reactions (i.e. modification of metabolic fluctuations), and on the other hand, it triggers the synthesis of stress proteins, stress metabolites, stress hormones, and activates gene expression and enzyme formation (Sharma *et al.*, 2024). The latter then regulate the minimum and maximum levels of stress resistance and further alter the metabolic reactions of the plants under stress. The relationship between metabolic reactions and gene expression is characterized by feedback controls (Engin, 2024). Many plant biology labs are currently conducting active research on these mechanisms, as well as the entire nature and chain of the signals perception and transduction processes under stress (Cao *et al.*, 2024).

6.4 Stress Detection by Using Chlorophyll Fluorescence Techniques

The traditional eco-physiological techniques of measuring rates of photosynthesis, contents of photosynthetic pigments (chlorophylls and carotenoids), and the concentration of stress metabolites can be used to identify stress-induced damages in plants (Sadhna *et al.*, 2023). The majority of stressors have an impact on the photosynthetic process, even if they have no direct effect on the makeup or operations of the photosynthetic systems (Farhan *et al.*, 2024a).

Heat emission and chlorophyll fluorescence significantly rise when the photosynthetic quantum conversion decreases under stress (Ahammed *et al.*, 2018). The red and far-red regions of the chlorophyll fluorescence have maxima around 685–690 nm and 730–740 nm, respectively (Verhoek, 2024). In the former case, both regions (Rfd 690 and Rfd 730) have the best Rfd-value measurements; the amplitude changes brought on by stress are greater in the 690 nm region than in the 740 nm region (Lichtenthaler *et al.*, 2005). The stress adaption index, which gives details on the stress exposure and strain on the photosynthetic machinery, can also be calculated using both chlorophyll fluorescence decrease ratio (Rfd) values (Hlihor *et al.*, 2022). The ratio of chlorophyll fluorescence emission in the red (F690) and far-red (F740) regions can be directly determined from the kinetics of chlorophyll fluorescence induction (Lichtenthaler, 2021). This ratio, F690/F740, is a measure of the amount of chlorophyll present in vivo (Verhoek, 2024). Despite the fact that they offered useful information about the stress conditions, various types of measurements of chlorophyll fluorescence still have fundamental drawbacks (Murchie and Lawson, 2013). In order to get an accurate picture of the functional state of photosynthetic system, it is essential to find the fluorescence signals of the leaf (Falcioni *et al.*, 2024) (Fig. 6.3).

6.5 Stress Detection by Fluorescence Imaging

The first high resolution laser-induced fluorescence (LIF) imaging system was created in collaboration with physicists. It enables the simultaneous imaging of chlorophyll fluorescence (F690 and F740) on a leaf surface. Plants exhibited a green (F520) and blue (F440) fluorescence emission when activated by UV-A light (Lichtenthaler, 2021). Unlike chlorophyll fluorescence, the blue and green fluorescence remain consistent during the light-induced fluorescence (Buschmann *et al.*, 2013). Phenolic compounds (i.e. hydroxycinnamic acids and flavonols) found in the cell walls of a plant are the main source of green and blue fluorescence (Otegui, 2021). Though not uniformly distributed across the entire leaf surface, the blue-green fluorescence emission is especially strong in the areas of leaf veins. In turn, vein-free leaf sections exhibit a strong emission of red and far-red chlorophyll florescence (Lichtenthaler, 2021). The high-resolution fluorescence imaging technology makes it simple to detect fluorescence gradients over the leaf surface (Buschmann *et al.*, 2013).

When it comes to identifying the effects of stress on plants, fluorescence imaging in the fluorescence bands established a new point of reference (Mangalraj and Cho, 2022). The blue/far-red (F440/F740), blue/red (F440/F690) and red/far-red (F690/F740) fluorescence ratios are very sensitive to the fluctuations in environment and plant growth conditions (Lichtenthaler, 2021). Accordingly, these ratios serve as early signs of alterations in the photosynthetic function of leaves, brought on by stress and strain (Lichtenthaler, 1996). Therefore, compared to point measurements of chlorophyll fluorescence used thus far, LIF imaging offers a very early, much better, and more accurate detection of stress and damage (Fig. 6.4.).

6.6 Stress-coping Mechanisms

Based on the type and intensity of stress, there are numerous stress-coping mechanisms that can be used by plants (Bondarchuk *et al.*, 2024) including proline, mannitol, and sorbitol amassing during salinity and drought; formation of heat shock proteins (HSPs) and radical scavenging compounds (ascorbate, glutathione, X-tocopherol); formation of UV-A and UV-B absorbing pigments in the epidermal layer; and an increase in superoxide dismutase (SOD) (Rai and Agrawal, 2017).

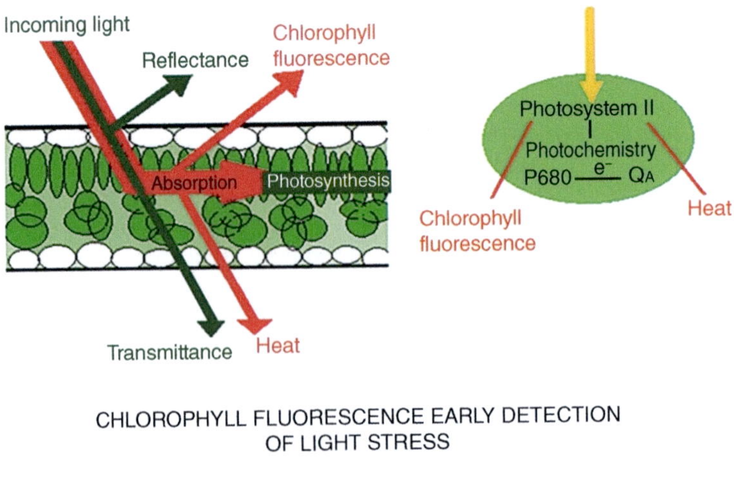

Fig. 6.3. Chlorophyll fluorescence: a silent whisper of stress, revealing how plants sense and respond to environmental challenges in real-time. (Figure author's own.)

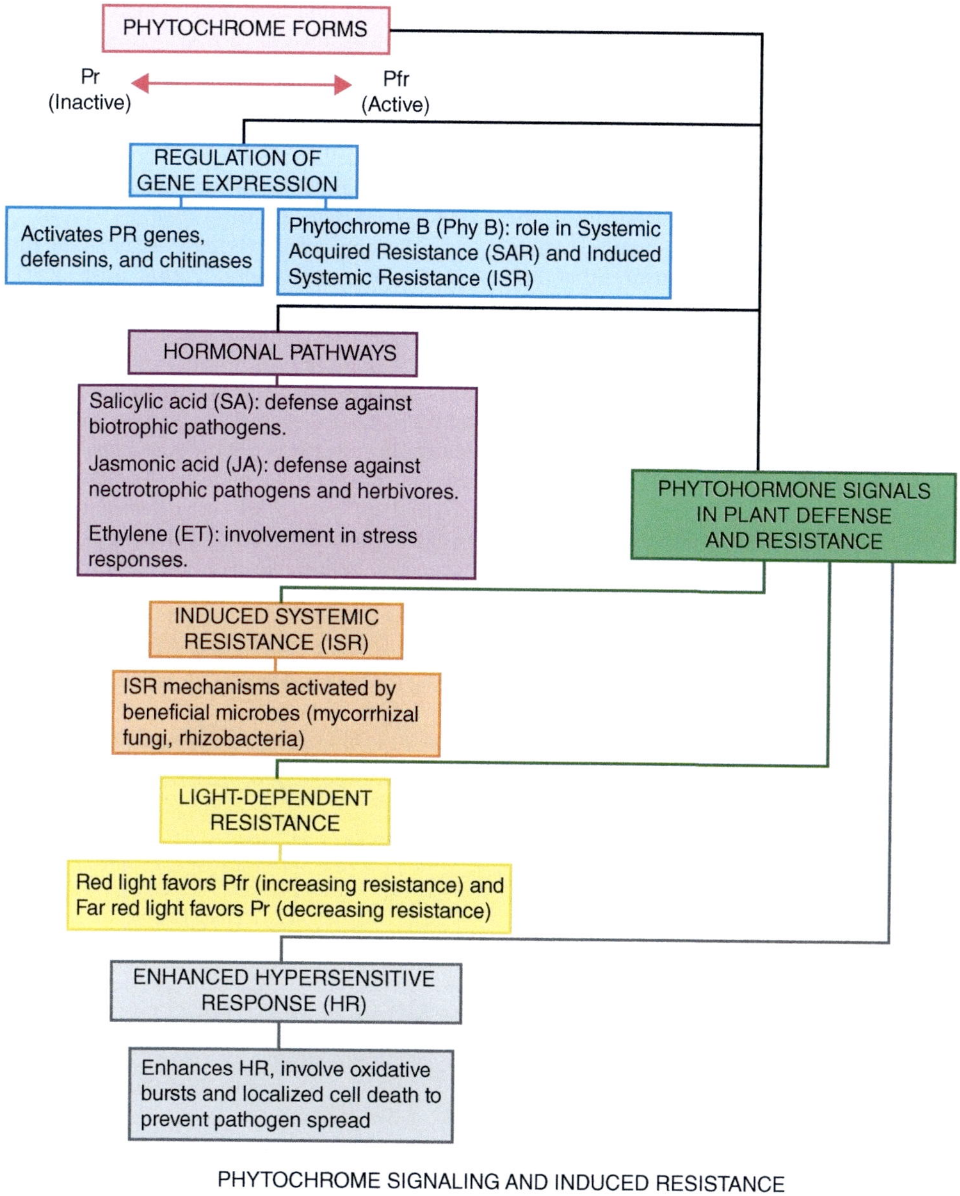

Fig. 6.4. Plants, resilient in nature, respond to stress through intricate signals with phytochromes, guiding their path to adaptation, and induced resistance. (Figure author's own.)

The carotenoid violaxanthin undergoes a rapid photo reduction (within minutes) in the thylakoids to zeaxanthin, which functions in the photosynthetic apparatus's photoprotection under high light conditions (Zia, 2010). After 05 hours of intense light exposure some plants, like *Tobacco aurea* mutant Su/su, can even triple their zeaxanthin levels through *de novo* production (Lichtenthaler, 2007).

The precise mechanism of zeaxanthin's photoprotective activity is yet unknown. However, zeaxanthin is not necessary for many chlorophyll fluorescence quenching mechanisms to occur (Dall'Osto *et al.*, 2014). A potential photoprotective mechanism of zeaxanthin includes an oxidation of zeaxanthin to violaxanthin either by the highly reactive oxygen species (ROS), formed at high-light conditions, or the epoxy groups formed at double bounds of thylakoid lipids (Kumar *et al.*, 2024b). The partial inactivation of photosystem-II centers by photo inhibition is a crucial mechanism. This process shields the photosystem-II centers from photo destruction. Thus, partial photo inhibition ensures that high enough CO_2 assimilation rates will be maintained under high light levels to support plant growth and development (Zivcak *et al.*, 2015).

6.7 Biotic Stress Responses

In order to prevent the loss of agricultural land and to fulfill the needs of the world population it is essential to find new ways to increase global food supply (Saini *et al.*, 2024).

Examples of biotic risks to food growth and transportation include bacteria, fungus, oomycetes, viruses, and insect pests (Balla *et al.*, 2021) (Fig. 6.5). Together, these factors cause up to 30% of global crop losses before and after harvest (Mesterhazy *et al.*, 2020).

All plant infections pose a threat to the plant's immune system. Plants have developed a defense mechanism to combat pathogen infection by either inhibiting or activating a wide range of genes (Roychowdhury *et al.*, 2024). In response to pathogen invasion, plants have a potential to recognize a variety of danger-associated molecular patterns (DAMPs) or pathogen-associated molecular patterns (PAMPs) (Patel, 2018). PAMPs or DAMPs are detected by plants using cell-surface localized pattern recognition receptors (PRRs). For example, chitin elicitor

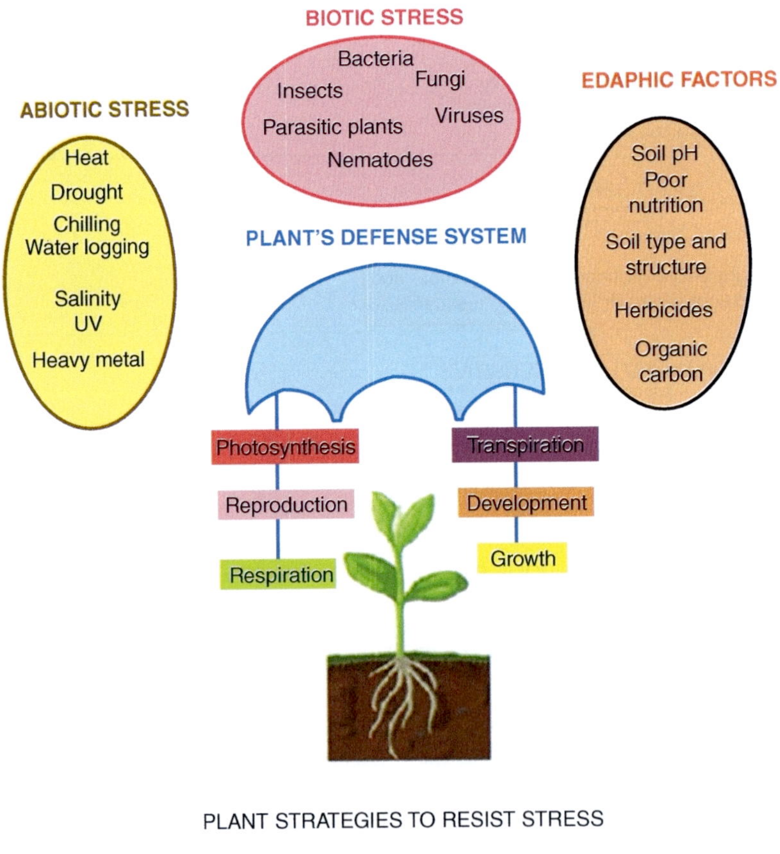

Fig. 6.5. Plants deploy a variety of survival strategies to thrive against stress, from deep roots to adaptive leaves, proving that even in adversity, growth finds a way. (Figure author's own.)

receptor kinase 1 (CERK1) and lysin motif receptor kinase 5 (LYK5) both detect fungi chitin, while elongation factor-TU (EF-Tu) receptor (EFR) and flagellin sensing 2 (FLS2) detect bacterial flagellin (Grabherr, 2011).

Salicylic acid (SA) buildup, ROS generation, callose deposition induction, and activation of pathogenesis-related (PR) genes are all components of PAMP-triggered immunity (PTI), which is activated by PAMP or DAMP (Tripathi *et al.*, 2024). Effector-triggered susceptibility (ETS) is the outcome of successful pathogens developing protein effectors that inhibit PTI (Naveed *et al.*, 2020). Effector-triggered immunity (ETI) is a secondary immunological response that plants have evolved in response (Yuan *et al.*, 2021). Resistance proteins, (which are able to identify and inhibit particular pathogen effectors) typically set off a more powerful and focused reaction, like the hypersensitive response (HR) (Wei *et al.*, 2024).

Small RNAs (sRNAs) are non-coding RNA molecules of about 20–30 nucleotides (nt) length (Röther and Meister, 2011). They control gene expressions in eukaryotes by a process known as RNA silencing (Almeida and Allshire, 2005). Small interfering RNA (siRNA) and microRNA (miRNA) are the two main types of sRNAs found in plants and are differentiated by their precursor and biogenesis mechanism (Singh *et al.*, 2018). Plants have a variety of siRNA subclasses, such as long siRNAs (lsiRNAs), heterochromatic siRNAs (hc-siRNAs), natural antisense transcript-derived siRNAs (nat-siRNAs) and trans-acting siRNAs (ta-siRNAs) (Javaid *et al.*, 2023).

The post-transcriptional gene silencing (PTGS) or transcriptional gene silencing (TGS) sRNAs affect the regulation of genes in both hosts and pathogens (Tan *et al.*, 2020). The TGS sRNA, which leads to histone modification, DNA methylation or chromatin modification, is typically mediated by siRNAs and certain specific miRNAs. While some of the processes in the biogenesis pathways of several sRNAs are common, many of them are complex and species-specific (Ponath *et al.*, 2022).

6.8 Responses Against Phytopathogens

The primary three types of phytopathogens are biotrophs, which feed on living tissue and gradually extract nutrients while maintaining the health and function of their host plants; necrotrophs, which overcome the plant's resistances by creating tissue-degrading enzymes and nourishing on dead tissues, thereby creating a loss of nutrients from host body; and hemibiotrophs, acting like a biotrophic during the early stages of infection but changing to necrotrophic mode in the later stages of disease cycle (Kashyap and Kumar, 2021).

The induced characteristic immunity in plants, sometimes referred to as basal resistance, provides the next line of active resistance after the physical defense barriers (Doughari, 2015). This come about at the plant's cell surface when unique trans-membrane proteins known as PRRs detect microbe-associated molecular patterns (MAMPs) on the host's cell surface (Kishore, 2010). These patterns include liposaccharides, β-glucans, pepetidoglycans, chitin, specific proteins, and bacterial flagellin, normally found in microbes (Kashyap and Kumar, 2021).

Non-host specific resistance that occurs at species level is caused by MAMP/ ETI. Physiological reactions to ETI, included increased Ca^{2+} ion concentration, generation of ROS, modifications of plant cell membranes and fortification of cell walls (Yang *et al.*, 2024). By injecting effector molecules, proteins or poisons produced by microbes can inactivate the regulators of the host immune system (Yin *et al.*, 2024). It has been demonstrated that some effectors, such as AvrPto1 from the *Pseudomonas syringae* strain, increase pathogen's virulence by deterring immune related proteins in tomato plants (Gowtham *et al.*, 2024). Plants have responded by developing resistance proteins, which are intracellular immune receptors that can initiate ETI (Liu and Tang, 2023). The HR can be induced in bacteria, viruses, fungi, and microscopic nematodes (Noman *et al.*, 2020). HR, linked to the synthesis of antimicrobial compounds and planned cell death at infection site, is a common characteristic of this second phase of protection.

In 1961, a phenomenon for tobacco plants infected with the Tobacco Mosaic Virus (TMV) was identified and named systemic-acquired resistance (SAR). SAR is a broad-spectrum, local HR immune response that mobilizes resources across the plant to respond quickly to any subsequent threats (Lapshina, 2024). SAR relies heavily on moving immunological signals. The infected tissue produces these signals, which are quickly spread to other plant components. Ethylene (ET), abscisic acid (ABA), SA and jasmonic acid (JA) are the main signaling phytohormones that accumulate with the onset of SAR (Kaya *et al.*, 2023). Generally, it has been observed that SA enhances biotic stress tolerance and ABA-mediated phytohormone signaling promotes abiotic stress tolerance (Faizan *et al.*, 2022).

6.9 sRNAs Role in Host PAMP-Triggered Immunity

In reaction to different pathogen attacks, plants have evolved complex defense mechanisms, and diseases have also evolved a variety of suppressors as counter defenses (Makandar, 2024). The proportional contribution of susceptibility and resistance variables determines the outcome of plant-pathogen interactions (Vanderplank, 2012). The host PAMP-triggered immunity (PTI) response is instantly triggered by the entry of pathogens (Wang *et al.*, 2022).

JA, auxin, ABA and SA are among the plant hormones that are modified by miRNAs and siRNAs, which are essential for PTI (Zhang *et al.*, 2011). *Arabidopsis* miRNA (miR393) was the first miRNA family, linked to PTI (Qin *et al.*, 2021). A peptide derived from flagellin (flg22) increased miR393 in response to an attack by bacterium (*Pseudomonas syringae)* (Li *et al.*, 2011).

While glucosinolates are antimicrobial compounds that help plants fight off pests and illnesses, SA is in charge of protection against biotrophic pathogens (Dutilloy *et al.*, 2022). Subsequent research showed that miR393's effect on auxin signaling and increased glucosinolate levels, decreased camalexin levels, hence improved *Arabidopsis* resistance to *P. syringae* (Das and Gutjahr, 2019). Additionally, in some other plants, e.g. *Oryza sativa*, miR393 works similarly to auxin signaling and causes early blooming and more tillering (Sihag *et al.*, 2024).

By focusing on genes involved in plant hormone production and signaling cascades, it was observed that bacterially-controlled miRNAs contributed significantly to plant defense (Spada *et al.*, 2024). *Candidatus Liberibacter asiaticus*(CLas), causes Huanglongbing (HLB), commonly referred to as a greening disease of *Citrus*, was the particular cause of miR399 in orange tree (*Citrus sinensis*) (Bilal *et al.*, 2024). Additional research indicates that phosphorus deficit brought on by HLB may be the reason of rise in miR399 (Zhong *et al.*, 2021). It has been confirmed that miR408 targets genes that encode plantacyanin-like proteins (PLCLs) in *Triticum* and *Arabidopsis* (Shao *et al.*, 2024). *Triticum aestivum* L. is more susceptible to stripe rust due to miR408, which also adversely affects the production of TaCLP1 (a chemocyanin like protein gene) (Zhou *et al.*, 2020). Fungal and oomycete infections have also been shown to involve miRNAs in PTI (Yang *et al.*, 2021). It was discovered that rice treated with fungal mycelia accumulated Osa-miR7695 (an miRNA) (Li *et al.*, 2021). Many miRNAs were generated in *T. aestivum* by a fungus (*Blumeria graminis*); among these miR1138, miR444, miR408, miR171, and miR167 are likely involved in PTI (Dube, 2017).

6.10 Pathogen Effectors and Host Defensive Mechanisms

Bacteria and fungi (oomycetes) have developed suppressors to slow down the host defensive mechanisms and promote infection in response to plant immunity (Tripathi *et al.*, 2024). Numerous effectors have been found. For example, *P. syringae* secretes about 30 effectors. By controlling the accumulation of sRNAs, pathogen effectors disrupt the RNA silencing system (Li *et al.*, 2024b). *Phytophthora* effectors of RNA silencing 1 and 2 were reported to decrease the accumulation of sRNAs in the oomycete *Phytophthora sojae* (Qiao *et al.*, 2021). By attaching itself to the conserved nuclear protein PSR1-interacting protein 1 (PINP1), PSR1 can reduce the amounts of endogenous siRNAs and miRNAs (Mencia *et al.*, 2023). According to Qiao *et al.* (2015), PINP1 is necessary for the amassing of different kinds of sRNAs by promoting the formation of complexes. PSR2 plays a role in reducing the accumulation of some trans-acting small-interfering RNAs (tasiRNAs). PSR2 does not impact miR390-mediated TAS3 tasiRNAs, but it can target miR173 to inhibit the production of ASRP1151 and ASRP255 tasiRNAs (Huang *et al.*, 2016).

Effectors can disrupt the RNA silencing process, in addition to inhibiting sRNA accumulation (Ye and Ma, 2016). Additionally, a recent study demonstrated that host PTI can be inhibited by using sRNAs, encoded by pathogens (Kunz *et al.*, 2024). Transferred from *Botrytis cinerea* to a host plant, Bc-sRNAs have the ability to take over the host argonaute-1 protein (AGO1) and hence inhibit RNA silencing. Broadly speaking, BcsiR3 and Bc-siR5 are both unique kinds of pathogen effectors (He *et al.*, 2023).

6.11 Host Effector-triggered Immunity to Defend Against Pathogens

Plants developed R gene-mediated immunity, also known as ETI, to combat pathogen effectors (Tailor and Bhatla, 2024). R proteins often produce an HR at the infection site by identifying pathogen effectors and inducing strong cellular alterations (Devi *et al.*, 2024). The nucleotide-binding site–leucine-rich repeat (NBS–LRR) gene family includes the majority of plant R genes (Yan *et al.*, 2024).

Plant genomes encode hundreds of different NBS–LRRs that enable the identification of several diseases (Zhao *et al.*, 2024). Under typical circumstances, R protein activity and quantity are kept low to conserve means for plant growth (Kaiser *et al.*, 2024). On the other hand, pathogen effectors can reduce PTI when plants are attacked, which causes plants' R genes to be upregulated and ultimately causes ETI (Yu *et al.*, 2024).

Directing the expression of R genes is the main function of sRNAs (Ribeiro *et al.*, 2024). Suppressor of NPR1-1, constitutive 1 (SNC1) and Recognition of *Peronospora parasitica* 4 (RPP4) are locus R genes in *Arabidopsis* that overexpress viral suppressors of RNA silencing (VSRs) (Henchiri, 2021). Furthermore, miRNAs are made to target the conserved area of NBS–LRRs in order to boost regulatory effectiveness (Liu *et al.*, 2024a). This means that a single miRNA can target several NBS–LRR genes and suppress their expression (Rodrigues *et al.*, 2024).

Furthermore, certain miRNAs have the ability to synthesize secondary siRNAs, which further strengthen the regulatory effect by directing the expression of targeted genes (Ali and Tang, 2024). Three miRNA (22 nt) families (miR2118, miR2109, and miR1507) have been identified in *Medicago truncatula*, targeting conserved regions in NBS-LRRs (Sudan *et al.*, 2020).

While miR482/miR2118 (a superfamily of miRNAs) are the main members that regulate R genes in tomatoes (*Solanum lycopersicum*), miR6019 and miR6020 (miRNAs conserved in solanaceae plant) regulate TIRNBS-LRR (TNL)-type (a large family of nucleotide binding site leucine rich repeats) genes in tobacco (*Nicotiana benthamiana*) (Rodrigues *et al.*, 2024). The cleavage induced by miRNAs in both situations leads to the creation of siRNAs, which strengthens the repression of R genes (Li *et al.*, 2024b). When resistant tomato cultivars are treated with the fungus *Fusarium oxysporum*, several miRNAs are downregulated, whereas miR398, miR482, and miR5300 (a number of 22 nt miRNAs) are increased (Ding *et al.*, 2024).

MiR5300/MiR482-mediated nucleotide binding (NB) gene regulation plays a significant role in tomato resistance to fungi, as evidenced by the induction of three additional NB domain genes (Shen *et al.*, 2024). miR472 in *Arabidopsis* has also been demonstrated to influence both ETI and PTI (Kumar *et al.*, 2024a). Furthermore, the regulation of Mildew locus, an allele encoded CCNBSLRR (N-terminal coiled-coil domain), receptors is suppressed in barley (*Hordeum vulgare* L.) by a regulation network formed by the miR9863 family (Ye and Ma., 2016). The overexpression of miR9863 decreased cell-death signaling and resistance to powdery mildew fungus caused by MLA1 (Ding *et al.*, 2024). The first plant endogenous siRNA, *Arabidopsis* natural antisense transcript-associated siRNA (nat-siRNAATGB2), (a bacterial infection-induced nat-siRNA) functions as a regulator in avrRpt2-triggered ETI (bacterial effector protein) (Seo *et al.*, 2013). Pentatricopeptide repeats (PPRL) are silenced by natsiRNAATGB2, which also stops PPRL's detrimental regulatory effects on the RPS2 resistance pathway (Huang *et al.*, 2016). According to a current study, miR393b is added to Argonaute 2 (AGO2) to inhibit MEMB12 (Member of Membrin Gene Family) gene expression. The plant's antibacterial activity is subsequently increased when MEMB12 is silenced because it increases PR1 exocytosis (small, secreted proteins and exocytosis regulators) (Zaheer *et al.*, 2024). The paired strand of miR393b, miR393, has been found to be loaded into AGO1 and participate in the auxin pathway (Rong *et al.*, 2024). As a result, two sRNAs produced from the same duplex promote the development of plant resistance via several AGOs pathways (Ortola and Daros., 2024).

6.12 sRNAs and Plant Immunity

Millions of insects depend on plants for their nourishment. Plants use trichomes, hairs, and wax as physical defenses against insect pest attacks. Aphids attack plants by sucking nutrients from the phloem using their specialized mouthparts, called stylets (Ali *et al.*, 2024b). Aphids release saliva while feeding as soon as they pierce the plant cell's plasma membrane. Aphid-secreted honeydew may also comprise compounds that change plant resistance mechanisms (Huang *et al.*, 2023). Aphids can add virus effectors into plants through these processes (Kumaraswamy and Huang, 2024). Herbivores (insects) may be poisoned by the accumulation of phytochemicals, which is caused by direction of genes involved in transmission of plant hormones (Yousuf *et al.*, 2024). Attacks by insect also change the amount of plant hormones (Singh *et al.*, 2024).

Latest studies indicate that plant RNA machinery, in addition to plant hormones, is crucial for plant protection against insect herbivores (Saakre *et al.*, 2024). Plant vulnerability to the moth *Manduca sexta* was markedly increased when RNA-dependent RNA polymerase-1 (RDR-1) in *Nicotiana attenuata* was silenced. The transcriptome of sRNAs in wild type (WT) and RDR-1-silenced plants of *N. attenuata* was characterized by both before and after oral secretion (OS) elicitation (Huang *et al.*, 2016). Numerous sRNAs were found to be up and downregulated in response to OS elicitation. These findings may be related to the extensive transcriptional alterations that follow herbivore attack (Chen *et al.*, 2024a; Singh *et al.*, 2024).

By feeding on photo assimilates and spreading numerous viruses, aphids are among the key herbivores that significantly reduced agricultural yields (Mostafa *et al.*, 2024). Aphid resistance has been demonstrated to be influenced by host surface-localized PRRs (Hao *et al.*, 2024). According to Prince *et al.* (2014), extracts from peach aphid (*Myzus persicae*) caused *Arabidopsis* to react similarly to PTI (Kumaraswamy *et al.*, 2024). Furthermore, R proteins may detect the effectors released by aphids and cause ETI (Devi *et al.*, 2024). Insect attack and plants have an active interaction that has been investigated at the molecular level. Secondary metabolites in plants are all regulated as a result of insect attack. (Panwar *et al.*, 2013; Farhan *et al.*, 2024b). In order to increase plant defense, it was discovered that aphids feeding on *Arabidopsis* force plants change one indole glucosinolate to another (Gloss *et al.*, 2022). Camalexin is another secondary metabolite that was discovered to support the defense response against aphids (Kumaraswamy *et al.*, 2024). Aphid feeding alters the amounts of plant hormones. For example, *Arabidopsis* was demonstrated to alter hormone level in response to *M. persicae* invasion (Yang *et al.*, 2024). It is remarkable to note that camalexin-related aphid resistance is likewise mediated by the *Arabidopsis* miRNA system (Chesnais *et al.*, 2022). The messenger (mRNA) level of peptidylarginine deiminase type 3 (PADI3), a marker for the camalexin biosynthesis pathway, significantly increased when they are exposed to aphids (Mechin *et al.*, 2010). It is yet unknown, nevertheless, whether insects produce sRNAs or double-stranded RNAs (dsRNAs) that enter into plant's body and control plant defense mechanisms (Verma and Modgil, 2024) (Fig. 6.6a, b and c).

6.13 Role of sRNA

Interaction between a pathogen and plant host is implied by sRNAs in plant immunity. Numerous plant pathogens, insect, parasite, and symbiotic interactions have recently been found to exhibit this communication (Mapuranga *et al.*, 2022; Huang *et al.*, 2024). In plants, sRNAs can be transported across phloem vascular tissue through plasmodesmata (Zainali *et al.*, 2024). The ability of transgenic plants to effectively silence genes in pathogens and pests through the expression of exogenous sRNAs/dsRNAs suggests that plants can transmit sRNAs as information to organisms (Ali *et al.*, 2024a). It is unclear how sRNA is transported between plants and pathogens, though. It has been demonstrated that both vesicular and non-vesicular transportation can transfer sRNAs into animal cells (Cai *et al.*, 2023). Sorting sRNAs into vesicles through fusing them with the plasma membrane

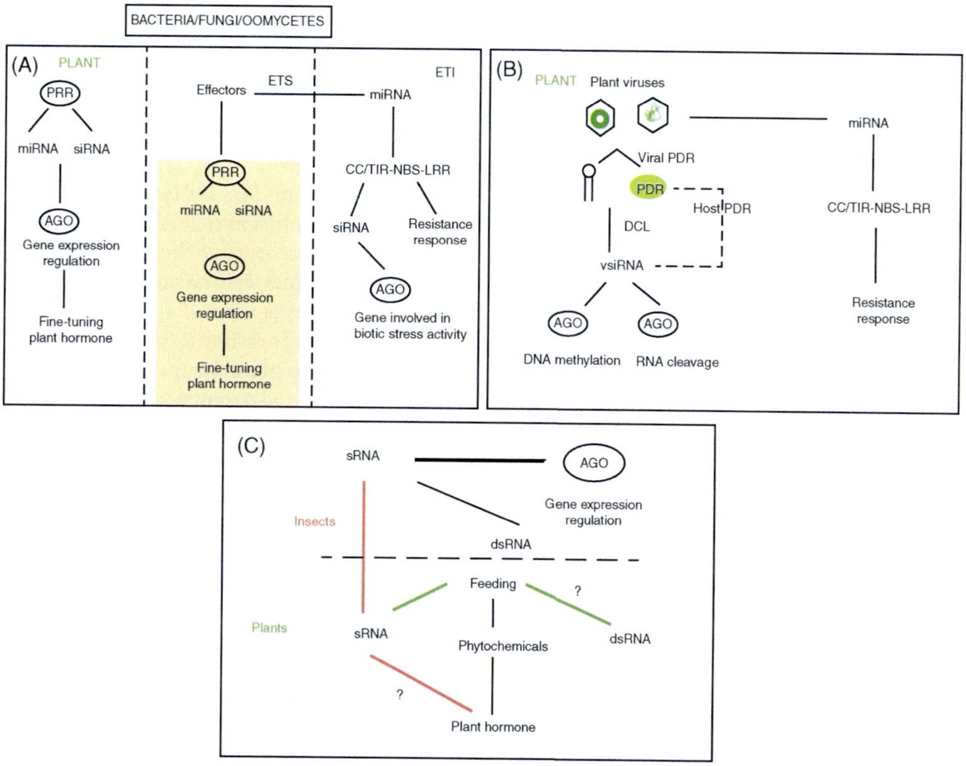

Fig. 6.6a,b,c. sRNAs play a crucial role in enhancing plant immunity by regulating defense pathways and targeting insect herbivores genes, enabling plants to effectively respond to herbivore attacks. (Figures author's own.)

and exosomes to liberate the sRNAs is known as vesicular transportation (Yadav *et al.*, 2024a). The direct uptake of RNA signals via RNA transporters is referred to as non-vesicular transportation (Stewart *et al.*, 2024). Systemic RNAi Defective 1 (SID1) and Systemic RNAi Defective 2 (SID2) are the two membrane-associated RNA transporters, which were found in *Caenorhabditis elegans* (Arellano, 2020). Certain fungi form specialized cells called haustoria to create an interface during their interactions with plants (Balios *et al.*, 2024). This interface is the site of numerous processes, such as food uptake, fungal effector protein production, enzyme/toxin provision to plant cells and cell surface sensor biogenesis (Li *et al.*, 2024a).

According to Koch *et al.* (2020) sRNAs may need membrane-associated receptors to migrate from the cytoplasm of plants to the haustorium through exosomes. Nevertheless, plant pathogens have not yet been found to possess membrane-associated RNA transporters (Chen *et al.*, 2024a). According to Kunz *et al.* (2024), sRNAs can move in both directions and are not only transferred from host to pathogen. It has been confirmed that the fungal pathogen *Botrytis cinerea* (Bc) may transfer sRNAs to plants and take control of the host RNA machinery (Meena, 2024). Following *B. cinerea* infection of *Arabidopsis* and tomato (*Solanum lycopersicum*), 832 Bc-RNAs were detected in both plants (Huang *et al.*, 2016). Additionally, Bc-sRNA expression in *Arabidopsis* helped to further describe the role of Bc-sRNAs (He *et al.*, 2023a). Bc-sRNA binds to *Arabidopsis* AGO1 and selectively silences host immunity genes. The transport of sRNAs from pathogen to host is suggested by the discovery of many RNA fragments (19–40 nt), derived from tRNA (Huang *et al.*, 2016). Small non-coding RNAs (snRNAs) from a bacterium *Wolbachia*, function as effectors in pathogen–animal interactions, modifying the expression of host genes in mosquitoes (Hussain and Asgari, 2014). Exosomes from the parasite *Leishmania* that were finally absorbed by host cells were found to contain sRNAs (Menezes and Tasca, 2024). It has also been demonstrated that exosomes secreted by the animal-parasitic worm *Heligmosomoides polygyrus* can transport miRNAs to cells (mammalian) (Vongai, 2020). These findings imply the possibility of a cross-kingdom RNA machinery as a knowledgeable mechanism (Xu *et al.*, 2024).

6.14 sRNAs Application in Plant Protection

Two main causes of biotic stress that restrict plant growth are pathogens and their diseases. During the struggle for existence and in order to protect themselves from pathogenic outbreak, plants developed sRNAs to suppress specific genes (Huang *et al.*, 2016). Sanford and Johnston introduced the idea of pathogen-derived resistance (PDR) prior to the discovery of RNA silencing by adding a pathogenic gene fragment into a host (Waheed *et al.*, 2021). By considering the characteristics of RNA silencing, scientists formulated a plan to produce plants that are more resilient to pathogens and insect herbivores (Sattar *et al.*, 2024). The development of transgenic papaya resistant to Ringspot Virus (RV) was the most effective use of PDR, extensively used against the virus (Hamim *et al.*, 2018).

Since RNA silencing was discovered, transgenic plants that have RNAi have been created to shield plants from a variety of pests and diseases (Cagliari *et al.*, 2019). The precursor of sRNA is essential for creating efficient PDR since it is closely linked to the effectiveness of silencing (Kryovrysanaki *et al.*, 2021). For example, transgenic plants expressing artificial miRNAs (amiRNAs) were most active in silencing the targeted gene and also improved herbivore resistance as compared to those plants expressing hairpin RNAs (hpRNAs) (Bally *et al.*, 2020). sRNAs derived from hairpin constructs were more effective at silencing green fluorescent protein (GFP) than those derived from sense/antisense constructs (Schumann *et al.*, 2013).

While choosing sRNAs, factors like off-target effects and sRNA perseverance are taken into account. It is vital to guarantee that PDR structures do not specifically target and adversely impact the host (Sharma *et al.*, 2023). Genes indispensable for the development of pathogens or those involved in plant pathogen interface should be considered promising PDR targets. PDR has been effectively used in a variety of major crops like rice, maize, cotton, wheat, and barley, as well as model plants like *Arabidopsis* and *Nicotiana benthamiana* (Johnson *et al.*, 2021).

Prior to the discovery of the RNAi mechanism, transgenic plants were created that targeted the viral DNA/RNA genome (Sajid *et al.*, 2021). To increase antiviral resistance, RNA interference (RNAi) was induced in transgenic plants through the manipulation and expression of viral transgene derived siRNAs, artificial miRNAs and viral-derived hpRNAs (Akbar *et al.*, 2022). Plant resistance to Cytomegalovirus (CMV) is enhanced by amiRNAs that contained sequences of CMV suppressor 2b, which effectively inhibits the production of the 2b gene (Wang *et al.*, 2011).

According to Ali *et al.* (2013) plants altered by a miRNA had enhanced resistance to the virus, e.g. Cotton Leaf Curl *Burewala* virus (CLCuBuV). PDR contrary to other pests necessitates the transmission of sRNAs to pathogens. PDR also controls the infection caused by fungal pathogens. It has been demonstrated that PDR to fungus, also known as host-induced gene silencing (HIGS), protects wheat and barley from fungal disease (*Blumeria graminis)* (Choudry *et al.*, 2024a). When fungal effector genes or fungal glucanosyl transferase genes are targeted by transgenic expression of dsRNAs, haustoria development is decreased, increasing resistance to *B. graminis* (Bakhat *et al.*, 2023). Fungal growth is also suppressed by inhibiting pathogenicity genes, such as those encoding a mitogen-activated protein kinase (MAPK), a cyclophilin and a calcineurin B in *Puccinia triticina* (Pt) (Panwar *et al.*, 2013).

Mechanisms similar to the precision delivery system (PDS) have been employed as a helpful tool in pest management in addition to defense against infections (Ahmad *et al.*, 2024). Lepidopteran and coleopteran insect pests are commonly managed with *Bacillus thuringiensis* insecticidal proteins (Yamamoto and Powell, 2024). However, a new strategy must be created to manage insect herbivores because transgenic plants that overexpress these *Bacillus thuringiensis* proteins have developed insect herbivore resistance (Nagaraj *et al.*, 2024).

In order to protect against insect herbivore attacks, transgenic plants that expressed dsRNAs are created (Saakre *et al.*, 2024). According to a research on cotton bollworm (*Helicoverpa armigera*), plants that expressed dsRNAs caused RNAi in the mid-gut of the moth, which inhibited CYP6AE14 appearance and slowed down the larval development (Khabbazi *et al.*, 2024). Cotton plants that expressed both plant cysteine proteases and CYP6AE14 dsRNAs showed greater resistance to the moth by increasing permeability to the mid-gut (Sood *et al.*, 2021). According to a recent study, the mid-gut cells of Colorado potato beetles and western corn rootworms can absorb dsRNAs but not siRNAs (Hussain, 2018). In order to control gene expression, the dsRNAs were converted into siRNAs (21 nt) by a variety of herbivores and amassed in cells (Parise *et al.*, 2024). These findings imply that RNAi can be used to lessen herbivore (insect) damage and that plants can create sRNAs or dsRNAs to fend against herbivore (insect) attack (Pappas *et al.*, 2020). Reduced development, a lower rate of reproduction, and vulnerability to insecticides have been seen in insect herbivores who consume transgenic plants containing RNAi constructs (Saakre *et al.*, 2024). By converting dsRNAs into siRNAs, the plant's Dicer and Dicer-like proteins stop a large number of dsRNAs from accumulating (Vaucheret and Voinnet, 2024). The effect in field cannot be accurately modeled by the effective use of PDR on plants, grown in greenhouses (Lopes, 2024). Given the prevalence of mixed infections, PDR is requisite to target genes in several different pathogens (Prentice *et al.*, 2024). It is also important to take into account how transgenic plants can withstand environmental challenges like salinity, drought, and temperature. To address these issues, more research is needed (Ngongolo and Mmbando, 2024).

6.15 Transgenerational Defense in Plants

There is a mark that a pathogenic attack can be used to trigger a specific defense response in plants known as priming, which can be transferred to the next generation (Tiwari and Singh, 2024). It has been found that homologous recombination frequency (HRF), callose deposition, PR1 expression and TMV resistance were all increased in the offspring of TMV-infected plants (Migicovsky and Kovalchuk, 2013). Furthermore, the next generation is more resistant to bacteria (*Pseudomonas syringae*) and oomycete (*Phytophthora nicotianae*) in addition to the virus (Zhu *et al.*, 2024). Apart from plant-virus interaction, *Arabidopsis* treated with *Pseudomonas syringae* pv. *tomato* DC3000, containing avrRpt2 (the effector gene) also showed transgenerational activation of defense (Abulfaraj, 2018). *Pseudomonas syringae* pv. *tomato* strain DC3000 (Pst DC3000) is a bacterium that causes Speck disease in tomatoes.

Another study found that even after one generation without an infection, the enhanced resistance persisted (Niu *et al.*, 2024). Insect herbivores are also capable of transgenerational resistance against pathogen stress (Mageroy *et al.*, 2023). The wild radish (*Raphanus raphanistrum*) produced radish plants with greater resistance as a result of transgenerational protection (Salzman *et al.*, 2021). Another indication of transgenerational resistance is the enhanced trichomes synthesis shown in the plants of yellow monkey flowers with injured leaves (Srivastava *et al.*, 2024). Furthermore, both species developed transgenerational resistance as a result of caterpillar herbivory on tomato and *Arabidopsis* (Shafi *et al.*, 2024).

According to recent research, sRNA accumulation, histone modification, and DNA methylation are the ways in which transgenerational resistance brought on by pathogen attack is transferred to the next generation (Farkas and Dobránszki, 2024). By infecting the plants with *Pseudomonas syringae* pv. *tomato* DC3000, Luna *et al.* (2012) produced *Arabidopsis* phenotypes that exhibited transgenerational resistance. The resistance phenotype was lacking in three DNA methyl transferases (DRM1 DRM2 CMT3 triple mutant of *Arabidopsis thaliana*) (Zhi and Chang, 2023). The transgenerational resistance phenotype was represented by hypo methylated DNA in the ddc mutant, suggesting that hypomethylated DNA is the means by which Pst DC3000 induces transgenerational resistance (Ashapkin *et al.*, 2020). By modulating the RNA-directed DNA methylation (RdDM) pathway, sRNAs seem to be crucial in this type of resistance inheritance (Li *et al.*, 2024c). The discovery that *Arabidopsis* dcl2/dcl3/dcl4 and nrpd2a/nrpd2b mutants, eliminated transgenerational resistance to caterpillar gives credibility to this concept (Yang *et al.*, 2016). According to Blevins *et al.* (2014) the resistance is transferred from one generation to the next generation during embryogenesis by some active DNA methyltransferases. Epigenetic inheritance has been demonstrated to be influenced by numerous proteins, involved in the production of sRNA (Ali and Tang, 2024).

Holeski *et al.* (2012) offered a sample of transgenerational induction in which sRNAs migrate from tissue (vegetative) to seeds as a form of stored information and chemical and physical defenses are persuaded in response to environmental signals (Ayoub *et al.*, 2024). It has not been proven, nevertheless, that sRNAs contain information for transgenerational protection. Plants have evolved extremely complex defense mechanisms against a variety of environmental stressors (Singh *et al.*, 2023). When the signaling pathways are activated, a series of stress responses are often triggered by a significant transcriptional reprogramming (Yuan *et al.*, 2024). In numerous epigenetic mechanisms, including histone modification, chromatin remodeling, and DNA methylation, non-coding RNAs are implicated in regulation of stress-responsive genes and transcriptional reprogramming (Abdulraheem *et al.*, 2024).

6.15.1 DNA methylation

DNA methylation is a form of chemical alteration of DNA that controls DNA stability, chromatin structure, and gene expression (Zhang and Zhu, 2024). One of the most frequent changes to DNA in eukaryotic cells is the bonding of methyl group to the cytosine-C5 site by DNA methyltransferase (Vilkaitis *et al.*, 2009). The most creative and prevalent methylation site in plants is CG, where C methylation is found in the setting of CG, CHH and CHG (H is a nucleotide other than G) (Rahman *et al.*, 2024).

Domains rearranged methyltransferase 1 (DRM1) and domains rearranged methyltransferase 2 (DRM2) have been found to catalyze DNA methylation in *Arabidopsis* at C site (Wickramasuriya *et al.*, 2024), whereas methyl transferase 1 (MET1) and chromo methylase 2/3 (CMT2/3) are necessary for maintaining DNA methylation during cell division (Zhang *et al.*, 2024e). According to a high-resolution profiling of DNA methylation in *Arabidopsis* and rice, DNA methylation can occur in a variety of chromatin locations, including gene promoters and intergenic transposable elements (TEs) (Wang *et al.*, 2017).

According to numerous studies DNA methylation in TEs is necessary to preserve genomic integrity (Nicu *et al.*, 2024). Additionally, methylation in DNA is linked to gene expression in *Arabidopsis*, while methylation in promoters typically suppresses gene expression (He *et al.*, 2022). DNA methylation is an epigenetic process that often controls how plants (wheat and barley etc.) react to the environmental stressors. According to a recent study, DNA methylation may control the expression of a gene in wheat that responds to salt (Afreen *et al.*, 2024). Two wheat cultivars, JN177 (salinity susceptible cultivar) and SR3 (salinity tolerant cultivar), showed a decrease in their DNA methylation level when exposed to salinity stress (Liang *et al.*, 2024). In contrast to the higher gene expression in cultivar SR3 as compared to cultivar JN177, the methylation level was lower in SR3 at the promoter of gene (stress-responsive) (Pandey *et al.*, 2022). This suggested that DNA methylation may show an active role in regulating wheat salinity tolerance (Hoseini *et al.*, 2024). Additionally, it has been demonstrated that in wheat and barley DNA methylation can modify the metal stress response (Zhang *et al.*, 2024e). The ATP binding cassette (ABC) transporters are crucial for heavy metal detoxifications. They are also involved in some other important plant processes, including pathogen response and development (Naaz *et al.*, 2024).

After being exposed to heavy metal stressors, Shafiq *et al.* (2019) discovered that the expression of ATP Binding Cassette (TaABCCs) and Heavy Metal ATpase2 (TaHMA2) was higher in wheat types that were resistant to heavy metals than in those that were susceptible. Additionally, during heavy metal stress, resistant varieties had low levels of DNA methylation at the promoter site of TaHMA2 and TaABCCs as compared to sensitive varieties (Fasani *et al.*, 2023). This indicated that DNA methylation is linked to the ability of wheat to withstand metal stress (Iqbal *et al.*, 2024). One of the key genes in barley that controls citrate efflux in the rhizosphere is Al-activated citrate transporter 1 (HvAACT1) (Qiu *et al.*, 2024). It is remarkable to note that the expression of HvAACT1 is linked to DNA methylation in a number of European barley accessions

(Guarino *et al.*, 2024). DNA demethylation led to increased HvAACT1 expression, particularly in the apical zone of root (Wu *et al.*, 2018a). In the meantime, low levels of HvAACT1 expression were linked to higher levels of DNA methylation, indicating that DNA methylation controls the expression of HvAACT1 (Kong *et al.*, 2020). In spite of a huge research done on how DNA methylation governs abiotic stress tolerance, little is known about how DNA methylation governs resistance against infections (Fan *et al.*, 2024).

According to a recent study into the wheat diploid ancestor *Aegilops tauschii*, DNA methylation, namely CHH methylation (where H is any base except G), has a role in regulating defense responses of *Blumeria graminis* f. sp. *tritici* (Doddavarapu *et al.*, 2024). CHH methylation was linked to a large number of differentially methylated regions (DMRs) following Bgt infection (Kong *et al.*, 2020). This suggested that DNA methylation plays an active role in controlling plant defense processes (Wang *et al.*, 2021). Furthermore, a number of cases showed that the sensitivity of transcription factors (TFs) to DNA methylation may influence how well the TF binds to chromatin (Zeng *et al.*, 2020). O'Malley *et al.* (2016) discovered that a particular DNA methyl transferase and the TF family may have a regulatory link in *Arabidopsis* (Talarico *et al.*, 2024). In order to balance the fine gene expression, DNA demethylation was used to eliminate 5-methyl cytosine and substituted with cytosine (Zheng *et al.*, 2024).

Passive and active demethylation are the two methods by which DNA demethylation takes place in plants (Fan *et al.*, 2024). Passive demethylation is the process by which methylated cytosine are swapped out for unmodified cytosine during DNA replication (Kisil *et al.*, 2024). Certain glycosylase hydrolyzed the N-glycosidic linkage between the ribose sugar and base, hence mediating active DNA demethylation (Fernandez-Rengel *et al.*, 2024). An active role of DNA demethylation is suggested by the considerably increased expression of HvDME (regulator gene) in barley seedlings to drought stress (Tiwari *et al.*, 2024).

6.15.2 Histone modification

Numerous histone modification methods, such as methylation, malonylation, succinylation acetylation, crotonylation, 2-hydroxyisobutyrylation, β-hydroxybutyrylation, butyrylation, and propionylation have been described as part of one of the most prevalent forms of epigenetic control (Qin *et al.*, 2024). The two most well-researched modifications (histone) are histone methylation/demethylation, and acetylation/deacetylation, controlled a variety of biological functions in plants (Barozai and Aziz, 2018). Specifically, H3K4me3, H3K36me3, H3K79me3, H3K9me2, and H3K27me3 are widely-conserved markers for gene expression (Bure *et al.*, 2022).

Histone methyltransferases (HMTs) and histone demethylases (HDMs) dynamically control histone methylation. For example, different DNA-binding proteins activate polycomb-repressive complex-1 (PRC1) and polycomb-repressive complex-2 (PRC2), which mediate the restrictive H3K27me3 modification (Ayad *et al.*, 2024).

Another kind of histone alteration are histone acetyltransferases (HATs) and histone deacetylases (HDACs) work antagonistically to maintain histone acetylation (Zhang *et al.*, 2024e). Acetylation encourages gene transcription by neutralizing lysine side chains on histones (positive charge) and decreasing their association with DNA backbone (negatively charged) (Brambilla *et al.*, 2024).

H3K4ac and H3K9ac are frequently linked to gene expression, which controls a variety of biological functions (Lepri *et al.*, 2023). According to evidence from *Arabidopsis* research, histone methylation is typically linked with histone acetylation (e.g. H3K4ac, H3K9ac) and simultaneously-regulated gene expression (Chaouche *et al.*, 2024). According to research, different HATs and HDACs alter plant gene expression in wheat and barley plants in response to stressors (Chen *et al.*, 2024b). Apart from abiotic stressors, biotic stressors like fungal infections also trigger plant reactions that are partially regulated by histone alterations (Balasubramani *et al.*, 2024). McCallum *et al.* (2024) suggested that *Puccinia triticina*, causes histone acetylation in wheat at the promoter's sites of genes (defense-related). These findings showed that histone acetylation have a role in controlling the wheat's reaction to *P. triticina* infection, however the exact mechanism is yet unknown (Gullner *et al.*, 2020).

It is interesting to note that in the model plant *Arabidopsis*, histone changes such histone methylation and acetylation were typically observed in conjunction with DNA methylation in response to stressors (Talarico *et al.*, 2024).

However, further research is needed to determine how histone changes interact with DNA methylation, to control wheat and barley's tolerance to environmental challenges (Dinkar *et al.*, 2024).

6.15.3 Chromatin material remodeling

In addition to histone modifications and DNA methylation, chromatin material remodeling is a process that alters particular DNA regions to transcription (Wong and Tremethick, 2024). In rice and *Arabidopsis*, chromatin modification has been shown to control plant growth and reaction to different stressors (Dinkar *et al.*, 2024; Jin *et al.*, 2024). The TaKCS6, along with TaKPAB1 (bHLH type TF 1), has been shown to interact with TaCHR729 (wheat chromatin remodeling factor) (Kong *et al.*, 2020). It is interesting to note that TaCHR729 has been shown to favorably regulate TaKCS6 transcription and enhanced H3K4me3 (Wang *et al.*, 2024). When TaCHR729 was silenced, the germination of Bgt conidia and the biosynthesis of cuticular wax (of wheat) were consistently reduced (Chang *et al.*, 2022). It was suggested that TaCHR729 controls the interaction of host and powdery mildew, regulating biosynthesis of cuticular wax by mediating histone methylation (Vassileva *et al.*, 2023). Relatively little is known about how wheat and barley's chromosomes change in reaction to stress. For example, in *Arabidopsis*, chromatin remodeling controls the ABA signal transduction that reacts to abiotic stressors such as cold, drought, and salt (Aizaz *et al.*, 2024).

Other research has demonstrated that SnRK2 kinase, can phosphorylate a number of chromatin regulators, included BRM SWI/SNF ATPase in the ABA signaling pathway (Bieluszewski *et al.*, 2023). These findings suggested that the phosphorylation and dephosphorylation states of SWI/SNF complexes may influence the ABA signaling pathway in response to environmental stressors. Other molecular pathways involving wheat and barley need to be investigated (Kong *et al.*, 2020).

6.16 Non-coding RNAs

Non-coding RNAs (ncRNAs) are significant epigenetic components that extensively control a variety of plant functions, such as growth, development, and even reactions to environmental stressors (Yadav *et al.*, 2024a). According to their size, ncRNAs can be classified as either long-chain or short-chain ncRNAs (Luszczki *et al.*, 2024).

Over the last few decades' large-scale research in plants has shown that non-coding RNAs, like siRNAs and miRNAs are involved in transcriptional and post-transcriptional of gene expression (Sarvari *et al.*, 2022). miRNAs and siRNAs are grouped as short RNAs (sRNAs) and typically include 18–24 nucleotides (Chakraborty *et al.*, 2024).

Long non-coding RNAs (lncRNAs) are commonly classified as those RNAs that are more than 200 nt long (Chen and Kim, 2024). Basic miRNAs, or pri-miRNAs (primary transcripts), are produced when miRNA-coding genes are transcribed in eukaryotic cells (Rengel *et al.*, 2024). Target specific post-transcriptional gene silencing (PTGS) in *Arabidopsis* is subsequently guided by ARGONAUTE (AGO) family proteins, which bind the mature miRNAs (Zaheer *et al.*, 2024).

In contrast to miRNA, RNA polymerase IV (RNAPIV) transcribes double-stranded RNAs with a length of 20–24 nt (Kong *et al.*, 2020). There are currently a number of siRNA subclasses known to exist, some of which are involved in transcriptional gene silencing (TGS) (Nien *et al.*, 2024).

LncRNAs are found in all species and take part in the control of gene expression (George *et al.*, 2024). LncRNAs are poly-adenylated and transcribed by RNA polymerase II (RNAPII) and they resemble mRNAs in structure and biogenesis (Tereshchenko, 2024). LncRNAs are prone to alternative splicing and have numerous exons, just like mRNAs. However, there is no noticeable coding potential in lncRNAs (Poliseno *et al.*, 2024). More recently, it has been discovered that lncRNAs play an active role in the key biological practices, such as growth, stress adaption, and vernalization in *Arabidopsis* (Numan *et al.*, 2024).

Stress-responsive gene expression in plants is controlled by siRNAs and lncRNAs through a variety of epigenetic mechanisms, such as histone modification, DNA methylation, and changes in genome topology etc. (Hassoun, 2023).

6.17 Beneficial Aspects of Biotic Stress

In nature plants growth and development is significantly subjective to biotic and abiotic factors (Atkinson and Urwin, 2012). Unlike animals, plants are sessile and continuously exposed to adverse ecological conditions (Singh and Roychoudhury, 2023). But in order to grow and survive plants have to develop some strategies to adapt themselves to the hostile conditions (Malinowski *et al.*, 2024).

Biotic interactions are not always harmful, they may be beneficial too (Potysz *et al.*, 2024). Interaction among plants and fungi (mycorrhizae), plants and bacteria (*Rhizobia*), plants and pollinators are the examples of mutual relationship in which both partners of the association benefit.

In legume rhizobia interactions, the host plant ensures the availability of nutrients and ecological niche, obligatory for their growth, while in return the bacterium fix nitrogen for the host plant. Similarly, some plants are unable to absorb nutrients especially phosphates (Santoro *et al.*, 2024). Hence they formed a beneficial symbiotic association with mycorrhizae (Duan *et al.*, 2024).

The biotic interactions among plants and other organisms may also be detrimental (Jiang *et al.*, 2024). These include interactions of plants with pathogens, pests, and parasites. Plant pathogens including viruses, bacteria, and fungi, spend a part of their life cycle inside the host plant (Jibril *et al.*, 2016). While plant pests are included, the herbivores (like nematodes, insects, and mammals) also caused harm to plants by eating them (Demis, 2024). It is an interesting fact that plants rising in a wild habitat very rarely developed diseases (Jeger, 2022).

However, the domestication of some economically-significant plant species leads to the expansion of monocultures, resulting in genetic uniformity, which makes them more vulnerable to various diseases (Mueller and Flachs, 2022).

Considering plant interactions with their environment prominently helps in dropping the usage of agrochemicals and hence facilitates reduction in pollution (Kiran *et al.*, 2024).

Another interaction known as "Allelopathic interaction" exists between various plants growing adjacent (Kong *et al.*, 2024). The growth and development of some plants is inhibited because of phytochemicals secreted by the adjacent plants (Hochma *et al.*, 2021).

6.18 Conclusion

Numerous biotic and abiotic stressors have a significant impact on plants and their physiological processes. Plants have evolved a wide range of intricate defense systems to help them endure and survive in all adverse environments. There are two types of defensive mechanisms: inducible, which activates defense when an attack is detected, and preformed, which stores poisonous secondary metabolites. After detecting biotic stressors, plants trigger their transcriptional or regulatory systems to produce the proper reaction. The interaction and influence of many signals to produce defense responses against biotic stress are still unclear, despite the fact that plant defense against pathogen attack is well understood. Usually, regulated growing conditions are used in laboratories to study how stresses affect plants. In contrast to controlled circumstances, the field environment is quite different and frequently exposes plants to a variety of challenges. Because the responses to the combined stresses are primarily regulated by distinct signalling pathways that may interact with and inhibit one another, the simultaneous occurrence of several stresses leads to a high degree of complexity in plant responses.

References

Abdulraheem, M.I., Xiong, Y., Moshood, A.Y., Cadenas-Pliego, G., Zhang, H. and Hu, J. (2024) Mechanisms of plant epigenetic regulation in response to plant stress: Recent discoveries and implications. *Plants* 13(2), 163.

Abulfaraj, A. A. (2018). Investigating the Role of the *Arabidopsis* Homologue of the Human G3BP in RNA Metabolism, Cellular Stress Responses and Innate Immunity.

Afreen, U., Mukhopadhyay, K. and Kumar, M. (2024) DNA methylation in wheat: Current understanding and future potential for enhancing biotic and abiotic stress tolerance. *Physiology and Molecular Biology of Plants* 13, 1–13.

Ahammed, G.J., Xu, W., Liu, A. and Chen, S. (2018) COMT1 silencing aggravates heat stress-induced reduction in photosynthesis by decreasing chlorophyll content, photosystem II activity, and electron transport efficiency in tomato. *Frontiers in Plant Science* 9, 998.

Ahmad, N., Hussain, H., Naeem, M., Rahman, S.U., Khan, K.A., Iqbal, B. and Umar, A.W. (2024) Metabolites-induced co-evolutionary warfare between plants, viruses, and their associated vectors: So close yet so far away. *Plant Science*, 346, 112165.

Ahmed, M., Tóth, Z. and Decsi, K. (2024) The impact of salinity on crop yields and the confrontational behavior of transcriptional regulators, nanoparticles, and antioxidant defensive mechanisms under stressful conditions: A review. *International Journal of Molecular Sciences* 25(5), 2654.

Aiello, I., Fedele, M.M., Román, F., Marpegan, L., Caldart, C., Chiesa, J.J. and Paladino, N. (2020) Circadian disruption promotes tumor-immune microenvironment remodeling favoring tumor cell proliferation. *Science Advances* 6(42), eaaz4530.

Aizaz, M., Lubna, J.R., Asaf, S., Bilal, S., Kim, K.M. and AL-Harrasi, A. (2024) Regulatory dynamics of plant hormones and transcription factors under salt stress. *Biology* 13(9), 673.

Akbar, S., Wei, Y. and Zhang, M.Q. (2022) RNA interference: Promising approach to combat plant viruses. *International Journal of Molecular Sciences* 23(10), 5312.

Ali, S. and Tang, Y. (2024) Non-coding RNA-mediated regulation of DNA methylation: Insights into plant epigenetic mechanisms. *Journal of Plant Growth Regulation*, 44, 1–16.

Ali, A., Shahbaz, M., Ölmez, F., Fatima, N., Umar, U.U.D., Ali, M.A. and Baloch, F.S. (2024a) RNA interference: A promising biotechnological approach to combat plant pathogens, mechanism and future prospects. *World Journal of Microbiology and Biotechnology* 40(11), 339.

Ali, J., Tonğa, A., Islam, T., Mir, S., Mukarram, M., Konôpková, A.S. and Chen, R. (2024b) Defense strategies and associated phytohormonal regulation in brassica plants in response to chewing and sap-sucking insects. *Frontiers in Plant Science* 15, 1376917.

Almeida, R. and Allshire, R.C. (2005) RNA silencing and genome regulation. *Trends in Cell Biology* 15(5), 251–258.

Arini, A., Muller, S., Coma, V., Grau, E., Sandre, O. and Baudrimont, M. (2023) Origin, exposure routes and xenobiotics impart nanoplastics with toxic effects on freshwater bivalves. *Environmental Science: Nano* 10(5), 1352–1371.

Ashapkin, V.V., Kutueva, L.I., Aleksandrushkina, N.I. and Vanyushin, B.F. (2020) Epigenetic mechanisms of plant adaptation to biotic and abiotic stresses. *International Journal of Molecular Sciences* 21(20), 7457.

Atkinson, N.J. and Urwin, P.E. (2012) The interaction of plant biotic and abiotic stresses: From genes to the field. *Journal of Experimental Botany* 63(10), 3523–3543.

Ayoub, L., Yaqoob, M., Zahoor, S., Wani, F.F., Irshad, S.S., Gull, A. and Sheikh, M.A. (2024) Role of induced resistance in insect-Pest management. In: *Plant Resistance to Insects in Major Field Crops*. Springer Nature Singapore, Singapore, pp. 249–277.

Balasubramani, S., Gayathiri, E., Arya, S.K., Gurusamy, D., Prakash, P., Appu, M. and Moola, A.K. (2024) Role of heat shock proteins in abiotic and biotic stress response in plants. In: *Defense-Related Proteins in Plants*. Academic Press, pp. 299–332, Elsevier: Cambridge, Massachusetts, United States.

Balios, V.A., Fischer, K., Bawin, T. and Krause, K. (2024) One organ to infect them all: The Cuscuta haustorium. *Annals of Botany*, 135, mcae208.

Balla, A., Silini, A., Cherif-Silini, H., Chenari Bouket, A., Moser, W.K., Nowakowska, J.A. and Belbahri, L. (2021) The threat of pests and pathogens and the potential for biological control in forest ecosystems. *Forests* 12(11), 1579.

Bally, J., Fishilevich, E., Doran, R.L., Lee, K., de Campos, S.B., German, M.A. and Waterhouse, P.M. (2020) Plin-amiR, a pre-microRNA-based technology for controlling herbivorous insect pests. *Plant Biotechnology Journal* 18(9), 1925–1932.

Barozai, M.Y.K. and Aziz, A.N. (2018) Recent plant growth and stress management-related significant advancements in epigenetics. *Annals of Agrarian Science* 16(4), 416–421.

Baseer, A.Q., Niazi, P., Monib, A.W., Hassand, M.H., Hejran, A.B., Sarwari, A. and Fahmawi, S.M.S. (2024) Lifecycle transitions in plant development: Ripening, senescence, and cell death. *Journal of Pharma Insights and Research* 2(2), 169–179.

Bhattacharya, A. and Bhattacharya, A. (2021) Mineral nutrition of plants under soil water deficit condition: A review. *Soil Water Deficit and Physiological Issues in Plants*, 287–391.

Bieluszewski, T., Prakash, S., Roulé, T. and Wagner, D. (2023) The role and activity of SWI/SNF chromatin remodelers. *Annual Review of Plant Biology* 74(1), 139–163.
Bilal, M.S., Paul, G., Yu, Z., Xu, L., Cheng, T., Cheng, B. and Zhao, H. (2024) Comparative transcriptome and sRNAome analysis suggest coordinated citrus immune responses against Huanglongbing disease. *Plants* 13(11), 1496.
Bonanno, G.A., Brewin, C.R., Kaniasty, K. and Greca, A.M.L. (2010) Weighing the costs of disaster: Consequences, risks, and resilience in individuals, families, and communities. *Psychological Science in the Public Interest* 11(1), 1–49.
Bondarchuk, O., Balakhtar, V., Pinchuk, N., Pustovalov, I. and Pavlenok, K. (2024) Coping with stressfull situations using coping strategies and their impact on mental health. *Multidisciplinary Reviews* 7.
Borowitzka, M.A. (2018) The 'stress' concept in microalgal biology—Homeostasis, acclimation and adaptation. *Journal of Applied Phycology* 30, 2815–2825.
Bure, I.V., Nemtsova, M.V. and Kuznetsova, E.B. (2022) Histone modifications and non-coding RNAs: Mutual epigenetic regulation and role in pathogenesis. *International Journal of Molecular Sciences* 23(10), 5801.
Buschmann, C., Konanz, S., Zhou, M., Lenk, S., Kocsányi, L. and Barócsi, A. (2013) Excitation kinetics of chlorophyll fluorescence during light-induced greening and establishment of photosynthetic activity of barley seedlings. *Photosynthetica* 51(2), 221–230.
Cagliari, D., Dias, N.P., Galdeano, D.M., Dos Santos, E.Á., Smagghe, G. and Zotti, M.J. (2019) Management of pest insects and plant diseases by non-transformative RNAi. *Frontiers in Plant Science* 10, 1319.
Cai, Q., Halilovic, L., Shi, T., Chen, A., He, B., Wu, H. and Jin, H. (2023) Extracellular vesicles: Cross-organismal RNA trafficking in plants, microbes, and mammalian cells. *Extracellular Vesicles and Circulating Nucleic Acids* 4(2), 262.
Cao, J., Bao, J., Lan, S., Qin, X., Ma, S. and Li, S. (2024) Research progress on low-temperature stress response mechanisms and mitigation strategies in plants. *Plant Growth Regulation* 104, 1–22.
Chakraborty, C., Bhattacharya, M. and Ranjan Sharma, A. (2024) miRNA, siRNA, and lncRNA: Recent development of bioinformatics tools and databases in support of combating different diseases. *Current Bioinformatics* 19(1), 39–60.
Challoumis, C. (2024) The future of money-exploring AI's role in finance and payments. In: *XVI International Scientific Conference*, pp. 158–189.
Chen, L.L. and Kim, V.N. (2024) Small and long non-coding RNAs: Past, present, and future. *Cell* 187(23), 6451–6485.
Chen, F., Zhao, Q., Yang, Z., Chen, R., Pan, H., Wang, Y. and Lan, L. (2024a) Citrate serves as a signal molecule to modulate carbon metabolism and iron homeostasis in Staphylococcus aureus. *PLoS Pathogens* 20(7), e1012425.
Chen, L., Chen, K., Xi, X., Du, X., Zou, X., Ma, Y. and Weining, S. (2024b) The evolution, expression patterns, and domestication selection analysis of the Annexin gene family in the barley Pan-genome. *International Journal of Molecular Sciences* 25(7), 3883.
Chesnais, Q., Golyaev, V., Velt, A., Rustenholz, C., Brault, V., Pooggin, M. and Drucker, M. (2022) Comparative plant transcriptome profiling of *Arabidopsis* and Camelina infested with Myzus persicae aphids acquiring circulative and non-circulative viruses reveals virus-and plant-specific alterations relevant to aphid feeding behavior and transmission. *bioRxiv*, 01.
Choudhary, V., Kumar, K.K., Sahu, B.H.I.M.E.S.H.W.A.R.I., Muthappa, S.K. and Das, A.D.A. (2021) Agronomic innovations in biotic stress management and its combined effect with abiotic stresses in crop production. *Indian Journal of Agronomy* 66(5), 237–257.
Choudry, M.W., Nawaz, P., Jahan, N., Riaz, R., Ahmad, B., Raza, M.H. and Afzal, S. (2024a) RNA based gene silencing modalities to control insect and fungal plant pests–challenges and future prospects. *Physiological and Molecular Plant Pathology* 102241.
Choudry, M.W., Riaz, R., Nawaz, P., Ashraf, M., Ijaz, B. and Bakhsh, A. (2024b) CRISPR-Cas9 mediated understanding of plants' abiotic stress-responsive genes to combat changing climatic patterns. *Functional and Integrative Genomics* 24(4), 132.
Cudjoe, D.K., Okyere, F.G., Clark, S.J., Virlet, N., Castle, M., Riche, A.B. and Hawkesford, M.J. (2024) Determination of optimal spectral reflectance indices for monitoring the nutritional status and agro-morpho-physiological responses in quinoa and cowpea under varying nitrogen and phosphorus availability using proximal sensing. *Rothamsted Repository Download* 87, 1.
Dall'Osto, L., Bassi, R. and Ruban, A. (2014) Photoprotective mechanisms: Carotenoids. *Plastid Biology* 393–435.
Das, C. (2024) A review on understanding the plant's secret language for communication and its application. *Haya Saudi J Life Sci* 9(1), 1–11.

Das, D. and Gutjahr, C. (2019) Role of phytohormones in arbuscular mycorrhiza development. *The Model Legume Medicago Truncatula* 485–500.

Demis, E. (2024) Mechanism of plant resistance to insects, weeds and pathogens. *Middle East Research Journal of Agriculture and Food Science* 4(2), 76–85.

Demmig-Adams, B., Stewart, J.J. and Adams, W.W.III (2014). Multiple feedbacks between chloroplast and whole plant in the context of plant adaptation and acclimation to the environment. *Philosophical Transactions of the Royal Society B: Biological Sciences* 369(1640), 20130244.

Devi, S., Dey, R., Dube, S.P. and Raghuwanshi, R. (2024) Effector Mediated Defense Mechanisms in Plants against Phytopathogens. In: *Plant Elicitor Peptides: New Tool for Sustainable Agriculture*. pp. 131–148, Springer: Germany.

Ding, T., Li, W., Li, F., Ren, M. and Wang, W. (2024) microRNAs: Key regulators in plant responses to abiotic and biotic stresses via endogenous and cross-kingdom mechanisms. *International Journal of Molecular Sciences* 25(2), 1154.

Dinkar, V., Pandey, S., Kumar, A., Shiv, A., Lal, D., Bharati, A. and Chandra, A.K. (2024) Epigenetic regulations under plant stress: A cereals perspective. *Environmental and Experimental Botany* 220, 105688.

Doddavarapu, B., Lata, C. and Shah, J.M. (2024) Epigenetic regulation influenced by soil microbiota and nutrients: Paving road to epigenome editing in plants. *Biochimica et Biophysica Acta (BBA)-General Subjects* 130580.

Doughari, J. (2015) An overview of plant immunity. *Journal of Plant Pathology and Microbiology* 6(11), 10–4172.

Du, B., Franzisky, B.L., Muhammad, W., Alfarraj, S., Geilfus, C.M. and Rennenberg, H. (2025) How to cope with stress in the desert—The date palm approach. *Plant, Cell and Environment* 48(1), 768–780.

Duan, S., Feng, G., Limpens, E., Bonfante, P., Xie, X. and Zhang, L. (2024) Cross-kingdom nutrient exchange in the plant–arbuscular mycorrhizal fungus–bacterium continuum. *Nature Reviews Microbiology* 22, 1–18.

Dube, G.Z. (2017) *MicroRNA Discovery in the Wheat-Russian Wheat Aphid Interaction* (Master's thesis). University of Johannesburg, South Africa.

Dutilloy, E., Oni, F.E., Esmaeel, Q., Clément, C. and Barka, E.A. (2022) Plant beneficial bacteria as bioprotectants against wheat and barley diseases. *Journal of Fungi* 8(6), 632.

Engin, A. (2024) Misalignment of circadian rhythms in diet-induced obesity. *Obesity and Lipotoxicity*, 27–71.

Faizan, M., Cheng, S.H., Tonny, S.H. and Robab, M.I. (2022) Specific roles of strigolactones in plant physiology and remediation of heavy metals from contaminated soil. *Plant Physiology and Biochemistry* 192, 186–195.

Falcioni, R., Oliveira, R.B.D., Chicati, M.L., Antunes, W.C., Demattê, J.A.M. and Nanni, M.R. (2024) Fluorescence and hyperspectral sensors for nondestructive analysis and prediction of biophysical compounds in the green and purple leaves of Tradescantia plants. *Sensors* 24(19), 6490.

Fan, Y., Sun, C., Yan, K., Li, P., Hein, I., Gilroy, E.M. and Bai, J. (2024) Recent advances in studies of genomic DNA methylation and its involvement in regulating drought stress response in crops. *Plants* 13(10), 1400.

Farhan, M., Sathish, M., Kiran, R., Mushtaq, A., Baazeem, A., Hasnain, A. and Moustafa, M. (2024a) Plant nitrogen metabolism: Balancing resilience to nutritional stress and abiotic challenges. *Phyton* 93(3), 0031–9457.

Farhan, M., Zhao, C., Akhtar, S., Ahmad, I., Jilong, P. and Zhang, S. (2024b) Assessment of Nano-formulated conventional insecticide-treated sugar baits on mosquito control and the effect on non-target Aphidophagous Coccinella septempunctata. *Insects* 15(1), 70.

Farkas, D. and Dobránszki, J. (2024) Vegetal memory through the lens of transcriptomic changes–recent progress and future practical prospects for exploiting plant transcriptional memory. *Plant Signaling and Behavior* 19(1), 2383515.

Fasani, E., Giannelli, G., Varotto, S., Visioli, G., Bellin, D., Furini, A. and DalCorso, G. (2023) Epigenetic control of plant response to heavy metals. *Plants* 12(18), 3195.

Fernandez-Rengel, I., Terrón-Camero, L.C. and Andrés-León, E. (2024) Inter-kingdom signalling: Exploring the relationships among microorganisms and MicroRNAs in human health. In: *miRNAs, Human Health and Diseases*. Springer Nature Switzerland, Cham, pp. 337–373.

Galviz, Y., Souza, G.M. and Lüttge, U. (2022) The biological concept of stress revisited: Relations of stress and memory of plants as a matter of space–time. *Theoretical and Experimental Plant Physiology* 34(2), 239–264.

Gechev, T.S., Van Breusegem, F., Stone, J.M., Denev, I. and Laloi, C. (2006) Reactive oxygen species as signals that modulate plant stress responses and programmed cell death. *BioEssays* 28(11), 1091–1101.

George, T.P., Subramanian, S. and Supriya, M.H. (2024) A brief review of non-coding RNA. *Egyptian Journal of Medical Human Genetics* 25(1), 98.

Gloss, A.D., Vergnol, A., Morton, T.C., Laurin, P.J., Roux, F. and Bergelson, J. (2022) Genome-wide association mapping within a local *Arabidopsis* thaliana population more fully reveals the genetic architecture for defensive metabolite diversity. *PhilosophicalTransactions of the Royal Society B* 377(1855), 20200512.

Gowtham, H.G., Murali, M., Shilpa, N., Amruthesh, K.N., Gafur, A., Antonius, S. and Sayyed, R.Z. (2024) Harnessing abiotic elicitors to bolster plant's resistance against bacterial pathogens. *Plant Stress*, 11, 100371.

Grabherr, H.M. (2011) *Characterisation of the Role of LysM Receptor-like Kinases and the CHIA Chitinase in the Perception of Peptidoglycan and in the Innate Immunity of Arabidopsis Thaliana*(Doctoral dissertation). Universität Tübingen.

Guarino, F., Cicatelli, A., Nissim, W.G., Colzi, I., Gonnelli, C., Basso, M.F. and Castiglione, S. (2024) Epigenetic effects induced by chronic and acute chromium stress treatments in *Arabidopsis* thaliana identified by the MSAP-Seq. *Chemosphere*, 362, 142642.

Haag, J. (2013) *Improving Photosynthetic Efficiency in Sports Turf*. Xlibris Corporation, Bloomington, Indiana, United States.

Hamim, I., Borth, W.B., Marquez, J., Green, J.C., Melzer, M.J. and Hu, J.S. (2018) Transgene-mediated resistance to papaya ringspot virus: Challenges and solutions. *Phytoparasitica* 46, 1–18.

Hao, Z.P., Sheng, L., Feng, Z.B., Fei, W.X. and Hou, S.M. (2024) Aphids may facilitate the spread of Sclerotinia stem rot in oilseed rape by carrying and depositing ascospores. *Journal of Fungi* 10(3), 202.

Hasanuzzaman, M., Nahar, K., Alam, M.M., Roychowdhury, R. and Fujita, M. (2013) Physiological, biochemical, and molecular mechanisms of heat stress tolerance in plants. *International Journal of Molecular Sciences* 14(5), 9643–9684.

Hassoun, Z.E.O. (2023) *A Novel Role of EWSR1 Protein in RNA Translation Regulation* (Doctoral dissertation). Universite de Liege, Belgium.

He, L., Huang, H., Bradai, M., Zhao, C., You, Y., Ma, J. and Zhu, J.K. (2022) DNA methylation-free *Arabidopsis* reveals crucial roles of DNA methylation in regulating gene expression and development. *Nature Communications* 13(1), 1335.

He, B., Cai, Q., Weiberg, A., Li, W., Cheng, A.P., Ouyang, S. and Jin, H. (2023) Botrytis cinerea small RNAs are associated with tomato AGO1 and silence tomato defense-related target genes supporting cross-kingdom RNAi. *Biorxiv*, 12.

Henchiri, H. (2021) *Description and Functional Analysis of the T-SNARE SNAP33 in Arabidopsis Thaliana* (Doctoral dissertation). Université Paris-Saclay.

Hlihor, R.M., Roşca, M., Hagiu-Zaleschi, L., Simion, I.M., Daraban, G.M. and Stoleru, V. (2022) Medicinal plant growth in heavy metals contaminated soils: Responses to metal stress and induced risks to human health. *Toxics* 10(9), 499.

Hochma, E., Yarmolinsky, L., Khalfin, B., Nisnevitch, M., Ben-Shabat, S. and Nakonechny, F. (2021) Antimicrobial effect of phytochemicals from edible plants. *PRO* 9(11), 2089.

Hofmann, G.E. and Todgham, A.E. (2010) Living in the now: Physiological mechanisms to tolerate a rapidly changing environment. *Annual Review of Physiology* 72(1), 127–145.

Holmgren, M. and Scheffer, M. (2010) Strong facilitation in mild environments: The stress gradient hypothesis revisited. *Journal of Ecology* 98(6), 1269–1275.

Hoseini, M., Arzani, A., Saeidi, G. and Araniti, F. (2024) Agro-physiological and DNA methylation responses to salinity stress in wheat (Triticum aestivum L.), Aegilops cylindrica host, and their introgressed lines. *Plants* 13(19), 2673.

Huang, J., Yang, M. and Zhang, X. (2016) The function of small RNAs in plant biotic stress response. *Journal of Integrative Plant Biology* 58(4), 312–327.

Huang, H.J., Wang, Y.Z., Li, L.L., Lu, H.B., Lu, J.B., Wang, X. and Zhang, C.X. (2023) Planthopper salivary sheath protein LsSP1 contributes to manipulation of rice plant defenses. *Nature Communications* 14(1), 737.

Hussain, T. (2018) Development of Potato Lines Resistant to Colorado Potato Beetle (CPB) using RNAi-Based Transformation Approach.

Hussain, M. and Asgari, S. (2014) MicroRNAs as mediators of insect host–pathogen interactions and immunity. *Journal of Insect Physiology* 70, 151–158.

Iqbal, B., Ahmad, N., Li, G., Jalal, A., Khan, A.R., Zheng, X. and Du, D. (2024) Unlocking plant resilience: Advanced epigenetic strategies against heavy metal and metalloid stress. *Plant Science* 349, 112265.

Issawi, M. (2018) *Plant Photodynamic Stress: Study of Molecular and Cellular Mechanisms in Plant and Plant Cells upon Porphyrin Treatment* (Doctoral dissertation). Université de Limoges.

Javaid, I., Faizullah, M.M., Sheikh, M.A. and Khan, K.A. (2023) Small RNAs: A new paradigm in plant microbe interactions. *Research and Reviews in Agriculture Science* III, 78.

Jeger, M.J. (2022) The impact of climate change on disease in wild plant populations and communities. *Plant Pathology* 71(1), 111–130.

Jena, J., Sahoo, S.K. and Dash, G.K. (2020) Chapter-9 An introduction to abiotic stress in plants. *Agronomy* 163, 1.

Jiang, F., Bennett, J.A., Crawford, K.M., Heinze, J., Pu, X., Luo, A. and Wang, Z. (2024) Global patterns and drivers of plant–soil microbe interactions. *Ecology Letters* 27(1), e14364.

Jibril, S.M., Jakada, B.H., Kutama, A.S. and Umar, H.Y. (2016) Plant and pathogens: Pathogen recognition, invasion and plant defense mechanism. *International Journal of Current Microbiology and Applied Sciences* 5(6), 247–257.

Jin, R., Yang, H., Muhammad, T., Li, X., Tuerdiyusufu, D., Wang, B. and Wang, J. (2024) Involvement of alfin-like transcription factors in plant development and stress response. *Genes* 15(2), 184.

Johnson, A.A., Gopal, D.S. and Sudhakar, C. (2021) GM crops for plant virus resistance: A review. *Genetically Modified Crops: Current Status, Prospects and Challenges* 2, 257–337.

Kaiser, E., Kusuma, P., Vialet-Chabrand, S., Folta, K., Liu, Y., Poorter, H. and Marcelis, L.F. (2024) Vertical farming goes dynamic: Optimizing resource use efficiency, product quality, and energy costs. *Frontiers in Science* 2, 1411259.

Kanaani, M. (ed.) (2022) *The Routledge Companion to Ecological Design Thinking: Healthful Ecotopian Visions for Architecture and Urbanism*. CRC Press: Florida, United States.

Kang, J.E., Graham-Engeland, J.E., Scott, S., Smyth, J.M. and Sliwinski, M.J. (2024) The relationship between loneliness and the experiences of everyday stress and stressor-related emotion. *Stress and Health* 40(2), e3294.

Kashyap, B. and Kumar, R. (2021) Sensing methodologies in agriculture for monitoring biotic stress in plants due to pathogens and pests. *Inventions* 6(2), 29.

Kaya, C., Ugurlar, F., Ashraf, M. and Ahmad, P. (2023) Salicylic acid interacts with other plant growth regulators and signal molecules in response to stressful environments in plants. *Plant Physiology and Biochemistry* 196, 431–443.

Kesineni, M., Yama, K.S., Lindgow, M.J. and Lakra, N. (2023) Multiple stresses are a big challenge for the development of tolerant varieties: Shared and unique physiological responses. In: *Salinity and Drought Tolerance in Plants: Physiological Perspectives*. Springer Nature Singapore, Singapore, pp. 29–45.

Khabbazi, S.D., Khabbazi, A.D. and Yaman, C. (2024) Different generations of genetically modified crops for insect resistance. In: *Plant Resistance to Insects in Major Field Crops*. Springer Nature Singapore, Singapore, pp. 279–299.

Kiran, P.S., Mandal, P., Jain, M., Ghosal, P.S. and Gupta, A.K. (2024) A comprehensive review on the treatment of pesticide-contaminated wastewater with special emphasis on organophosphate pesticides using constructed wetlands. *Journal of Environmental Management* 368, 122163.

Kishore, U. (ed.) (2010) *Target Pattern Recognition in Innate Immunity*. Vol. 653, Springer Science and Business Media: Germany.

Kisil, O., Sergeev, A., Bacheva, A. and Zvereva, M. (2024) Methods for detection and mapping of methylated and Hydroxymethylated cytosine in DNA. *Biomolecules* 14(11), 1346.

Koch, A., Höfle, L., Jelonek, L., Busche, T., Möbus, A. and Hardt, M. (2020) Short Distance Spreading of Non-coding RNAs: Vesicle Trafficking and Exosome Secretion.

Kong, C.H., Li, Z., Li, F.L., Xia, X.X. and Wang, P. (2024) Chemically mediated plant–plant interactions: Allelopathy and allelobiosis. *Plants* 13(5), 626.

Koyro, H.W. and Breckle, S.W. (2024) Ecophysiological constraints under salinity stress: Halophytes versus non-halophytes. In: *Halophytes Vis-à-Vis Saline Agriculture: Perspectives and Opportunities for Food Security*. Springer Nature Singapore, Singapore, pp. 179–229.

Kranner, I., Minibayeva, F.V., Beckett, R.P. and Seal, C.E. (2010) What is stress? Concepts, definitions and applications in seed science. *New Phytologist* 188(3), 655–673.

Kryovrysanaki, N., James, A., Tselika, M., Bardani, E. and Kalantidis, K. (2021) RNA silencing pathways in plant development and defense. *The International Journal of Developmental Biology* 66(1-2-3), 163–175.

Kuderov, P., Dzhivelikian, E. and Panov, A.I. (2024) Hebbian spatial encoder with adaptive sparse connectivity. *Cognitive Systems Research* 88, 101277.

Kumar, P., Sharma, K., Saini, L. and Dey, S.R. (2021) Role and behavior of microbial volatile organic compounds in mitigating stress. In: *Volatiles and Metabolites of Microbes*. pp. 143–161, Elsevier: Cambridge, Massachusetts, United States.

Kumar, A., Dhansu, P. and Mann, A. (eds) (2023) *Salinity and Drought Tolerance in Plants: Physiological Perspectives*. Springer Nature: Germany.

Kumar, K., Mandal, S.N., Neelam, K. and de los Reyes, B. (2024a) *Regulation of Host Plant Defenses against Pathogens and Herbivores by miRNA: Balancing Resistance and Productivity Trade-Offs*. Authorea Preprints.

Kumar, P., Banik, S.P., Ohia, S.E., Moriyama, H., Chakraborty, S., Wang, C.K. and Bagchi, D. (2024b) Current insights on the Photoprotective mechanism of the macular carotenoids, lutein and zeaxanthin: Safety, efficacy and bio-delivery. *Journal of the American Nutrition Association*, 43, 1–14.

Kumaraswamy, S. and Huang, Y. (2024) Molecular interactions between plants and aphids: Recent advances and future perspectives. *Insects* 15(12), 935.

Kunz, L., Poretti, M., Praz, C.R., Müller, M.C., Wyler, M., Keller, B. and Bourras, S. (2024) *High-Copy Transposons from a Pathogen Give Rise to a Conserved sRNA Family with a Novel Host Immunity Target. Molecular Plant-Microbe Interactions, (Ja).*

Lamalakshmi Devi, E., Kumar, S., Basanta Singh, T., Sharma, S.K., Beemrote, A., Devi, C.P. and Wani, S.H. (2017) Adaptation strategies and defence mechanisms of plants during environmental stress. *Medicinal plants and environmental challenges*, 359–413.

Larcher, W. (2003) *Physiological Plant Ecology: Ecophysiology and Stress Physiology of Functional Groups*. Springer Science and Business Media: Germany.

Lepri, A., Longo, C., Messore, A., Kazmi, H., Madia, V.N., Di Santo, R. and Vittorioso, P. (2023) Plants and small molecules: An up-and-coming synergy. *Plants* 12(8), 1729.

Li, Y., Wang, W. and Zhou, J.M. (2011) Role of small RNAs in the interaction between *Arabidopsis* and pseudomonas syringae. *Frontiers in Biology* 6, 462–467.

Li, P., Lu, Y.J., Chen, H. and Day, B. (2020) The lifecycle of the plant immune system. *Critical Reviews in Plant Sciences* 39(1), 72–100.

Li, J., Zhang, H., Yang, R., Zeng, Q., Han, G., Du, Y. and Luo, Q. (2021) Identification of miRNAs contributing to the broad-spectrum and durable blast resistance in the Yunnan local rice germplasm. *Frontiers in Plant Science* 12, 749919.

Li, F., Lu, Y., Xi, K., Li, Y., Chen, X., Wang, P. and Huang, X. (2024a) Interkingdom communication via extracellular vesicles: Unraveling plant and pathogen interactions and its potential for next-generation crop protection. *Microorganisms* 12(12), 2392.

Li, H., Chen, Y., Dai, Y., Yang, L. and Zhang, S. (2024b) Genome-wide identification and expression analysis of histone deacetylase and histone acetyltransferase genes in response to drought in poplars. *BMC Genomics* 25(1), 657.

Li, Q., Wang, Y., Sun, Z., Li, H. and Liu, H. (2024c) The biosynthesis process of small RNA and its pivotal roles in plant development. *International Journal of Molecular Sciences* 25(14), 7680.

Liang, X., Li, J., Yang, Y., Jiang, C. and Guo, Y. (2024) Designing salt stress-resilient crops: Current progress and future challenges. *Journal of Integrative Plant Biology* 66(3), 303–329.

Lichtenthaler, H.K. (1996) Vegetation stress: An introduction to the stress concept in plants. *Journal of Plant Physiology* 148(1–2), 4–14.

Lichtenthaler, H.K. (2007) Biosynthesis, accumulation and emission of carotenoids, α-tocopherol, plastoquinone, and isoprene in leaves under high photosynthetic irradiance. *Photosynthesis Research* 92, 163–179.

Lichtenthaler, H.K. (2021) Multi-colour fluorescence imaging of photosynthetic activity and plant stress. *Photosynthetica* 59(2), 4–20.

Lichtenthaler, H.K., Buschmann, C. and Knapp, M. (2005) How to correctly determine the different chlorophyll fluorescence parameters and the chlorophyll fluorescence decrease ratio R Fd of leaves with the PAM fluorometer. *Photosynthetica* 43, 379–393.

Linke, J.A., Munn, L.L. and Jain, R.K. (2024) Compressive stresses in cancer: Characterization and implications for tumour progression and treatment. *Nature Reviews Cancer* 24, 1–24.

Liu, J. and Tang, D. (2023) Plant immunity research in China. *Phytopathology Research* 5(1), 37.

Liu, F., Tang, J., Li, T. and Zhang, Q. (2024a) The microRNA miR482 regulates NBS-LRR genes in response to ALT1 infection in apple. *Plant Science* 343, 112078.

Liu, S., Xu, H., Wang, G., Jin, B., Cao, F. and Wang, L. (2025) Tree longevity: Multifaceted genetic strategies and beyond. *Plant, Cell and Environment* 48(1), 244–259.

Lopes, A.M. (2024) Policy Coherence in Lao PDR: A Climate, Land-Use, Energy and Water Systems Model.

Luszczki, E., Wyszynska, J., Dymek, A., Drożdż, D., González-Ramos, L., Hartgring, I. and Alvarez-Pitti, J. (2024) The effect of maternal diet and lifestyle on the risk of childhood obesity. *Metabolites* 14(12), 655.

Mageroy, M.H., Nagy, N.E., Steffenrem, A., Krokene, P. and Hietala, A.M. (2023) Conifer defences against pathogens and pests—Mechanisms, breeding, and management. *Current Forestry Reports* 9(6), 429–443.

Makandar, R. (2024) Plant immunity inducers: Strategies to identify and isolate them to boost Defense responses in plants. In: *Plant Elicitor Peptides: New Tool for Sustainable Agriculture*. Springer Nature Singapore, Singapore, pp. 283–317.

Malinowski, R., Singh, D., Kasprzewska, A., Blicharz, S. and Basińska-Barczak, A. (2024) Vascular tissue–boon or bane? How pathogens usurp long-distance transport in plants and the defence mechanisms deployed to counteract them. *New Phytologist* 243(6), 2075–2092.

Mangalraj, P. and Cho, B.K. (2022) Recent trends and advances in hyperspectral imaging techniques to estimate solar induced fluorescence for plant phenotyping. *Ecological Indicators* 137, 108721.

Márquez, D.A., Gardner, A. and Busch, F.A. (2025) Navigating Challenges in Interpreting Plant Physiology Responses through Gas Exchange Results in Stressed Plants.

McKersie, B.D. and Lesheim, Y. (2013) *Stress and Stress Coping in Cultivated Plants*. Springer Science and Business Media: Germany.

Mechin, M.C., Coudane, F., Adoue, V., Arnaud, J., Duplan, H., Charveron, M. and Simon, M. (2010) Deimination is regulated at multiple levels including auto-deamination of peptidylarginine deiminases. *Cellular and Molecular Life Sciences* 67, 1491–1503.

Meena, M.K. (2024) Regulating postharvest pathogen infection and decay using biotechnological tools. In: *Bioengineered Fruit and Vegetables*, pp. 135–161. Apple Academic Press: Florida, United States.

Mencia, R., Gonzalo, L., Tossolini, I. and Manavella, P.A. (2023) Keeping up with the miRNAs: Current paradigms of the biogenesis pathway. *Journal of Experimental Botany* 74(7), 2213–2227.

Menezes, S.A. and Tasca, T. (2024) Extracellular vesicles in parasitic diseases–from pathogenesis to future diagnostic tools. *Microbes and Infection*, 26, 105310.

Mesterházy, Á., Oláh, J. and Popp, J. (2020) Losses in the grain supply chain: Causes and solutions. *Sustainability* 12(6), 2342.

Migicovsky, Z. and Kovalchuk, I. (2013) Changes to DNA methylation and homologous recombination frequency in the progeny of stressed plants. *Biochemistry and Cell Biology* 91(1), 1–5.

Mishra, R.C. and Bae, H. (2019) Plant cognition: Ability to perceive 'touch' and 'sound'. *Sensory Biology of Plants*, 137–162.

Morales, A. and Kaiser, E. (2020) Photosynthetic acclimation to fluctuating irradiance in plants. *Frontiers in Plant Science* 11, 268.

Mueller, N.G. and Flachs, A. (2022) Domestication, crop breeding, and genetic modification are fundamentally different processes: Implications for seed sovereignty and agrobiodiversity. *Agriculture and Human Values* 39(1), 455–472.

Murchie, E.H. and Lawson, T. (2013) Chlorophyll fluorescence analysis: A guide to good practice and understanding some new applications. *Journal of Experimental Botany* 64(13), 3983–3998.

Naaz, S., Ahmad, N. and Qureshi, M.I. (2024) ATP binding cassette (ABC) transporters in plant development and Defense. In: *Molecular and Physiological Insights into Plant Stress Tolerance and Applications in Agriculture-Part 2*. Bentham Science Publishers, pp. 251–269.

Nagaraj, S., Rajasekaran, R., Palaniappan, J., Rangasamy, S., Narayanasamy, C. and Narayanan, M.B. (2024) Emerging technological developments to address pest resistance in Bt cotton. *Journal of Cotton Research* 7(1), 30.

Naveed, Z.A., Wei, X., Chen, J., Mubeen, H. and Ali, G.S. (2020) The PTI to ETI continuum in Phytophthora-plant interactions. *Frontiers in Plant Science* 11, 593905.

Ngongolo, K. and Mmbando, G.S. (2024) Harnessing biotechnology and breeding strategies for climate-resilient agriculture: Pathways to sustainable global food security. *Discover Sustainability* 5(1), 431.

Nicu, A.T., Ionel, I.P., Stoica, I., Burlibasa, L. and Jinga, V. (2024) Recent advancements in research on DNA methylation and testicular germ cell Tumors: Unveiling the intricate relationship. *Biomedicine* 12(5), 1041.

Nien, Y.C., Vanek, A. and Axtell, M.J. (2024) Trans-species mobility of RNA interference between plants and associated organisms. *Plant and Cell Physiology* 65(5), 694–703.

Niu, H., Gu, J. and Zhang, Y. (2024) Bacterial persisters: Molecular mechanisms and therapeutic development. *Signal Transduction and Targeted Therapy* 9(1), 174.

Noman, A., Aqeel, M., Qari, S.H., Al Surhanee, A.A., Yasin, G., Alamri, S. and Al-Saadi, A.M. (2020) Plant hypersensitive response vs pathogen ingression: Death of few gives life to others. *Microbial Pathogenesis* 145, 104224.

Numan, M., Sun, Y. and Li, G. (2024) Exploring the emerging role of long non-coding RNAs (lncRNAs) in plant biology: Functions, mechanisms of action, and future directions. *Plant Physiology and Biochemistry* 212, 108797.

Ortola, B. and Daros, J.A. (2024) RNA interference in insects: From a natural mechanism of gene expression regulation to a biotechnological crop protection promise. *Biology* 13(3), 137.

Otegui, M.S. (2021) Imaging polyphenolic compounds in plant tissues. *Recent Advances in Polyphenol Research* 7, 281–295.

Pandey, A., Prasad, S.M. and Kumar, J. (2022) Genomics (characterization of genes) and molecular aspects of metal tolerance and Hyperaccumulation. In: *Heavy Metals in Plants*, pp. 310–322. CRC Press: Florida, United States.

Panwar, V., McCallum, B. and Bakkeren, G. (2013) Endogenous silencing of Puccinia triticina pathogenicity genes through in planta-expressed sequences leads to the suppression of rust diseases on wheat. *The Plant Journal* 73(3), 521–532.

Pappas, M.L., Baptista, P., Broufas, G.D., Dalakouras, A., Djobbi, W., Flors, V. and Papadopoulou, K. (2020) Biological and molecular control tools in plant defense. *Plant Defence: Biological Control*, 3–43.

Parise, C., Galetto, L., Abbà, S., Bodino, N., Marzachì, C. and Bosco, D. (2024) RNA interference protocols for gene silencing in the spittlebug Philaenus spumarius, vector of Xylella fastidiosa. *Scientific Reports* 14(1), 25812.

Patel, S. (2018) Danger-associated molecular patterns (DAMPs): The derivatives and triggers of inflammation. *Current Allergy and Asthma Reports* 18, 1–12.

Poliseno, L., Lanza, M. and Pandolfi, P.P. (2024) Coding, or non-coding, that is the question. *Cell Research* 34(9), 609–629.

Ponath, F., Hör, J. and Vogel, J. (2022) An overview of gene regulation in bacteria by small RNAs derived from mRNA 3′ ends. *FEMS Microbiology Reviews* 46(5), fuac017.

Potysz, A., Kierczak, J. and van Hullebusch, E.D. (2024) Metallurgical slags in the environment and treatment systems: Insights regarding biotic interactions. In: *Geomicrobiology: Natural and Anthropogenic Settings*. Springer Nature Switzerland, Cham, pp. 215–239.

Prasad, L., Katoch, S. and Shahid, S. (2022) Microbial interaction mediated programmed cell death in plants. *3 Biotech* 12(2), 43.

Qiao, Y., Xia, R., Zhai, J., Hou, Y., Feng, L., Zhai, Y. and Ma, W. (2021) Small RNAs in plant immunity and virulence of filamentous pathogens. *Annual Review of Phytopathology* 59(1), 265–288.

Qiao, M., Hong, C., Jiao, Y., Hou, S. and Gao, H. (2024) Impacts of drought on photosynthesis in major food crops and the related mechanisms of plant responses to drought. *Plants* 13(13), 1808.

Qin, L., Zhao, L., Wu, C., Qu, S. and Wang, S. (2021) Identification of microRNA transcriptome in apple response to Alternaria alternata infection and evidence that miR390 is negative regulator of defense response. *Scientia Horticulturae* 289, 110435.

Qiu, C.W., Ma, Y., Gao, Z.F., Sreesaeng, J., Zhang, S., Liu, W. and Wu, F. (2024) Genome-wide profiling of genetic variations reveals the molecular basis of aluminum stress adaptation in Tibetan wild barley. *Journal of Hazardous Materials* 461, 132541.

Quan, Y. (2025). Plant-Insect Interactions in Grassland Ecosystems: Assessing Consequences of Landscape Change and Environmental Stressors.

Rahman, A., Rahman, M.H.S., Uddin, M.S., Sultana, N., Akhter, S., Nath, U.K. and Hossain, A. (2024) Advances in DNA methylation and its role in cytoplasmic male sterility in higher plants. *Journal of Integrative Agriculture* 23(1), 1–19.

Rai, K. and Agrawal, S.B. (2017) Effects of UV-B radiation on morphological, physiological and biochemical aspects of plants: An overview. *Journal of Scientific Research* 61, 87–113.

Ribeiro, C.A.M., de Oliveira Cerqueirae Costa, M., Farias, A.B., dos Reis Ribeiro, R., de Lima, Y.F.A. *et al.* (2024) Regulatory small RNAs as antimicrobial drug targets. In: *Structure-Based Drug Design*. Springer International Publishing, Cham, pp. 31–79.

Rodrigues, J.C.M., Carrijo, J., Anjos, R.M., Cunha, N.B., Grynberg, P., Aragão, F.J.L. and Vianna, G.R. (2024) The role of microRNAs in NBS-LRR gene expression and its implications for plant immunity and crop development. *Transgenic Research* 1–16.

Rong, F., Lv, Y., Deng, P., Wu, X., Zhang, Y., Yue, E. and Wu, L. (2024) Switching action modes of miR408-5p mediates auxin signaling in rice. *Nature Communications* 15(1), 2525.

Röther, S. and Meister, G. (2011) Small RNAs derived from longer non-coding RNAs. *Biochimie* 93(11), 1905–1915.

Roychowdhury, R., Hada, A., Biswas, S., Mishra, S., Prusty, M.R., Das, S.P. and Sarker, U. (2024) Jasmonic acid (JA) in plant immune response: Unravelling complex molecular mechanisms and networking of defence signalling against pathogens. *Journal of Plant Growth Regulation* 1-26.

Saakre, M., Jaiswal, S., Rathinam, M., Raman, K.V., Tilgam, J., Paul, K. and Pattanayak, D. (2024) Host-delivered RNA interference for durable pest resistance in plants: Advanced methods, challenges, and applications. *Molecular Biotechnology* 66(8), 1786–1805.

Sadhna, P.B.D., Tripathi, S., Devi, R.S., Singh, R. and Bhadouria, R. (2023) 7 Ecophysiological Responses of Tropical Plants to Rising Air Pollution. In: Ecophysiology of Tropical Plants: Recent Trends and Future Perspectives, CRC Press: Florida, United States.

Saini, L., Sharma, K. and Kumar, P. (2024) Water efficiency in agriculture as a vital approach toward water management in rural areas. In: *Water Resources Management for Rural Development*, pp. 127–138. Elsevier: Cambridge, Massachusetts, United States.

Santoro, V., Schiavon, M. and Celi, L. (2024) Role of soil abiotic processes on phosphorus availability and plant responses with a focus on strigolactones in tomato plants. *Plant and Soil* 494(1), 1–49.

Sarvari, P., Sarvari, P., Ramírez-Díaz, I., Mahjoubi, F. and Rubio, K. (2022) Advances of epigenetic biomarkers and epigenome editing for early diagnosis in breast cancer. *International Journal of Molecular Sciences* 23(17), 9521.

Sattar, M.N., Iftikhar, S., Gul, A., Aliya, K., El-Ganainy, S.M. and Al-Khayri, J.M. (2024) RNAi technology utilization in plant breeding. In: *Plant Molecular Breeding in Genomics Era: Applications*. Springer Nature Switzerland, Cham, pp. 263–314.

Schumann, U., Smith, N.A., Kazan, K., Ayliffe, M. and Wang, M.B. (2013) Analysis of hairpin RNA transgene-induced gene silencing in fusarium oxysporum. *Silence* 4, 1–16.

Seo, J.K., Wu, J., Lii, Y., Li, Y. and Jin, H. (2013) Contribution of small RNA pathway components in plant immunity. *Molecular Plant-Microbe Interactions* 26(6), 617–625.

Shafi, I., Gautam, M. and Kariyat, R. (2024) Integrating ecophysiology and omics to unlock crop response to drought and herbivory stress. *Frontiers in Plant Science* 15, 1500773.

Shafiq, S., Zeb, Q., Ali, A., Sajjad, Y., Nazir, R., Widemann, E., and Liu, L. (2019). Lead, cadmium and zinc phytotoxicity alter DNA methylation levels to confer heavy metal tolerance in wheat. *International Journal of Molecular Sciences*, 20(19), 4676.

Shahid, M., Shafi, Z., Ilyas, T., Singh, U.B. and Pichtel, J. (2024) Crosstalk between phytohormones and pesticides: Insights into unravelling the crucial roles of plant growth regulators in improving crop resilience to pesticide stress. *Scientia Horticulturae* 338, 113663.

Shao, M., Feng, Y., Yang, S., Feng, T., Zeng, F., Lu, S. and Mao, J. (2024) Molecular evolution of Phytocyanin gene and analysis of expression at different coloring periods in apple (Malus domestica). *BMC Plant Biology* 24(1), 374.

Sharma, S., Kumar, A., Kumari, N. and Walia, A. (2023) RNAi as a tool to enhance crop yield and biotic stress management in the plants. *Plant Cell, Tissue and Organ Culture (PCTOC)* 152(3), 437–454.

Sharma, K., Jalota, K., Agarwal, C., Pal, P. and Jindal, S. (2024) Biochemical responses of plants to individual and combined abiotic stresses. *Plant Secondary Metabolites and Abiotic Stress*, 1–34.

Shen, Q., Hasegawa, K., Oelerich, N., Prakken, A., Tersch, L.W., Wang, J. and Maekawa, T. (2024) Cytoplasmic calcium influx mediated by plant MLKLs confers TNL-triggered immunity. *Cell Host and Microbe* 32(4), 453–465.

Shi, W., Huang, S., Zhang, K., Liu, B., Liu, D., Huang, Q. and Chao, L. (2022) Quantifying the superimposed effects of drought-flood abrupt alternation stress on vegetation dynamics of the Wei River basin in China. *Journal of Hydrology* 612, 128105.

Sihag, N., Singh, T., Sheoran, S., Singh, O., Malik, R., Kumar, L. and Singh, J. (2024) Role of RNA interference in drought stress management: Physiological, biochemical and molecular approach. Crop and pasture. *Science* 75(1), NULL-NULL.

Silveira, J.A. and Sousa, R.H. (2024) Looking for a systemic concept and physiological diagnosis of a plant stress state. *Theoretical and Experimental Plant Physiology*, 36, 1–11.

Singh, A. and Roychoudhury, A. (2023) Abscisic acid in plants under abiotic stress: Crosstalk with major phytohormones. *Plant Cell Reports* 42(6), 961–974.

Singh, A., Gautam, V., Singh, S., Sarkar Das, S., Verma, S., Mishra, V. and Sarkar, A.K. (2018) Plant small RNAs: Advancement in the understanding of biogenesis and role in plant development. *Planta* 248, 545–558.

Singh, A., Rajput, V.D., Sharma, R., Ghazaryan, K. and Minkina, T. (2023) Salinity stress and nanoparticles: Insights into antioxidative enzymatic resistance, signaling, and defense mechanisms. *Environmental Research*, 235, 116585.

Singh, A., Yajnik, K.N., Mogilicherla, K. and Singh, I.K. (2024) Deciphering the role of growth regulators in enhancing plant immunity against herbivory. *Physiologia Plantarum* 176(6), e14604.
Sood, M., Kapoor, D., Kumar, V., Kalia, N., Bhardwaj, R., Sidhu, G.P. and Sharma, A. (2021) Mechanisms of plant defense under pathogen stress: A review. *Current Protein and Peptide Science* 22(5), 376–395.
Spada, M., Pugliesi, C., Fambrini, M. and Pecchia, S. (2024) Challenges and opportunities arising from host–Botrytis cinerea interactions to outline novel and sustainable control strategies: The key role of RNA interference. *International Journal of Molecular Sciences* 25(12), 6798.
Srivastava, S., Fuentes, F. and Bhargava, A. (2024) Insights in plant epigenetics. In: *Integrative Approaches to Biotechnology*, pp. 3–28. CRC Press: Florida, United States.
Stewart, A.F., Fulton, S.L., Durand-de Cuttoli, R., Thompson, R.E., Chen, P.J. *et al.* (2024). Hippocampal γCaMKII dopaminylation promotes synaptic-to-nuclear signaling and memory formation. bioRxiv.
Sudan, A.K., Grewal, S. and Vakhlu, J. (2020) Plant growth regulation by small RNA-mediated plant-biotic interactions. In: *Plant Small RNA*, pp. 321–351. Elsevier: Cambridge, Massachusetts, United States.
Tailor, A. and Bhatla, S.C. (2024) R gene-mediated resistance in the management of plant diseases. *Journal of Plant Biochemistry and Biotechnology* 33(1), 5–23.
Talarico, E., Zambelli, A., Araniti, F., Greco, E., Chiappetta, A. and Bruno, L. (2024) Unravelling the epigenetic code: DNA methylation in plants and its role in stress response. *Epigenomes* 8(3), 30.
Tan, H., Li, B. and Guo, H. (2020) The diversity of post-transcriptional gene silencing mediated by small silencing RNAs in plants. *Essays in Biochemistry* 64(6), 919–930.
Tereshchenko, M. (2024) *Investigation of Biomolecular Condensate Formation in the Cellular Response to Genotoxic Stress* (Doctoral dissertation). University of Toronto, Canada.
Tiwari, M. and Singh, P. (2024) Guardians of wheat: Unleashing transgenerational immune priming with Trichoderma against spot blotch. *South African Journal of Botany* 169, 242–254.
Tiwari, M., Devi, B., Sinha, S., Yadav, N. and Singh, P. (2024) Intergenerational priming by Trichoderma alleviates drought stress in barley. *Environmental and Experimental Botany* 223, 105772.
Tripathi, A., Pandey, V.K., Jha, A.K., Srivastava, S., Jakhar, S., Singh, G. and Choudhary, P. (2024) Intricacies of plants' innate immune responses and their dynamic relationship with fungi: A review. *Microbiological Research* 285, 127758.
Vanderplank, J.E. (2012) *Host-Pathogen Interactions in Plant Disease*. Elsevier: Cambridge, Massachusetts, United States.
Vassileva, V., Georgieva, M., Zehirov, G. and Dimitrova, A. (2023) Exploring the genotype-dependent toolbox of wheat under drought stress. *Agriculture* 13(9), 1823.
Vaucheret, H. and Voinnet, O. (2024) The plant siRNA landscape. *The Plant Cell* 36(2), 246–275.
Velini, E.D., Trindade, M.L., Barberis, L.R.M. and Duke, S.O. (2010) Growth regulation and other secondary effects of herbicides. *Weed Science* 58(3), 351–354.
Verhoek, A. E. (2024). Exploring Laser-Induced Fluorescence An Experimental Approach Towards the Remote Application of Laser.
Verma, K. and Modgil, M. (2024) RNA interference (RNAi) mediated technique for combating plant diseases: Harnessing nanoparticles for effective delivery and enhanced efficacy. *Plant Cell, Tissue and Organ Culture (PCTOC)* 157(3), 53.
Vongai, C. (2020) *Identification and Characterization of PAMPS/NAMPS Released by Nematodes* (Doctoral dissertation). Ghent University.
Waheed, S., Anwar, M., Saleem, M.A., Wu, J., Tayyab, M. and Hu, Z. (2021) The critical role of small RNAs in regulating plant innate immunity. *Biomolecules* 11(2), 184.
Wang, S., Yan, W., Yang, X., Zhang, J. and Shi, Q. (2021) Comparative methylome reveals regulatory roles of DNA methylation in melon resistance to Podosphaera xanthii. *Plant Science* 309, 110954.
Wang, Y., Pruitt, R.N., Nuernberger, T. and Wang, Y. (2022) Evasion of plant immunity by microbial pathogens. *Nature Reviews Microbiology* 20(8), 449–464.
Wei, J., Li, Y., Chen, X., Tan, P., Muhammad, T. and Liang, Y. (2024) Advances in understanding the interaction between Solanaceae NLR resistance proteins and the viral effector Avr. *Plant Signaling and Behavior* 19(1), 2382497.
Weigelt, O., Siestrup, K. and Prem, R. (2021) Continuity in transition: Combining recovery and day-of-week perspectives to understand changes in employee energy across the 7-day week. *Journal of Organizational Behavior* 42(5), 567–586.
Weyers, J.D. and Paterson, N.W. (2001) Plant hormones and the control of physiological processes. *New Phytologist* 152(3), 375–407.
Wickramasuriya, A.M., Sanahari, W.M.A., Weeraman, J.W.J.K., Karunarathne, M.A.U., Hendalage, D.P.B., Sandumina, K.H.N. and Artigala, A.D.T.S. (2024) DNA-(cytosine-C5) methyltransferases

and demethylases in Theobroma cacao: Insights into genomic features, phylogenetic relationships, and protein–protein interactions. *Tree Genetics and Genomes* 20(2), 7.

Wong, L.H. and Tremethick, D.J. (2024) Multifunctional histone variants in genome function. *Nature Reviews Genetics* 26, 1–23.

Wu, L., Yu, J., Shen, Q., Huang, L., Wu, D. and Zhang, G. (2018a) Identification of microRNAs in response to aluminum stress in the roots of Tibetan wild barley and cultivated barley. *BMC Genomics* 19, 1–14.

Wu, W., Ma, B.L. and Whalen, J.K. (2018b) Enhancing rapeseed tolerance to heat and drought stresses in a changing climate: Perspectives for stress adaptation from root system architecture. *Advances in Agronomy* 151, 87–157.

Xu, T., Zhu, Y., Lin, Z., Lei, J., Li, L., Zhu, W. and Wu, D. (2024) Evidence of cross-kingdom gene regulation by plant MicroRNAs and possible reasons for inconsistencies. *Journal of Agricultural and Food Chemistry* 72(9), 4564–4573.

Yamamoto, T. and Powell, G.K. (2024) Bacillus thuringiensis crystal proteins: Recent advances in understanding its insecticidal activity. *Advanced Engineered Pesticides* 3–42.

Yan, L., Su, B.X., Li, J.J., Li, Y.Y., Chen, S.Y., Feng, C.Y. and Zhang, Q.H. (2024) Genome-wide identification of nucleotide-binding site–leucine-rich repeat gene family in Cymbidium ensifolium and expression profiles in response to fusarium wilt infection. *Horticulturae* 10(6), 634.

Yang, X., Zhang, L., Yang, Y., Schmid, M. and Wang, Y. (2021) miRNA mediated regulation and interaction between plants and pathogens. *International Journal of Molecular Sciences* 22(6), 2913.

Yang, X., Zhang, L., Li, Y., Liu, X., Chen, C., Deng, Y. and Chen, X. (2024) Fortifying crop defenses: Unraveling the molecular arsenal against aphids. *Horticulture Advances* 2(1), 22.

Ye, W. and Ma, W. (2016) Filamentous pathogen effectors interfering with small RNA silencing in plant hosts. *Current Opinion in Microbiology* 32, 1–6.

Yin, R., Cheng, J. and Lin, J. (2024) The role of the type VI secretion system in the stress resistance of plant-associated bacteria. *Stress Biology* 4(1), 16.

Yousuf, P., Razzak, S., Parvaiz, S., Rather, Y.A. and Lone, R. (2024) Role of plant phenolics in the resistance mechanism of plants against insects. In: *Plant Phenolics in Biotic Stress Management*. Springer Nature Singapore, Singapore, pp. 191–215.

Yu, X.Q., Niu, H.Q., Liu, C., Wang, H.L., Yin, W. and Xia, X. (2024) PTI-ETI synergistic signal mechanisms in plant immunity. *Plant Biotechnology Journal*, 22, 2113–2128.

Yuan, M., Ngou, B.P.M., Ding, P. and Xin, X.F. (2021) PTI-ETI crosstalk: An integrative view of plant immunity. *Current Opinion in Plant Biology* 62, 102030.

Yuan, B., Wang, W.B., Wang, Y.T. and Zhao, X.Q. (2024) Regulatory mechanisms underlying yeast chemical stress response and development of robust strains for bioproduction. *Current Opinion in Biotechnology* 86, 103072.

Zaheer, U., Munir, F., Salum, Y.M. and He, W. (2024) Function and regulation of plant ARGONAUTE proteins in response to environmental challenges: A review. *PeerJ* 12, e17115.

Zainali, N., Alizadeh, H. and Delavault, P. (2024) Gene silencing in broomrapes and other parasitic plants of the Orobanchaceae family: Mechanisms, considerations, and future directions. *Journal of Experimental Botany* 76 erae388.

Zeng, Y., Gong, M., Lin, M., Gao, D. and Zhang, Y. (2020) A review about transcription factor binding sites prediction based on deep learning. *IEEE Access* 8, 219256–219274.

Zhang, H. and Zhu, J.K. (2024) Epigenetic gene regulation in plants and its potential applications in crop improvement. *Nature Reviews Molecular Cell Biology*, 26, 1–17.

Zhang, W., Gao, S., Zhou, X., Chellappan, P., Chen, Z., Zhou, X. and Jin, H. (2011) Bacteria-responsive microRNAs regulate plant innate immunity by modulating plant hormone networks. *Plant Molecular Biology* 75, 93–105.

Zhang, Y., Chen, C., Jin, Z., Yang, Z. and Li, Y. (2022) Leaf anatomy, photosynthesis, and chloroplast ultrastructure of Heptacodium miconioides seedlings reveal adaptation to light environment. *Environmental and Experimental Botany* 195, 104780.

Zhang, C., Chen, L. and Hou, S. (2024a) The emerging roles of clathrin-mediated endocytosis in plant development and stress responses. *Journal of Plant Physiology* 295, 154189.

Zhang, Q., Wang, Z., Gao, R. and Jiang, Y. (2024b) Sugars, lipids and more: New insights into plant carbon sources during plant–microbe interactions. *Plant, Cell and Environment* 48, 1656–1673.

Zhang, W., Stelinski, L.L., Mohamed, A., Wang, G., Tettamanti, G., Chen, M. and Zhao, Q. (2024c) *Unlocking Agro-Ecosystem Sustainability: Exploring the Bottom-up Effects of Microbes, Plants, and Insect Herbivores*. Integrative Zoology. Elsevier: Cambridge, Massachusetts, United States.

Zhang, X., Zhou, Y., Liu, Y., Li, B., Tian, S. and Zhang, Z. (2024d) Research progress on the mechanism and function of histone acetylation regulating the interaction between pathogenic fungi and plant hosts. *Journal of Fungi* 10(8), 522.

Zhao, K., Li, Y., Li, Z., Cao, Z., Ma, X., Ren, R. and Yin, D. (2024) Genome-wide analysis of AhCN genes reveals the AhCN34 involved in bacterial wilt resistance in peanut. *Journal of Integrative Agriculture*.

Zheng, K., Lyu, Z., Chen, J. and Chen, G. (2024) 5-Hydroxymethylcytosine: Far beyond the intermediate of DNA demethylation. *International Journal of Molecular Sciences* 25(21), 11780.

Zhong, Y., Cheng, C., Moniruzzaman, M., Jiang, B., Jiang, N. and Zhong, G. (2021) Expression of miRNAs and their target genes in roots of 'Sanhu'tangerine (Citrus reticulata Blanco cv.'Sanhu') in response to Candidatus Liberibacter asiaticus infection. *Journal of Plant Diseases and Protection* 128, 407–420.

Zhou, Z., Cao, Y., Li, T., Wang, X., Chen, J., He, H. and Zhang, H. (2020) MicroRNAs are involved in maize immunity against fusarium verticillioides ear rot. *Genomics, Proteomics and Bioinformatics* 18(3), 241–255.

Zhu, F., Cao, M.Y., Zhang, Q.P., Mohan, R., Schar, J., Mitchell, M. and Fu, Z.Q. (2024) Join the green team: Inducers of plant immunity in the plant disease sustainable control toolbox. *Journal of Advanced Research* 57, 15–42.

Zia, A . (2010). Study of factors controlling the photoprotection capacity of the thylakoid membrane (Doctoral dissertation).

Zivcak, M., Brestic, M., Kunderlikova, K., Olsovska, K. and Allakhverdiev, S.I. (2015) Effect of photosystem I inactivation on chlorophyll a fluorescence induction in wheat leaves: Does activity of photosystem I play any role in OJIP rise? *Journal of Photochemistry and Photobiology B: Biology* 152, 318–324.

7 Abiotic Stress Responses: Drought, Salinity and Temperature

Muhammad Ahsan Altaf[1,2], Naveed Mushtaq[1,2] and Riaz Ahmad[3*]

[1]Key Laboratory for Quality Regulation of Tropical Horticultural Crops of Hainan Province, School of Breeding and Multiplication (Sanya Institute of Breeding and Multiplication), Hainan University, Sanya 572025, China; [2]Key Laboratory for Quality Regulation of Tropical Horticultural Crops of Hainan Province, School of Tropical Agriculture and Forestry, Hainan University, Haikou 570228, China; [3]Department of Horticulture, The University of Agriculture, Dera Ismail Khan 29111, Pakistan

Abstract

Abiotic stress refers to the adverse effects of non-living environmental factors on plants, significantly influencing their growth and development. Key abiotic stressors include drought, salinity, extreme temperatures, flooding, and nutrient deficiencies. These stresses can lead to various physiological and biochemical responses, ultimately affecting crop productivity and sustainability. Plants have evolved complex mechanisms to sense and respond to various types of environmental stress, such as drought, salinity, and temperatures. These responses involve intricate signaling pathways and physiological adaptations. Plant hormones, or phytohormones, play a crucial function in enhancing abiotic stress tolerance in plants. These signaling molecules regulate various physiological processes and help plants adapt to abiotic stresses. Understanding the mechanisms by which phytohormones operate can lead to innovative agricultural practices that improve crop resilience in the face of climate change.

Keywords: salinity, oxidative damage, photosynthesis, drought, antioxidants, temperature

7.1 Introduction

Climate change has already led to an increase in temperatures due to global warming and other factors. Currently, the scientific community faces one of the most challenging scenarios (Shahid *et al.*, 2021; Kumar *et al.*, 2023). A wide range of agricultural systems are impacted by climatic conditions, so a lack of adaptation and mitigation measures may have a huge impact on them (Dos Santos *et al.*, 2022; Mushtaq *et al.*, 2024).

Plants respond to abiotic stresses such as drought, salinity, extreme temperatures, and nutrient deficiencies through complex mechanisms involving signaling and transcription (Zhu, 2016; Zhang *et al.*, 2023). Plants exhibit changes in growth, development, and physiology to adapt to stress conditions (Dresselhaus and Hückelhoven, 2018). Plants alter their metabolic pathways to cope with stress, such as increasing the production of osmoprotectants and antioxidants (Kumar, 2020). Abiotic stresses can severely impact

* Email: riazahmadbzu@gmail.com

 Molecular Biology of Plant Stress Responses (Muhammad Waseem and Pingwu Liu eds)
DOI: 10.1079/9781800628823.0007

plant physiology, leading to reduced growth, development, and productivity. Abiotic stresses increase respiration rates, leading to greater energy expenditure and reduced biomass accumulation (Li *et al.*, 2021; Koza *et al.*, 2022).

Drought, salinity, and extreme temperatures can impair photosynthetic efficiency by reducing CO_2 uptake, damaging chloroplasts, and disrupting electron transport chains (ETCs) (Khan *et al.*, 2021). Ion transport mechanisms are modulated to maintain ion homeostasis under salt stress (Amin *et al.*, 2021). Salinity toxicity can interfere with nutrient availability in the soil and impair root function, leading to nutrient deficiencies (Basu *et al.*, 2021). In addition, salinity causes ion toxicity and imbalances, disrupting nutrient homeostasis. Furthermore, drought alters the production and sensitivity of plant hormones like abscisic acid (ABA), auxin, cytokinin (CK), and ethylene (ET), disrupting growth regulation (Rahman *et al.*, 2016; Ali *et al.*, 2021). Salinity significantly damages plant physiology through various mechanisms, leading to physiological, biochemical, and morphological changes. Salinity creates an osmotic imbalance that hinders water uptake by the roots. The high concentration of salts in the soil solution reduces the water potential, making it difficult for plants to absorb sufficient water. Drought triggers ABA accumulation, leading to stomatal closure and growth inhibition (Tombesi *et al.*, 2015). Plant growth and development are significantly affected by salinity (Safdar *et al.*, 2019), drought (Ahanger *et al.*, 2016), and extreme temperature (Adams *et al.*, 2001). Guilioni *et al.* (2003) reported that plant growth can be decreased by high temperatures or water deficits in field pea, resulting in lower yield productivity.

Abiotic stresses increase the production of reactive oxygen species (ROS), causing oxidative damage to proteins, lipids, and DNA (Sachdev *et al.*, 2021). On the hand, antioxidant systems are upregulated to mitigate ROS, but severe stress can overwhelm these defenses (Altaf *et al.*, 2022). Extreme temperatures and drought can disrupt membrane structure and function, leading to ion leakage and cellular damage (Bhattacharjee, 2012). Different abiotic stresses such as drought, salinity, and temperature stress, effectively reduced seed germination, impaired seedling health index, hindered mineral nutrient accumulation, restricted water availability, cell death, imbalanced ROS homeostasis, led to excessive malondialdehyde (MDA) concentration, altered antioxidant defense mechanisms, and resulted in changes in cell wall, which in turn affected photosynthesis, water relations, oxidative status, and membrane integrity (Fig. 7.1) (Balfagón *et al.*, 2020; Dos Santos *et al.*, 2022). Understanding these stress responses is crucial for developing stress-tolerant crops and sustainable agricultural practices (Kumar *et al.*, 2019; Yadav *et al.*, 2020).

7.2 Response Mechanism to Abiotic Stresses in Plants

Abiotic stresses are often encountered by plants. Globally, abiotic stresses (salinity, drought, and temperature) adversely affect survival, biomass production, and yield of staple food crops by up to 70%, thereby undermining food security (Cramer *et al.*, 2011). Plant growth and productivity are limited by abiotic stresses such as drought, salinity, and temperatures (Chen and Murata, 2008). Importantly, the rise in atmospheric CO_2 may

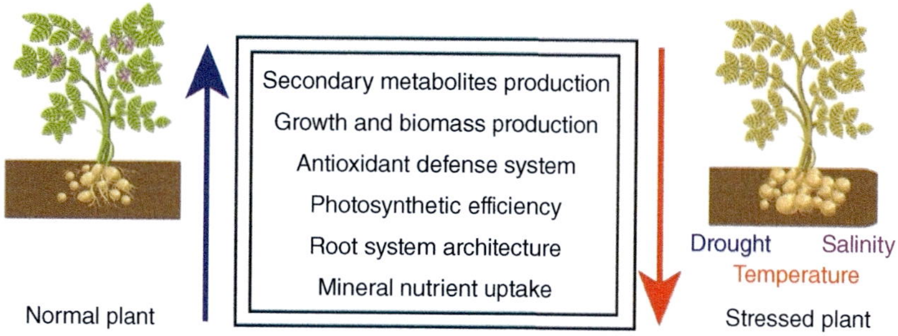

Fig. 7.1. Plant response to different abiotic stresses (salinity, drought, and temperature). (Figure author's own.)

stimulate alterations in the net photosynthetic rate of plants, leading to variations in their growth rate. This often has a beneficial effect on total biomass, but it can also result in a decrease in nutritional quality (Uddling *et al.*, 2018). Plants exhibit specific alterations in their growth and physiological processes in response to stress-induced stimuli.

7.3 Physiological and Molecular Mechanisms of Stress Tolerance

When plants are exposed to high salt levels, they activate an osmotic adjustment process to preserve the clarity of their cells, which leads to the reduced development of stressed plants (Table 7.1). Meanwhile, the modifications caused by salinity stress in plants can vary depending on the species, developmental stage, severity, and period of the stress (Liang *et al.*, 2018). The limiting of plant development caused by salinity is believed to be a result of the osmotic impact caused by an imbalance of ions in the early phase, and a direct influence of the ions themselves in the latter phase of mild to moderate stress (Mishra *et al.*, 2021).

Cellular dehydration is the result of the loss of intracellular water from the cell as a result of salinity, drought, and temperature (Kaushal and Wani, 2016). The antioxidant enzymes found in plant cells counteract the detrimental effects of ROS (Balasubramaniam *et al.*, 2023). The MDA and electrolyte leakage (EL) levels were reduced due to a significant increase in antioxidant enzyme production (Shilev, 2020). When plants are exposed to temperature, salinity, and drought, they seem to experience osmotic and oxidative stress (Saberi Riseh *et al.*, 2021). Mozafariyan *et al.* (2016) reported that salinity toxicity negatively damaged the photosynthetic capacity, hindered water availability, disturbed root system architecture, and decreased plant growth and development.

Plants have the ability to activate several physiological and biochemical systems in order to cope with stress (Fig. 7.2). In addition, these mechanisms include changes in the plant's structure, water balance, photosynthesis, hormone levels, distribution of ions, and metabolic adjustments (Ashraf and Harris, 2013).

7.4 Drought Response in Plants

The water scarcity issue has garnered significant attention and prompted increased research efforts into the underlying science of plant

Table 7.1. Salinity and drought stress response in plants. (Table author's own.)

Stress	Crop	Response	Reference
Salinity	Maize	Decreased grain size and leaf area, lowered photosynthesis and antioxidant enzyme activity	Ali *et al.* (2021)
Drought	Frankenia	Decreased leaf water potential, hindered osmotic adjustment, reduced mineral nutrient uptake	Chegah *et al.* (2013)
Heat	Tomato	Reduced growth, hindered photosynthesis, declined fruit yield and quality	Adams *et al.* (2001)
Salinity	Potato	Reduced leaf size area, secondary metabolites production, and osmolytes production	Chourasia *et al.* (2021)
Drought	Wheat	Reduced leaf size, increased enzymes activity, increased oxidative damage	Saqib *et al.* (2013)
Heat	Rice	Reduced enzyme activity and proline activity, and enhanced oxidative damage in rice leaves	Bhattacharjee (2012)
Salinity	Tomato	Increased sodium (Na^+) concentration, lowered potassium (K^+) content, and caused oxidative damage	Flores *et al.* (2001)
Drought	Pea	Reduced grain size, lowered leaf water potential, increased oxidative damage	Guilioni *et al.* (2003)
Heat	Pepper	Increased electrolyte leakage (EL), lowered water potential, altered antioxidant enzyme activity	Gisbert-Mullor *et al.* (2021)

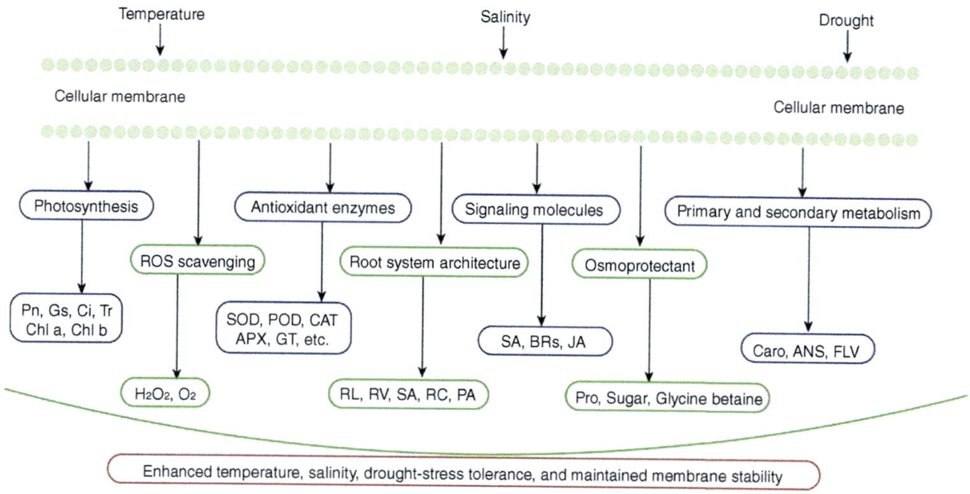

Fig. 7.2. Abiotic stresses (drought, salinity, and temperature) damage physiological pathways. (Figure author's own.)

drought resistance, as well as the application of this knowledge for the development of water-efficient and drought-resistant crops (Chaves *et al.*, 2003; Fang and Xiong, 2015). In recent years, many nations and international organizations have initiated research initiatives to investigate the drought-resistance and water-conservation systems of plants in order to find crucial genes or methods for enhancing plant drought resilience (Yang *et al.*, 2021).

For over 20 years, researchers have undertaken extensive investigations, including studies on morphological characteristics and the structural, physiological, biochemical, and molecular approaches to elucidate the processes of drought responses in plants (Cattivelli *et al.*, 2008; Ilyas *et al.*, 2021). When plants encounter drought or, more frequently, water deficit, they respond to the stress by employing a variety of adaptive mechanisms and responses at the morphological, physiological, and molecular levels. However, different plant species or genotypes within a species often exhibit significant variations in the utilization of these mechanisms (Carraro and Di Iorio, 2022). In the first phase of water scarcity, plants often possess the capacity to effectively extract water from the soil via their root systems, partly obscure stomata to decrease transpiration, and modify their metabolism so it corresponds with the available carbon resources (Farooq *et al.*, 2009). Plants have developed many drought-resistance techniques that enable them to adapt to certain settings, enhancing their growth and development (Waraich *et al.*, 2011). Drought resistance refers to the comprehensive capacity of plants to respond to and adapt to the severe circumstances induced by drought stress (Seleiman *et al.*, 2021).

Plant roots continually absorb water and nutrients from the soil. Consequently, the root system is crucial in responding to water deficiency stress (Anjum *et al.*, 2011). Some kinds of plants possess a strong capacity to enhance root development during the first phase of drought stress to get water from deeper soil layers. Root morphological traits are considerably damaged by drought (Hasanuzzaman *et al.*, 2013b). Root development is also affected by the hydration or nutritional availability of the plant's aerial parts (Siddique *et al.*, 2016).

It is essential to understand how leaves respond morphologically and physiologically to drought stress in order to minimize water loss and increase water usage efficiency (Ahluwalia *et al.*, 2021). Plant growth is primarily influenced by photosynthesis, which is mostly regulated by stomata for CO_2 and water exchange, as well as by the photosynthetic activity inside mesophyll cells (Jaleel *et al.*, 2009; Kaur and Asthir, 2017).

In conditions of acute water scarcity, the elongation of cells in higher plants may be impeded by the disruption of water transport from the xylem to the adjacent elongating cells (Ozturk *et al.*, 2021). The capacity of agricultural plants to adapt to varying environments is directly or indirectly linked to their photosynthetic acclimation, which subsequently influences biochemical and physiological processes, hence affecting overall plant growth and output (Chegah *et al.*, 2013). Drought stress significantly impeded the gas exchange elements of agricultural plants, likely attributable to reduced leaf growth, compromised photosynthetic apparatus and stomatal closure, and caused oxidative damage and changes in pigments concentration.

Chlorophyll is a key component of chloroplasts essential for photosynthesis, and its relative abundance correlates positively with the photosynthetic rate (Okunlola *et al.*, 2017). The reduction in chlorophyll concentration under drought stress is regarded as a characteristic indicator of oxidative stress, perhaps resulting from pigment photo-oxidation and chlorophyll a breakdown (Harb *et al.*, 2010). Photosynthetic pigments are crucial for plants primarily for light absorption and the generation of reducing power.

Drought stress significantly reduces seed germination and impedes seedling growth in plants (Anjum *et al.*, 2011). It also severely hinders gas exchange in agricultural species, primarily due to stomatal closure, reduced leaf expansion, and damage to the photosynthetic apparatus (Anjum *et al.*, 2011). These effects are compounded by oxidative damage and a reduction in photosynthetic pigment concentrations (Dhanda *et al.*, 2004; Beck *et al.*, 2007), which collectively disrupt the antioxidant defense system (AOS) and lower leaf water potential (Beck *et al.*, 2007). In potato cultivation, drought stress effectively decreases leaf area size and tuber yield (Nasir and Toth, 2022). One study recorded a 40% reduction in yield, which was attributed to decreased leaf area, lower leaf count, reduced photosynthetic capacity, and diminished tuber size and quality (Saravia *et al.*, 2016).

The authors described that antioxidant enzymes activity, osmolytes production, secondary metabolites accumulation, leaf photosynthesis, and growth were damaged in rapeseed plants under drought treatment. In addition, drought treatment caused oxidative damage by excessive ROS accumulation in rapeseed plants (Khan *et al.*, 2019). Nahar *et al.* (2016) reported that drought showed considerable improvement in the oxidative stress parameters such as lipid peroxidation production, and superoxide ion uptake, and caused oxidative damage in rice (Khan *et al.*, 2019).

7.5 Salinity Response in Plants

Soil salinity is a significant abiotic factor that influences agricultural productivity, prompting extensive research to cultivate plants with enhanced salt tolerance. Root-zone salinization poses a significant barrier to plant production (Volkmar *et al.*, 1998). Soil salinity is a worldwide issue that influences around 20% of irrigated land and markedly diminishes agricultural output (Negrão *et al.*, 2017). The physiological reactions of plants to salt are complicated and wide-ranging, complicating the design and interpretation of research (Balasubramaniam *et al.*, 2023). The impact of salinity on water relations is significant, leading to stomatal closure and the suppression of leaf growth (Zhang and Dai, 2019). The presence of salt ions induces hyperionic salinity toxicity, which adversely affects plant cells. Stunted growth is an adaptation strategy for survival that enables plants to withstand salinity toxicity (Zhao *et al.*, 2020). Salinity toxicity may diminish the expression of essential regulatory genes associated with cell cycle development, resulting in a reduction of cell numbers in the meristem and inhibiting growth, so impairing the plant's capacity to effectively absorb mineral nutrient accumulation and water (Botella *et al.*, 2005).

Salinity effectively damages the plant growth, seedling health, seed germination, fruiting, flowering, secondary metabolites production, antioxidant defense mechanism, root growth, and mineral nutrient accumulation (Bernstein, 2019). The most well known and essential process in plants, photosynthesis, turns energy from the Sun into chemical form (Hernández, 2019). The decrease in chlorophyll content has been documented in salt-stress circumstances, attributed to enhanced oxidation and degradation of chlorophyll caused by a formation of ROS, with chlorophyll loss decreasing with salinity levels (Angon *et al.*, 2022).

Plant development in nutrient-deficient root zones may be hindered by either a shortage or toxicity of nutrients. The challenge of acquiring mineral nutrients during salinity toxicity is mostly due to reduced nutrient availability resulting from competition with dominant sodium (Na^+) and chloride (Cl^-) ions (Rivero *et al.*, 2014). The accelerated acquisition of ions results in their formation inside plant cells, adversely influencing plant-water connections (Hasegawa *et al.*, 2000).

Salinity stress results in a greater severe osmotic potential in plant cells owing to elevated soil salt concentrations, creating an osmotic gradient that facilitates water loss from the cells and reduces turgor pressure (Dos Santos *et al.*, 2022). Salinity toxicity considerably reduces the mineral nutrient content, metabolites production, increased Na^+ accumulation, and reduced potassium (K^+) uptake in tomato (Flores *et al.*, 2001). Na^+ salts result in inadequate root development, chlorosis, leaf curling and leaf burning in rice. In addition, salinity also triggers stomatal closure, impedes photosynthesis, modifies cellular metabolism, produces oxidative damage in the rice crop, affects spikelet sterility, and impacts grain production (Rodríguez Coca *et al.*, 2023). The reduction in growth and production of bread wheat due to salinity was mostly attributed to decreased water and turgor potentials, diminished photosynthetic assimilation rate, damaged root growth pattern, reduced stomatal conductance, lowered leaf K^+ levels, and elevated leaf Na^+ concentrations (Saqib *et al.*, 2013). Furthermore, the development of the potato plant is impeded by salt salinity toxicity, resulting from osmotic stress-induced ion toxicity. Osmotic stress induced by salinity results in physiological alterations in the plant, including nutritional imbalance, compromised detoxification of ROS, cellular membrane damage, and diminished photosynthetic activity (Chourasia *et al.*, 2021).

Salinity toxicity induces an ionic imbalance that affects several facets of plant growth and development (Cabot *et al.*, 2014). Under a salinity stress environment, there is a decreased absorption of mineral nutrients, which disrupts photosynthesis, root system architecture, antioxidant enzymes accumulation, cellular metabolism, and metabolites production (Tuteja, 2007).

In order to deal with salinity toxicity, plants have evolved to regulate ion homeostasis and its compartmentalization in order to preserve intracellular ion homeostasis (Safdar *et al.* 2019). Salinity toxicity hinders the growth of tomato (GulRukh and Khan, 2020), damages root growth in rice (Puvanitha and Mahendran, 2017) and hinders leaf photosynthesis activity in wheat (Hossain *et al.*, 2021).

7.6 Heat Response in Plants

Global warming affects agricultural output and risks food security. In addition, global warming is adversely influencing farming practices (Hemantaranjan *et al.*, 2014). Heat stress is a significant abiotic barrier faced by plants (Qu *et al.*, 2013). Heat stress is one of the major threats that this temperature rise is providing to sustainable agricultural production in many regions across the world (Kotak *et al.*, 2007). Heat stress considerably decreased plant growth, seedling growth, seed germination, secondary metabolites production, and flowering. It also caused leaf damage and oxidative damage, hindered photosynthesis, and lowered fruit yield and quality (Baniwal *et al.*, 2004; Kan *et al.*, 2023). High temperatures reduce crop productivity due to a shorter life cycle and expedited senescence across various agro-climatic zones (Sun and Guo, 2016). Heat stress significantly influences the yields of several economically vital crops, including tomato, wheat, rice, potato, and pepper (Ohama *et al.*, 2017).

The primary mechanism behind carbon uptake, plant development, and biomass productivity is photosynthesis. Elevated ambient temperature impairs plant production by disrupting photosynthesis (Hasanuzzaman *et al.*, 2013a). The reduction of the photosynthetic rate at moderately elevated temperatures is reversible; however, exposure to very high temperatures may result in irreversible damage to the completely photosynthetic system (Zhou *et al.*, 2022). In addition, heat stress effectively decreased the chlorophyll content and pigment concentration, and lowered photosynthesis levels in plants (Fig. 7.3) (Wahid *et al.*, 2012; Anfoka *et al.*, 2016).

In recent decades, sustaining development and agricultural yield against severe environmental stressors, such as water scarcity, has become the primary problem in contemporary agriculture (Janni *et al.*, 2020). Heat stress causes cellular membrane damage, increases EL, MDA production, and triggers oxidative damage in plants (Prerostova, and Vankova, 2023).

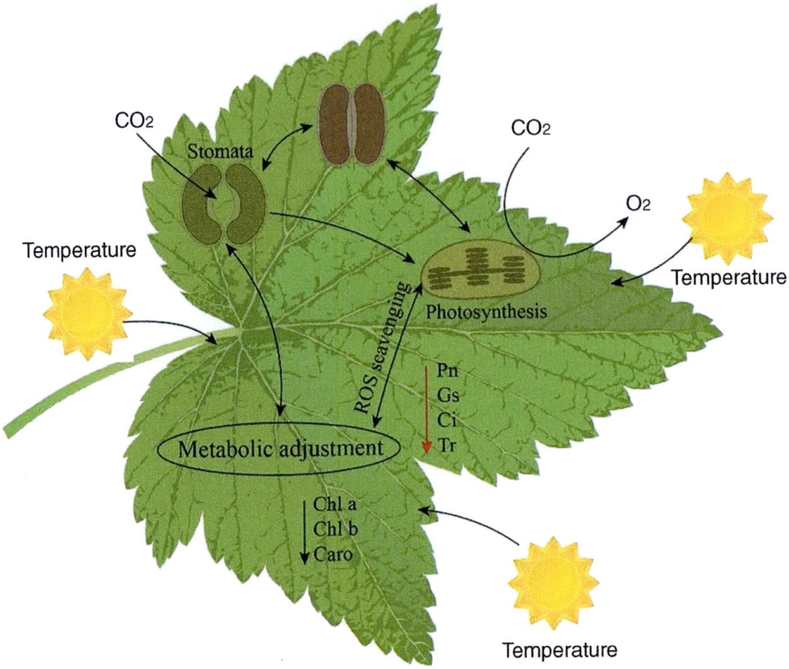

Fig. 7.3. Heat stress hinders the photosynthetic apparatus in plants. (Figure author's own.)

Plants modify their metabolism in multiple ways in response to heat stress, notably by synthesizing secondary metabolites that organize proteins production and cellular membrane structures, preserving cell turgor through osmotic adjustment, and maintaining the AOS to restore cellular redox balance and ion homeostasis (Jagadish *et al.*, 2021).

Plant responses to heat stress differ based on temperature intensity, duration, and species. Heat stress significantly impairs plant development across multiple species and growth stages. It considerably decreases seed germination, seedling growth, and overall plant biomass (Baniwal *et al.*, 2004; Singh *et al.*, 2017; Kan *et al.*, 2023). The physiological impacts are severe, including damage to the photosynthetic apparatus, reduced chlorophyll content, impaired photosynthesis, and oxidative stress (Akter and Rafiqul, 2017; Wang *et al.*, 2023). These effects are often accompanied by altered antioxidant enzyme activity and increased EL (Gisbert-Mullor *et al.*, 2021; Wang *et al.*, 2023). Morphologically, heat stress leads to reduced leaf area, size, and number, as well as a decline in leaf water content, which alters leaf ultrastructure (Shrestha *et al.*, 2022; Wang *et al.*, 2023). This stress also negatively impacts reproductive processes, reducing flowering and fruit yield while compromising fruit quality (Baniwal *et al.*, 2004; Alsamir *et al.*, 2021). Furthermore, it disrupts metabolic functions, altering glucose metabolism and decreasing the production of secondary metabolites (Baniwal *et al.*, 2004; Guo *et al.*, 2020). The cumulative effect of these disruptions is a significant limitation on the growth, yield, and quality of major crops such as tomato, chili, water spinach, rice, wheat, and pepper (Alsamir *et al.*, 2021; Utami and Aryanti, 2021; Ullah *et al.*, 2022; Wu *et al.*, 2022).

7.7 Conclusion and Future Prospects

The scientific community underscores the need of enhancing agricultural productivity to meet rising food demand and population growth. Nevertheless, there has been a notable overlook in the proper use of biotechnological technologies to mitigate the detrimental impacts of abiotic stressors on morphological, physiological, biochemical, and molecular processes. In light of the challenges faced by modern agriculture, the critical inquiries should examine what information should be used and how to effectively implement the insights derived from these strategies to really alleviate these challenges.

This chapter compiles the primary mechanisms through which major agricultural stressors have impacted crop production in recent decades. The aim is to enhance and consolidate understanding of these critical processes.

Further research is required to clarify plant defense mechanisms under various stressors, particularly in conditions when heat, drought, salt, and increased CO_2 co-exist, while also accounting for biotic stresses. Local ecotypes and landraces must be meticulously evaluated to find individuals capable of withstanding environmental stresses, with the objective of elucidating plant defense mechanisms.

References

Adams, S.R., Cockshull, K.E. and Cave, C.R.J. (2001) Effect of temperature on the growth and development of tomato fruits. *Annals of Botany* 88(5), 869–877.

Ahanger, M.A., Morad-Talab, N., Abd-Allah, E.F., Ahmad, P. and Hajiboland, R. (2016) Plant growth under drought stress: Significance of mineral nutrients. *Water Stress and Crop Plants: A Sustainable Approach* 2, 649–668.

Ahluwalia, O., Singh, P.C. and Bhatia, R. (2021) A review on drought stress in plants: Implications, mitigation and the role of plant growth promoting rhizobacteria. *Resources, Environment and Sustainability* 5, 100032.

Akter, N. and Rafiqul Islam, M. (2017) Heat stress effects and management in wheat. A review. *Agronomy for Sustainable Development* 37, 1–17.

Ali, M., Afzal, S., Parveen, A., Kamran, M., Javed, M.R. *et al.* (2021) Silicon mediated improvement in the growth and ion homeostasis by decreasing Na^+ uptake in maize (Zea mays L.) cultivars exposed to salinity stress. *Plant Physiology and Biochemistry* 158, 208–218.

Alsamir, M., Mahmood, T., Trethowan, R. and Ahmad, N. (2021) An overview of heat stress in tomato (*Solanum lycopersicum* L.). *Saudi Journal of Biological Sciences* 28(3), 1654–1663.

Altaf, M.A., Shahid, R., Ren, M.X., Naz, S., Altaf, M.M. *et al.* (2022) Melatonin improves drought stress tolerance of tomato by modulating plant growth, root architecture, photosynthesis, and antioxidant defense system. *Antioxidants* 11(2), 309.

Akter, N. and Rafiqul Islam, M. (2017) Heat stress effects and management in wheat. A review. *Agronomy for Sustainable Development* 37, pp. 1–17.

Amin, I., Rasool, S., Mir, M.A., Wani, W., Masoodi, K.Z. and Ahmad, P. (2021) Ion homeostasis for salinity tolerance in plants: A molecular approach. *Physiologia Plantarum* 171(4), 578–594.

Anfoka, G., Moshe, A., Fridman, L., Amrani, L., Rotem, O.R. *et al.* (2016) Tomato yellow leaf curl virus infection mitigates the heat stress response of plants grown at high temperatures. *Scientific Reports* 6(1), 19715.

Angon, P.B., Tahjib-Ul-Arif, M., Samin, S.I., Habiba, U., Hossain, M.A. and Brestic, M. (2022) How do plants respond to combined drought and salinity stress?—a systematic review. *Plants* 11(21), 2884.

Anjum, S.A., Xie, X.Y., Wang, L.C., Saleem, M.F., Man, C. and Lei, W. (2011) Morphological, physiological and biochemical responses of plants to drought stress. *African Journal of Agricultural Research* 6(9), 2026–2032.

Ashraf, M.H.P.J.C. and Harris, P.J. (2013) Photosynthesis under stressful environments: An overview. *Photosynthetica* 51, 163–190.

Balasubramaniam, T., Shen, G., Esmaeili, N. and Zhang, H. (2023) Plants' response mechanisms to salinity stress. *Plants* 12(12), 2253.

Balfagón, D., Zandalinas, S.I., Mittler, R. and Gómez-Cadenas, A. (2020) High temperatures modify plant responses to abiotic stress conditions. *Physiologia Plantarum* 170(3), 335–344.

Baniwal, S.K., Bharti, K., Chan, K.Y., Fauth, M., Ganguli, A. *et al.* (2004) Heat stress response in plants: a complex game with chaperones and more than twenty heat stress transcription factors. *Journal of Biosciences* 29, 471–487.

Basu, S., Kumar, A., Benazir, I. and Kumar, G. (2021) Reassessing the role of ion homeostasis for improving salinity tolerance in crop plants. *Physiologia Plantarum* 171(4), 502–519.

Beck, E.H., Fettig, S., Knake, C., Hartig, K. and Bhattarai, T. (2007) Specific and unspecific responses of plants to cold and drought stress. *Journal of Biosciences* 32, 501–510.

Bernstein, N. (2019) Plants and salt: Plant response and adaptations to salinity. In: *Model Ecosystems in Extreme Environments*. pp. 101–112, Academic Press: Cambridge.

Bhattacharjee, S. (2012) An inductive pulse of hydrogen peroxide pretreatment restores redox-homeostasis and oxidative membrane damage under extremes of temperature in two rice cultivars. *Plant Growth Regulation* 68, 395–410.

Botella, M.A., Rosado, A., Bressan, R.A. and Hasegawa, P.M. (2005) Plant adaptive responses to salinity stress. In: Jenks, M.A. and Hasegawa, P.M. (eds) *Plant Abiotic Stress*. Blackwell Publishing: Iowa, USA, pp. 37–70.

Cabot, C., Sibole, J.V., Barceló, J. and Poschenrieder, C. (2014) Lessons from crop plants struggling with salinity. *Plant Science* 226, 2–13.

Carraro, E. and Di Iorio, A. (2022) Eligible strategies of drought response to improve drought resistance in woody crops: a mini-review. *Plant Biotechnology Reports* 16(3), 265–282.

Cattivelli, L., Rizza, F., Badeck, F.W., Mazzucotelli, E., Mastrangelo, A.M. *et al.* (2008). Drought tolerance improvement in crop plants: an integrated view from breeding to genomics. *Field Crops Research* 105(1–2), 1–14.

Chaves, M.M., Maroco, J.P. and Pereira, J.S. (2003) Understanding plant responses to drought—from genes to the whole plant. *Functional Plant Biology* 30(3), 239–264.

Chegah, S., Chehrazi, M. and Albaji, M. (2013) Effects of drought stress on growth and development frankenia plant (Frankenia Leavis). *Bulgarian Journal of Agricultural Science* 19(4), 659–666.

Chen, T.H. and Murata, N. (2008) Glycinebetaine: an effective protectant against abiotic stress in plants. *Trends in Plant Science* 13(9), 499–505.

Chourasia, K.N., Lal, M.K., Tiwari, R.K., Dev, D., Kardile, H.B. *et al.* (2021) Salinity stress in potato: Understanding physiological, biochemical and molecular responses. *Life* 11(6), 545.

Cramer, G.R., Urano, K., Delrot, S., Pezzotti, M. and Shinozaki, K. (2011) Effects of abiotic stress on plants: A systems biology perspective. *BMC Plant Biology* 11, 1–14.

Dhanda, S.S., Sethi, G.S. and Behl, R.K. (2004) Indices of drought tolerance in wheat genotypes at early stages of plant growth. *Journal of Agronomy and Crop Science* 190(1), 6–12.

Dos Santos, T.B., Ribas, A.F., de Souza, S.G.H., Budzinski, I.G.F. and Domingues, D.S. (2022) Physiological responses to drought, salinity, and heat stress in plants: a review. *Stresses* 2(1), 113–135.

Dresselhaus, T. and Hückelhoven, R. (2018) Biotic and abiotic stress responses in crop plants. *Agronomy* 8(11), 267.

Fang, Y. and Xiong, L. (2015) General mechanisms of drought response and their application in drought resistance improvement in plants. *Cellular and Molecular Life Sciences* 72, 673–689.

Farooq, M., Wahid, A., Kobayashi, N.S.M.A., Fujita, D.B.S.M.A. and Basra, S.M. (2009) Plant drought stress: Effects, mechanisms and management. *Sustainable Agriculture* 153–188.

Flores, P., Carvajal, M., Cerdá, A. and Martínez, V. (2001) Salinity and ammonium/nitrate interactions on tomato plant development, nutrition, and metabolites. *Journal of Plant Nutrition* 24(10), 1561–1573.

Gisbert-Mullor, R., Padilla, Y.G., Martínez-Cuenca, M.R., López-Galarza, S. and Calatayud, Á. (2021) Suitable rootstocks can alleviate the effects of heat stress on pepper plants. *Scientia Horticulturae* 290, 110529.

Guilioni, L., Wéry, J. and Lecoeur, J. (2003) High temperature and water deficit may reduce seed number in field pea purely by decreasing plant growth rate. *Functional Plant Biology* 30(11), 1151–1164.

GulRukh, E. and Khan, A.S. (2020) Effects of salinity on growth of tomato. *American Academic Scientific Research Journal for Engineering, Technology, and Sciences* 74, 124–129.

Guo, R., Wang, X., Han, X., Chen, X. and Wang-Pruski, G. (2020) Physiological and transcriptomic responses of water spinach (Ipomoea aquatica) to prolonged heat stress. *BMC Genomics* 21, 1–15.

Harb, A., Krishnan, A., Ambavaram, M.M. and Pereira, A. (2010) Molecular and physiological analysis of drought stress in Arabidopsis reveals early responses leading to acclimation in plant growth. *Plant Physiology* 154(3), 1254–1271.

Hasanuzzaman, M., Nahar, K., Alam, M.M., Roychowdhury, R. and Fujita, M. (2013a) Physiological, biochemical, and molecular mechanisms of heat stress tolerance in plants. *International Journal of Molecular Sciences* 14(5), 9643–9684.

Hasanuzzaman, M., Nahar, K., Gill, S.S. and Fujita, M. (2013b) Drought stress responses in plants, oxidative stress, and antioxidant defense. *Climate Change and Plant Abiotic Stress Tolerance* 209–250.

Hasegawa, P.M., Bressan, R.A., Zhu, J.K. and Bohnert, H.J. (2000) Plant cellular and molecular responses to high salinity. *Annual Review of Plant Biology* 51(1), 463–499.

Hemantaranjan, A., Bhanu, A.N., Singh, M.N., Yadav, D.K., Patel, P.K., Singh, R. and Katiyar, D.J.A.P.A.R. (2014) Heat stress responses and thermotolerance. *Advances in Plants and Agriculture Research* 1(3), 1–10.

Hernández, J.A. (2019) Salinity tolerance in plants: Trends and perspectives. *International Journal of Molecular Sciences* 20(10), 2408.

Hossain, A., Skalicky, M., Brestic, M., Maitra, S., Ashraful Alam, M. *et al.* (2021) Consequences and mitigation strategies of abiotic stresses in wheat (Triticum aestivum L.) under the changing climate. *Agronomy* 11(2), 241.

Ilyas, M., Nisar, M., Khan, N., Hazrat, A., Khan, A.H. *et al.* (2021). Drought tolerance strategies in plants: a mechanistic approach. *Journal of Plant Growth Regulation* 40, 926–944.

Jagadish, S.K., Way, D.A. and Sharkey, T.D. (2021) Plant heat stress: Concepts directing future research. *Plant, Cell and Environment* 44(7), 1992–2005.

Jaleel, C.A., Manivannan, P., Wahid, A., Farooq, M., Al-Juburi, H.J., Somasundaram, R. and Panneerselvam, R. (2009) Drought stress in plants: A review on morphological characteristics and pigments composition. *International Journal of Agriculture and Biology* 11(1), 100–105.

Janni, M., Gullì, M., Maestri, E., Marmiroli, M., Valliyodan, B., Nguyen, H.T. and Marmiroli, N. (2020) Molecular and genetic bases of heat stress responses in crop plants and breeding for increased resilience and productivity. *Journal of Experimental Botany* 71(13), 3780–3802.

Kan, Y., Mu, X.R., Gao, J., Lin, H.X. and Lin, Y. (2023) The molecular basis of heat stress responses in plants. *Molecular Plant* 16(10), 1612–1634.

Kaur, G. and Asthir, B. (2017) Molecular responses to drought stress in plants. *Biologia Plantarum* 61, 201–209.

Kaushal, M. and Wani, S.P. (2016) Rhizobacterial-plant interactions: strategies ensuring plant growth promotion under drought and salinity stress. *Agriculture, Ecosystems and Environment* 231, 68–78.

Khan, M.N., Zhang, J., Luo, T., Liu, J., Ni, F. *et al.* (2019) Morpho-physiological and biochemical responses of tolerant and sensitive rapeseed cultivars to drought stress during early seedling growth stage. *Acta Physiologiae Plantarum* 41, 1–13.

Khan, N., Ali, S., Shahid, M.A., Mustafa, A., Sayyed, R.Z. and Curá, J.A. (2021) Insights into the interactions among roots, rhizosphere, and rhizobacteria for improving plant growth and tolerance to abiotic stresses: a review. *Cells* 10(6), 1551.

Kotak, S., Larkindale, J., Lee, U., von Koskull-Döring, P., Vierling, E. and Scharf, K.D. (2007) Complexity of the heat stress response in plants. *Current Opinion in Plant Biology* 10(3), 310–316.

Koza, N.A., Adedayo, A.A., Babalola, O.O. and Kappo, A.P. (2022) Microorganisms in plant growth and development: Roles in abiotic stress tolerance and secondary metabolites secretion. *Microorganisms* 10(8), 1528.

Kumar, S. (2020) Abiotic stresses and their effects on plant growth, yield and nutritional quality of agricultural produce. *The International Journal of Agriculture and Food Science* 4(4), 367–378.

Kumar, A., Patel, J.S., Meena, V.S. and Ramteke, P.W. (2019) Plant growth-promoting rhizobacteria: strategies to improve abiotic stresses under sustainable agriculture. *Journal of Plant Nutrition* 42(11–12), 1402–1415.

Kumar, R., Lal, M.K., Tiwari, R.K., Kumar, A., Behera, B. *et al.* (2023) Traditional and emerging climate-resilient agricultural practices for enhancing food production and nutritional quality. In: *New Frontiers in Plant-Environment Interactions: Innovative Technologies and Developments*. Springer Nature Switzerland, Cham, pp. 551–570.

Li, S., Zheng, H., Lin, L., Wang, F. and Sui, N. (2021) Roles of brassinosteroids in plant growth and abiotic stress response. *Plant Growth Regulation* 93, 29–38.

Liang, W., Ma, X., Wan, P. and Liu, L. (2018) Plant salt-tolerance mechanism: A review. *Biochemical and Biophysical Research Communications* 495(1), 286–291.

Mishra, P., Mishra, J. and Arora, N.K. (2021) Plant growth promoting bacteria for combating salinity stress in plants–Recent developments and prospects: A review. *Microbiological Research* 252, 126861.

Mozafariyan, M., Kamelmanesh, M.M. and Hawrylak-Nowak, B. (2016) Ameliorative effect of selenium on tomato plants grown under salinity stress. *Archives of Agronomy and Soil Science* 62(10), 1368–1380.

Mushtaq, N., Altaf, M.A., Ning, J., Shu, H., Fu, H. *et al.* (2024) Silicon improves the drought tolerance in pepper plants through the induction of secondary metabolites, GA biosynthesis pathway, and suppression of chlorophyll degradation. *Plant Physiology and Biochemistry* 214, 108919.

Nahar, S., Kalita, J., Sahoo, L. and Tanti, B., 2016. Morphophysiological and molecular effects of drought stress in rice. *Ann Plant Sci* 5(9), pp. 1409–1416.

Nasir, M.W. and Toth, Z. (2022) Effect of drought stress on potato production: A review. *Agronomy* 12(3), 635.

Negrão, S., Schmöckel, S.M. and Tester, M. (2017) Evaluating physiological responses of plants to salinity stress. *Annals of Botany* 119(1), 1–11.

Ohama, N., Sato, H., Shinozaki, K. and Yamaguchi-Shinozaki, K. (2017) Transcriptional regulatory network of plant heat stress response. *Trends in Plant Science* 22(1), 53–65.

Okunlola, G.O., Olatunji, O.A., Akinwale, R.O., Tariq, A. and Adelusi, A.A. (2017) Physiological response of the three most cultivated pepper species (Capsicum spp.) in Africa to drought stress imposed at three stages of growth and development. *Scientia Horticulturae* 224, 198–205.

Ozturk, M., Turkyilmaz Unal, B., García-Caparrós, P., Khursheed, A., Gul, A. and Hasanuzzaman, M. (2021) Osmoregulation and its actions during the drought stress in plants. *Physiologia Plantarum* 172(2), 1321–1335.

Prerostova, S. and Vankova, R. (2023) Phytohormone-mediated regulation of heat stress response in plants. In: *Plant Hormones and Climate Change.* Springer Nature Singapore, Singapore, pp. 167–206.

Puvanitha, S. and Mahendran, S. (2017) Effect of salinity on plant height, shoot and root dry weight of selected rice cultivars. *Scholars Journal of Agriculture and Veterinary Sciences* 4(4), 126–131.

Qu, A.L., Ding, Y.F., Jiang, Q. and Zhu, C. (2013) Molecular mechanisms of the plant heat stress response. *Biochemical and biophysical research communications*, 432(2), pp. 203–207.

Rahman, A., Hossain, M.S., Mahmud, J.A., Nahar, K., Hasanuzzaman, M. and Fujita, M. (2016) Manganese-induced salt stress tolerance in rice seedlings: regulation of ion homeostasis, antioxidant defense and glyoxalase systems. *Physiology and Molecular Biology of Plants* 22, 291–306.

Rivero, R.M., Mestre, T.C., Mittler, R., Rubio, F., Garcia-Sanchez, F. and Martinez, V. (2014) The combined effect of salinity and heat reveals a specific physiological, biochemical and molecular response in tomato plants. *Plant, Cell and Environment* 37(5), 1059–1073.

Rodríguez Coca, L.I., García González, M.T., Gil Unday, Z., Jiménez Hernández, J., Rodríguez Jáuregui, M.M. and Fernández Cancio, Y. (2023) Effects of sodium salinity on rice (Oryza sativa L.) cultivation: A review. *Sustainability* 15(3), 1804.

Saberi Riseh, R., Ebrahimi-Zarandi, M., Tamanadar, E., Moradi Pour, M. and Thakur, V.K. (2021) Salinity stress: toward sustainable plant strategies and using plant growth-promoting rhizobacteria encapsulation for reducing it. *Sustainability* 13(22), 12758.

Sachdev, S., Ansari, S.A., Ansari, M.I., Fujita, M. and Hasanuzzaman, M. (2021) Abiotic stress and reactive oxygen species: Generation, signaling, and defense mechanisms. *Antioxidants* 10(2), 277.

Safdar, H., Amin, A., Shafiq, Y., Ali, A., Yasin, R. *et al.* (2019) A review: Impact of salinity on plant growth. *Natural Sciences* 17(1), 34–40.

Saqib, M., Akhtar, J., Abbas, G. and Nasim, M. (2013) Salinity and drought interaction in wheat (Triticum aestivum L.) is affected by the genotype and plant growth stage. *Acta Physiologiae Plantarum* 35, 2761–2768.

Saravia, D., Farfán-Vignolo, E.R., Gutiérrez, R., De Mendiburu, F., Schafleitner, R., Bonierbale, M. and Khan, M.A. (2016) Yield and physiological response of potatoes indicate different strategies to cope with drought stress and nitrogen fertilization. *American Journal of Potato Research* 93, 288–295.

Seleiman, M.F., Al-Suhaibani, N., Ali, N., Akmal, M., Alotaibi, M. *et al.* (2021). Drought stress impacts on plants and different approaches to alleviate its adverse effects. *Plants* 10(2), 259.

Shahid, R., Shijie, L., Shahid, S., Altaf, M.A. and Shahid, H. (2021) Determinants of reactive adaptations to climate change in semi-arid region of Pakistan. *Journal of Arid Environments* 193, 104580.

Shilev, S. (2020) Plant-growth-promoting bacteria mitigating soil salinity stress in plants. *Applied Sciences* 10(20), 7326.

Shrestha, S., Mahat, J., Shrestha, J., Madhav, K.C. and Paudel, K. (2022) Influence of high-temperature stress on rice growth and development. A review. *Heliyon* 8(12), e12651.

Siddique, Z., Jan, S., Imadi, S.R., Gul, A. and Ahmad, P. (2016) Drought stress and photosynthesis in plants. *Water Stress and Crop Plants: A Sustainable Approach* 1, 1–11.

Singh, A.K., Singh, M.K., Singh, V., Singh, R., Raghuvanshi, T. and Singh, C. (2017) Debilitation in tomato (*Solanum lycopersicum* L.) as result of heat stress. *Journal of Pharmacognosy and Phytochemistry* 6(6), 1917–1922.

Sun, A.Z. and Guo, F.Q. (2016) Chloroplast retrograde regulation of heat stress responses in plants. *Frontiers in Plant Science* 7, 398.

Tombesi, S., Nardini, A., Frioni, T., Soccolini, M., Zadra, C. *et al.* (2015) Stomatal closure is induced by hydraulic signals and maintained by ABA in drought-stressed grapevine. *Scientific Reports* 5(1), 12449.

Tuteja, N. (2007) Mechanisms of high salinity tolerance in plants. *Methods in Enzymology* 428, 419–438.

Uddling, J., Broberg, M.C., Feng, Z. and Pleijel, H. (2018) Crop quality under rising atmospheric CO2. *Current Opinion in Plant Biology* 45, 262–267.

Ullah, A., Nadeem, F., Nawaz, A., Siddique, K.H. and Farooq, M. (2022) Heat stress effects on the reproductive physiology and yield of wheat. *Journal of Agronomy and Crop Science* 208(1), 1–17.

Utami, D. and Aryanti, E. (2021) Impact of heat stress on germination and seedling growth of chili pepper (*Capsicum annuum* L.). In: *IOP Conference Series: Earth and Environmental Science*. IOP Publishing, p. 012032.

Volkmar, K.M., Hu, Y. and Steppuhn, H. (1998) Physiological responses of plants to salinity: A review. *Canadian Journal of Plant Science* 78(1), 19–27.

Wahid, A., Farooq, M., Hussain, I., Rasheed, R. and Galani, S. (2012) Responses and management of heat stress in plants. In: Ahmad, P. and Prasad, M.N.V. (eds) *Environmental Adaptations and Stress Tolerance of Plants in the Era of Climate Change*, Springer: New York, Dordrecht, Heidelberg, London, pp. 135–157.

Wang, X., Altaf, M.A., Hao, Y., Wang, Z. and Zhu, G. (2023) Effect of heat stress on root architecture, photosynthesis, and antioxidant profile of water spinach (Ipomoea aquatica Forsk) seedlings. *Horticulturae* 9(8), 923.

Waraich, E.A., Ahmad, R. and Ashraf, M.Y. (2011) Role of mineral nutrition in alleviation of drought stress in plants. *Australian Journal of Crop Science* 5(6), 764–777.

Wu, C., Cui, K. and Fahad, S. (2022) Heat stress decreases rice grain weight: Evidence and physiological mechanisms of heat effects prior to flowering. *International Journal of Molecular Sciences* 23(18), 10922.

Yadav, S., Modi, P., Dave, A., Vijapura, A., Patel, D. and Patel, M. (2020) Effect of abiotic stress on crops. *Sustainable Crop Production* 3(17), 5–16.

Yang, X., Lu, M., Wang, Y., Wang, Y., Liu, Z. and Chen, S. (2021) Response mechanism of plants to drought stress. *Horticulturae* 7(3), 50.

Zhang, Q. and Dai, W. (2019) Plant response to salinity stress. In: *Stress Physiology of Woody Plants*. pp. 155–173, CRC Press: Boca Raton, Floride, USA.

Zhang, Y., Xu, J., Li, R., Ge, Y., Li, Y. and Li, R. (2023) Plants' response to abiotic stress: Mechanisms and strategies. *International Journal of Molecular Sciences* 24(13), 10915.

Zhao, C., Zhang, H., Song, C., Zhu, J.K. and Shabala, S. (2020) Mechanisms of plant responses and adaptation to soil salinity. *The Innovation* 1(1), 100017.

Zhou, Y., Xu, F., Shao, Y. and He, J. (2022) Regulatory mechanisms of heat stress response and thermomorphogenesis in plants. *Plants* 11(24), 3410.

Zhu, J.K. (2016) Abiotic stress signaling and responses in plants. *Cell* 167(2), 313–324.

8 Molecular Responses to Biotic Stress in Plants

Muhammad Waseem[1]*, Sana Basharat[1], Iffat Shaheen[2], Wajid Saeed[1], Pingwu Liu[1]*, Guangsheng Yang[1] and Waleed Hussain[3]

[1]*College of Tropical Agriculture and Forestry, Hainan University, Danzhou, Hainan, China;* [2]*Department of Agriculture, Bahauddin Zakariya University, Multan, Pakistan;* [3]*Department of Botany, University of Narowal, Punjab, Pakistan*

Abstract

Plants are constantly challenged by biotic stress factors, including pathogens such as fungi, bacteria, viruses, and pests like insects and nematodes, which significantly affect crop productivity. This chapter provides an in-depth overview of plant responses to biotic stress, detailing the types of stress and their impact on plants. It discusses plant defense mechanisms, including both pre-formed barriers and inducible responses triggered by pathogen recognition and signaling cascades, such as pattern-triggered immunity (PTI) and effector-triggered immunity (ETI). The chapter also covers pest resistance mechanisms, illustrating how plants detect and respond to herbivores. The roles of key defense hormones, including salicylic acid (SA), jasmonic acid (JA), and ethylene (ET), in coordinating stress responses are examined. Systemic acquired resistance (SAR), a plant-wide, long-term immune response that protects against future attacks, and the Gene-for-Gene Hypothesis, explaining the interaction between plant resistance (R) genes and pathogen avirulence (Avr) genes, are central themes. The chapter concludes by addressing the ongoing challenges in biotic stress management, such as rapid pathogen evolution and pest adaptation, and emphasizes the need for innovative solutions. Future directions, including genetic engineering, breeding for durable resistance, and integrated pest management strategies, are discussed to ensure sustainable crop protection.

Keywords: plant defense mechanisms, pathogen resistance, pest resistance, systemic acquired resistance (SAR), gene-for-gene hypothesis, hormonal regulation

8.1 Introduction

Biotic stress refers to damage caused by living organisms, including pathogens such as bacteria, fungi, viruses, and oomycetes, as well as pests like herbivorous insects, nematodes, and parasitic plants (Dhiman *et al.*, 2024). These biotic agents can infect plant tissues, disrupt metabolic processes, and, in severe cases, kill the plant. In agriculture, biotic stress is one of the primary challenges faced by farmers and crop producers globally (Adetuyi and Adetunji, 2024). It drastically reduces crop yields and quality, posing a major threat to food security and agricultural sustainability (Wani *et al.*, 2024). Crop losses due to biotic stress can range from minor yield reductions to complete destruction of crops, depending on the severity of the infestation or infection (Sunani *et al.*, 2024).

The economic impact of biotic stress is profound. In developing regions, where access to

* Email: m.waseem.botanist@gmail.com

* Email: hnulpw@hainanu.edu.cn

DOI: 10.1079/9781800628823.0008

chemical treatments and advanced agricultural technologies may be limited, biotic stress can devastate food supplies (Ahlawat *et al.*, 2024). Globally, farmers spend billions on pesticides, fungicides, and other chemical interventions to mitigate the effects of pathogens and pests (Venugopal *et al.*, 2024). However, these solutions are not only costly but can also lead to environmental degradation, development of resistance in pests and pathogens, and potential health risks for humans and animals (Onyeaka *et al.*, 2024). As a result, the need for more sustainable, effective methods of managing biotic stress is more pressing than ever (Biswas and Das, 2024).

Plants have evolved a wide array of defense mechanisms to cope with biotic stress. These include preformed defenses, such as thick cell walls, waxy cuticles, and secondary metabolites that act as toxins or deterrents against herbivores and pathogens (Muhammad *et al.*, 2024). When these preformed defenses are breached, plants can activate a second line of defense known as induced defenses. These involve complex signaling pathways that help plants recognize the presence of a pathogen or pest and mount a targeted response (Riseh and Vazvani, 2024). For instance, when a pathogen invades, plants can recognize specific molecules called pathogen-associated molecular patterns (PAMPs) through pattern recognition receptors (PRRs), initiating a defense response known as PAMP-triggered immunity (Yu *et al.*, 2024). However, many pathogens have evolved to overcome PTI by producing effectors that interfere with plant defenses. In response, plants have developed effector-triggered immunity (ETI), a more robust and specific defense mechanism governed by resistance (R) genes (Nabi *et al.*, 2024). Similarly, plants defend against pests like insects and nematodes through chemical deterrents, physical structures such as trichomes or thorns, and proteins that interfere with pest digestion (Ayoub *et al.*, 2024). Hormonal pathways involving molecules like salicylic acid (SA), jasmonic acid (JA), and ethylene (ET) are central to coordinating these defenses, with different hormones playing key roles in responses to biotrophic pathogens (those that require living host tissue) and necrotrophic pathogens (those that kill host tissue) (Ghosh and Roychoudhury, 2024).

Despite these natural defense mechanisms, pathogens and pests continuously evolve new strategies to circumvent plant defenses, leading to an ongoing evolutionary "arms race". For example, herbivorous insects may develop resistance to plant toxins, or pathogens may evolve new effector molecules that can bypass recognition by plant immune systems (Mathur *et al.*, 2024). This dynamic interaction highlights the need for ongoing research to understand plant defense mechanisms at the molecular level and to develop new strategies for enhancing crop resistance (Ali *et al.*, 2023).

In modern agriculture, biotechnological interventions have become critical tools in managing biotic stress. Genetic engineering and selective breeding allow for the development of crop varieties with enhanced resistance to pathogens and pests (Sahu *et al.*, 2023). For instance, transgenic plants expressing specific resistance genes or proteins that inhibit pest feeding can offer more durable resistance. Additionally, techniques such as RNA interference (RNAi) are being used to silence genes in pests or pathogens, reducing their ability to harm crops (Saakre *et al.*, 2024). The use of microbial biocontrol agents, which involve beneficial microorganisms that either outcompete or directly inhibit harmful pathogens and pests, is another promising approach to reduce reliance on chemical pesticides (Lee *et al.*, 2023).

In the context of climate change, the challenges posed by biotic stress are expected to intensify. Warmer temperatures, changing precipitation patterns, and shifts in growing seasons can alter the range and behavior of pests and pathogens, leading to new outbreaks in previously unaffected regions (Nitta *et al.*, 2024). This highlights the need for adaptive, integrated approaches to managing biotic stress, combining traditional agricultural practices with cutting-edge scientific advances in plant biology and biotechnology (Şimşek *et al.*, 2024). This chapter will explore the complex interaction between plants and their biotic stressors, delving into the molecular and physiological mechanisms underlying plant resistance. It will also examine the role of biotechnological innovations in improving crop resilience, with a focus on sustainable agricultural practices. By understanding how plants respond to biotic stress, we can better equip crops to withstand these challenges, ultimately contributing to food security and reducing the environmental impact of agriculture.

8.2 Types of Biotic Stress

Pathogen stress is caused by infections from various organisms, including bacteria, fungi, viruses, and oomycetes. These pathogens invade plant tissues and disrupt cellular functions, leading to a wide range of diseases (Ray, 2024). For instance, bacterial pathogens like *Xanthomonas* can induce bacterial blight, characterized by water-soaked lesions and leaf drop, while fungi such as *Fusarium* can cause wilt by blocking vascular tissues, leading to rapid plant decline (Somashetty *et al.*, 2023). Oomycetes, like those responsible for late blight, thrive in moist environments and can rapidly decimate crops, particularly potatoes and tomatoes, by causing extensive tissue rot. These pathogens attack plants by producing toxins and enzymes that degrade cell walls, leading to symptoms such as stunted growth, tissue necrosis, and significantly reduced yields. In severe cases, infections can result in complete crop loss (Choudhury, 2023).

On the other hand, herbivory stress is caused by pests such as insects, nematodes, and other herbivores that feed on various plant parts, including leaves, stems, roots, and fruits (Dhiman *et al.*, 2024). Insects like aphids and caterpillars damage plants through their feeding habits, which can lead to decreased photosynthesis, nutrient depletion, and overall plant vigor. For example, aphids not only suck sap but also transmit viral diseases, exacerbating the stress on the plant (Poonia *et al.*, 2024). Nematodes, microscopic roundworms, invade roots and cause galls or lesions that disrupt nutrient and water uptake, leading to stunted growth and yellowing of leaves (Gilbert and Parker, 2023). Other herbivores, such as spider mites, create stippling effects on leaves and can produce silk webs that further inhibit plant growth. The physical damage from these herbivores can also create entry points for secondary infections by pathogens, compounding the stress on the plant (Mathur *et al.*, 2024).

Both pathogen and herbivory stress can severely impact crop productivity and quality, posing significant challenges for farmers worldwide. The economic ramifications can be substantial, with billions lost annually due to reduced yields and increased management costs (Ayanwale *et al.*, 2024). Effective management strategies are essential for mitigating these stresses and sustaining agricultural output. Approaches include breeding disease-resistant crop varieties that can withstand specific pathogens, employing biocontrol methods such as introducing beneficial insects (like ladybugs for aphid control) or microbial agents (like *Bacillus thuringiensis* for caterpillar pests), and implementing integrated pest management (IPM) practices (Singh *et al.*, 2023a). IPM combines cultural practices, such as crop rotation and intercropping, mechanical controls like traps and barriers, biological controls, and judicious use of chemical pesticides to minimize damage while protecting beneficial organisms (Singh *et al.*, 2023a). Moreover, advancements in biotechnology offer promising avenues for enhancing plant resistance to biotic stress. Techniques such as genetic engineering and CRISPR gene editing allow for the precise introduction of resistance traits from other species, potentially providing crops with enhanced defenses against both pathogens and pests (Manzoor *et al.*, 2024).

Understanding the intricate dynamics of plant-pathogen and plant-pest interactions is crucial for developing sustainable agricultural practices that can improve crop resilience, ensure food security, and reduce reliance on chemical pesticides. By leveraging these strategies, the agricultural sector can better equip itself to face the challenges posed by biotic stress, ultimately contributing to more sustainable food production systems (Rathore *et al.*, 2024; Singh *et al.*, 2024).

8.3 Plant Defense Mechanisms

Plants have evolved a sophisticated array of defense mechanisms to protect themselves from biotic stress caused by pathogens and pests, which can be broadly categorized into structural barriers and induced responses (Bhatla and Lal, 2023). Structural defenses act as the first line of defense, limiting pathogen entry and pest damage. The cuticle, a waxy layer covering the leaves and stems, plays a critical role in preventing water loss while also serving as a hydrophobic barrier that impedes the infiltration of waterborne pathogens (Raza *et al.*, 2024). Beneath the cuticle, the cell wall, composed of cellulose, hemicellulose, and lignin, provides mechanical support and resistance to physical invasion. In response to pathogen attack, plants can enhance their cell walls by depositing additional cellulose, lignin, and other reinforcing compounds, which create a more formidable barrier against pathogens. Furthermore, many plants produce specialized structures like trichomes and thorns to deter herbivores (Dhiman *et al.*, 2024). Trichomes, which are hair-like projections, can physically obstruct feeding and may produce secondary metabolites that are toxic or unpalatable to herbivores. Thorns and spines serve as physical deterrents, making it difficult for larger herbivores to access the plant and feed on its tissues (Vukeya, 2021; Pittaway, 2023).

When pathogens or pests breach these structural defenses, plants initiate induced responses to mount an effective defense. The hypersensitive response (HR) is a rapid and localized reaction to pathogen invasion, characterized by the programmed cell death of cells surrounding the infection site (Nadarajah, 2024). This strategic sacrifice of host tissue restricts the pathogen's access to nutrients and effectively limits its spread within the plant. The HR is often accompanied by the production of reactive oxygen species (ROS), highly reactive molecules that serve dual roles: they directly damage pathogens by disrupting cellular structures and act as signaling molecules that trigger further defense mechanisms (Thiruvengadam *et al.*, 2024). In addition to local responses, plants can activate systemic acquired resistance (SAR) after an HR, which results in a long-lasting immune response that provides enhanced protection against a broad spectrum of pathogens (Liu *et al.*, 2024). This systemic response involves the production of signaling molecules, particularly SA, which spread throughout the plant and primes uninfected tissues to respond more rapidly and effectively to future infections.

In addition to these responses, plants can produce a diverse range of chemical defenses (Parihar *et al.*, 2024). Secondary metabolites such as phenolics, alkaloids, and terpenoids are synthesized in response to herbivory and pathogen attack. These compounds often serve multiple purposes: some act as toxins that disrupt the digestive systems of herbivores, while others may interfere with the growth and reproduction of pathogens. For example, alkaloids like nicotine are toxic to many insects, while phenolics can provide antimicrobial properties that help inhibit pathogen growth (Xu *et al.*, 2024). Additionally, some plants release volatile organic compounds (VOCs) that attract beneficial insects, such as predators or parasitosis, which can help control pest populations. Together, these structural and induced defense mechanisms enable plants to enhance their resilience against biotic stress (Adetuyi and Adetunji, 2024). By effectively reinforcing physical barriers and activating complex signaling pathways upon recognizing pathogens or pests, plants improve their chances of survival and reproduction. Understanding these intricate defense mechanisms is crucial for developing strategies to enhance crop resistance and ensure sustainable agricultural practices, ultimately contributing to food security and ecosystem health (Khatri *et al.*, 2024).

8.4 Pathogen Recognition and Signalling

Plants have developed sophisticated mechanisms to recognize and respond to pathogen attacks, primarily through the detection of PAMPs. These are conserved molecular features found on a wide range of pathogens, including proteins, lipopolysaccharides, and chitin (van Hensbergen and Hu, 2024). Plants utilize PRRs, which are typically located on the cell surface, to identify these PAMPs. The recognition of PAMPs triggers a defense response known as PAMP-triggered immunity (PTI). This process involves a series of biochemical and physiological changes within the plant, including the activation of defense-related genes, production of antimicrobial compounds, and reinforcement of cell walls (DeFalco and Zipfel, 2021). Key components of PTI include the generation of ROS and the activation of mitogen-activated protein kinase (MAPK) pathways, both of which contribute to strengthening the plant's defenses and limiting pathogen growth (Yang *et al.*, 2022).

However, many pathogens have evolved sophisticated strategies to overcome PTI by delivering effector proteins directly into plant cells. These effectors can manipulate the plant's cellular machinery to suppress immune responses and facilitate infection (Rufián *et al.*, 2023). For instance, some effector proteins can inhibit the activity of PRRs or interfere with the signaling pathways that activate PTI. In response to this threat, plants have developed specialized R proteins that recognize these effectors, often through a direct interaction or by sensing the effects of the effector proteins on plant cellular processes (Saraiva *et al.*, 2023). This interaction activates a more robust immune response known as ETI, which is typically characterized by a rapid and localized cell death response, known as the HR. The HR restricts the pathogen's ability to spread within the plant.

The signaling pathways involved in plant defense are complex and interconnected, primarily involving key molecules such as SA, JA, and ET (Sutaoney *et al.*, 2023). SA plays a central role in mediating responses to biotrophic pathogens, which rely on living host tissue for nutrition. SA is crucial for establishing SAR, a state of heightened defense that protects uninfected tissues following localized infection. It promotes the expression of defense-related genes and enhances the plant's overall immune capacity by facilitating the production of pathogenesis-related (PR) proteins (Walker, 2023).

Conversely, JA is primarily involved in responses to necrotrophic pathogens, which kill host tissue, and to herbivorous pests. JA signaling activates the production of defensive proteins and secondary metabolites that deter herbivores and inhibit the growth of pathogens (Shah and Gupta, 2023). For example, the activation of JA-responsive genes can lead to the synthesis of proteinase inhibitors that make it harder for herbivores to digest the plant material.

ET, another critical signaling molecule, often works in conjunction with SA and JA to fine-tune the plant's response to various biotic stresses (Song *et al.*, 2023). ET signaling is especially important during fruit ripening and in the plant's response to mechanical stress and pathogens. Moreover, the interplay between these signaling pathways allows plants to mount appropriate defenses based on the type of threat they face. For instance, some pathogens can trigger both SA and JA pathways, necessitating a coordinated response (Chin *et al.*, 2024). This integration of signals helps the plant differentiate between types of attacks and adjust its defense strategy accordingly, which is vital for maximizing survival.

In summary, the recognition of PAMPs and effectors, along with the intricate signaling pathways involving SA, JA, and ET, forms a robust defense network in plants (Al-Bader *et al.*, 2023). This ability to distinguish between different pathogens and activate appropriate immune responses is essential for maintaining plant health and resilience. Understanding these mechanisms not only enhances our knowledge of plant biology but also has significant implications for developing disease-resistant crop varieties, ultimately contributing to more sustainable agricultural practices and improved food security (Islam *et al.*, 2023a).

8.5 Pest Resistance Mechanisms

Plants have developed a multifaceted array of pest resistance mechanisms to defend themselves against herbivorous pests, encompassing chemical, protein-based, and physical defenses that work synergistically to enhance survival (Khan *et al.*, 2023). Chemical defenses play a crucial role in this protective strategy, with plants producing a diverse range of secondary metabolites that deter herbivores (Al-Khayri *et al.*, 2023). Notable examples include alkaloids, such as nicotine, caffeine, and morphine, which exhibit potent toxic properties that can incapacitate or kill herbivores. Terpenoids, including essential oils and resins, not only serve as repellents but can also alter the feeding behavior of pests (Doggett and Lee, 2023). For instance, compounds like limonene and menthol have been shown to effectively deter insect herbivores. Phenolics, such as flavonoids and tannins, contribute to plant defense by binding proteins and inhibiting digestion, making the plant less palatable and nutritious. Tannins, in particular, can also interfere with the growth and reproduction of herbivores, providing a double layer of protection (Singh *et al.*, 2023b).

In addition to these chemical strategies, plants also employ protein-based defenses. Protease inhibitors are particularly significant, as they disrupt the digestive enzymes in herbivores, making it difficult for them to break down proteins in the plant tissue. This inhibition can lead to reduced growth rates and lower survival chances for the pests (Divekar *et al.*, 2023). Furthermore, lectins, which are carbohydrate-binding proteins, can impair the digestive processes of insects by binding to gut epithelial cells, leading to nutritional deficiencies and toxicity. This targeted approach not only deters herbivory but also demonstrates the plant's capacity for active defense (Xiang *et al.*, 2023).

Complementing these biochemical defenses are physical barriers that enhance the plant's ability to withstand herbivore pressure. Thorns and spines serve as effective deterrents against larger herbivores, inflicting pain or injury and making feeding difficult (La Vergata, 2023). Meanwhile, trichomes, which are hair-like structures on the surfaces of leaves and stems, provide multifaceted protection. These structures can create physical obstructions that hinder insect access and feeding, and some trichomes secrete sticky or toxic substances that can trap or harm pests (Yousuf *et al.*, 2024). Additionally, the presence of a waxy cuticle on plant surfaces not only minimizes water loss but also acts as a slippery barrier, making it less appealing for herbivores to feed. Together, these mechanisms form a comprehensive defense system that significantly enhances a plant's resilience against pest damage (Whitehill *et al.*, 2023).

The interplay of chemical, protein-based, and physical defenses allows plants to adapt to varying pest pressures, effectively reducing the likelihood of herbivory and supporting overall plant health. Understanding these diverse strategies is crucial for developing sustainable pest management practices in agriculture, as well as improving crop resistance to herbivory, ultimately contributing to food security and ecosystem stability (Nikpay *et al.*, 2023). By harnessing knowledge of these natural defenses, researchers can explore innovative approaches, such as breeding programs that enhance pest resistance or the application of biopesticides derived from plant metabolites, thereby promoting sustainable agricultural practices (Arora *et al.*, 2024).

8.6 Hormonal Regulation in Defense Responses

Hormonal regulation, particularly through SA, JA, and ET, plays a crucial role in coordinating plant defenses against various biotic stresses, with SA mainly defending against biotrophic pathogens that feed on living host tissue (Hewedy *et al.*, 2023). When a plant is attacked by such pathogens, including certain bacteria and fungi, SA levels increase significantly. This rise in SA triggers a cascade of signaling events that activate a variety of defense-related genes, leading to the production of PR proteins and other antimicrobial compounds (Islam *et al.*, 2023b). These proteins play a crucial role in hindering pathogen growth and preventing the spread of infection. In addition, SA is instrumental in establishing SAR, a long-lasting immune state that prepares uninfected tissues for potential future attacks (Riseh and Vazvani, 2024). This systemic response enhances the plant's overall resilience by allowing it to respond more effectively to subsequent infections, thereby improving its chances of survival in environments where biotic threats are prevalent (Mishra *et al.*, 2024).

In contrast, JA and ET serve as key regulators of plant defense against necrotrophic pathogens, which kill host tissue and herbivorous pests. JA is synthesized in response to various stress signals, including tissue damage, herbivory, and the presence of necrotrophic pathogens. This hormone activates a wide range of defense responses, including the production of protease inhibitors and other defensive proteins that deter herbivores and inhibit the growth of pathogens (Hu *et al.*, 2024). For example, JA is crucial for the synthesis of secondary metabolites such as alkaloids and terpenoids, which not only repel pests but also have antimicrobial properties. Furthermore, JA signaling interacts with other hormonal pathways, allowing the plant to fine-tune its response based on the specific type of threat it faces. This interaction is vital for ensuring that the plant allocates resources effectively to combat the most pressing threats (Sun *et al.*, 2024). ET plays a complementary role to JA, particularly in response to herbivores and necrotrophic pathogens. This hormone is involved in regulating plant responses to mechanical stress and is crucial for the coordination of defense responses initiated by JA. ET enhances the production of defensive compounds in the presence of pests or pathogens, working in concert with JA to amplify the plant's defensive capabilities (Nadarajah, 2024). Moreover, ET is also implicated in various developmental processes, such as ripening and senescence, which can influence how a plant reallocates resources in response to stress. The interplay between JA and ET can lead to synergistic effects that enhance the plant's overall defense, allowing it to mount a more robust response to combined biotic threats (Mishra *et al.*, 2024).

The integration of these hormonal pathways is essential for ensuring that plants can mount appropriate and effective defenses against a variety of biotic stresses. The balance and interaction between SA, JA, and ET determine not only the type of defense response activated but also the plant's overall fitness and survival (Balasubramani *et al.*, 2024). By understanding the roles of these hormones in plant defense, researchers can develop strategies to enhance crop resistance, improve agricultural sustainability, and ensure food security in the face of increasing biotic threats. This knowledge can lead to innovative approaches, such as genetic modifications or selective breeding aimed at improving hormone signaling pathways, ultimately resulting in crops that are more resilient to pests and diseases, thereby contributing to more sustainable agricultural practices (Jagadeesh Chandra Bose *et al.*, 2024).

8.7 Systemic Acquired Resistance (SAR) and Gene-for-Gene Hypothesis

SAR represents a remarkable defense mechanism in plants, providing them with a robust and long-lasting immunity following a localized infection. When a plant is initially attacked by a pathogen, it activates SAR as a "whole-plant" response, essentially turning on a defense network that prepares the entire organism for potential future assaults (Li, 2024). This response is often initiated when the plant recognizes PAMPs or when specific pathogens breach the plant's first line of defense. Once activated, SAR triggers a cascade of signaling events that lead to the production of an array of defensive compounds and proteins, including PR proteins (Roychowdhury *et al.*, 2024). PR proteins are integral to the SAR response; they are synthesized throughout the plant and play diverse roles in bolstering defenses. These proteins can degrade pathogen cell walls, produce antimicrobial compounds, and reinforce plant tissues to prevent further invasion. In addition to PR proteins, the establishment of SAR involves the production of signaling molecules, most notably SA. This hormone spreads the defense signals to uninfected tissues, allowing the plant to mount a rapid response to subsequent infections (Ray, 2024). The result is a heightened state of readiness, enabling the plant to fend off a wide spectrum of pathogens more effectively. For instance, research has shown that plants exhibiting SAR can quickly respond to a variety of pathogens, from bacteria to fungi, showcasing the versatility and efficiency of this defense mechanism (Rai *et al.*, 2024).

On the other hand, the Gene-for-Gene Hypothesis offers a fascinating insight into the specificity of plant-pathogen interactions. This hypothesis posits that for every plant R gene, there exists a corresponding pathogen avirulence (Avr) gene. The interaction between these resistance and Avr genes plays a crucial role in determining the plant's ability to resist specific pathogens. When a pathogen possesses an Avr gene that corresponds to an resistance gene in the plant, the plant can recognize the threat and activate a strong immune response, often culminating in localized cell death known as the HR (Agarrwal *et al.*, 2024). This programmed cell death restricts the pathogen's access to nutrients and effectively limits its spread within the plant. However, the evolutionary arms race between plants and pathogens is relentless. Pathogens continually adapt by altering their Avr genes, making it crucial for plants to evolve new resistance genes to maintain their defense capabilities. This dynamic interplay drives the diversity of plant immune systems and underscores the complexity of plant-pathogen interactions (Kamble *et al.*, 2024). For example, researchers have identified resistance genes in crops like rice and wheat that confer resistance to specific strains of pathogens, allowing for targeted breeding programs that enhance crop resilience.

Together, SAR and the Gene-for-Gene Hypothesis illuminate the intricate and adaptive nature of plant defense mechanisms. SAR provides a broad-spectrum, systemic response that equips plants to respond to a range of biotic threats, while the Gene-for-Gene Hypothesis highlights the specificity of immune responses tailored to combat particular pathogens (Castel *et al.*, 2024). Understanding these mechanisms is essential not only for advancing our knowledge of plant biology but also for developing innovative strategies to enhance crop resilience in agriculture. By harnessing the principles of SAR and the Gene-for-Gene Hypothesis, researchers and agronomists can create disease-resistant crop varieties that contribute to sustainable agricultural practices and ensure food security in an ever-changing environment (Bijlwan *et al.*, 2024). The ongoing exploration of these defense strategies promises exciting developments in the quest for resilient crops capable of thriving amidst the challenges posed by evolving pathogens (Table 8.1) (Chaffai *et al.*, 2024).

8.8 Challenges and Future Directions

Despite the remarkable advances in plant defense mechanisms and agricultural technologies, there are significant "challenges" that must be addressed to ensure the long-term sustainability of food production systems. Chief among these challenges is the evolution of pathogens and pests, which poses a constant threat to crop resilience. As pathogens and pests are exposed to plant defenses or chemical treatments, they can adapt and evolve, developing resistance over time. This arms race between crops and their attackers necessitates continuous research and development of new resistant crop varieties. For example, while transgenic crops like Bt cotton have proven highly effective against pests such as the cotton bollworm, these pests have started to evolve resistance to the Bt toxin. Similarly, pathogens such as fungi and bacteria can evolve mechanisms to bypass plant immune responses or resist fungicides, making it crucial for breeders and scientists to stay ahead by continually identifying new genes for resistance or by developing multi-layered defense strategies.

The changing climate adds another layer of complexity to managing biotic stress. Rising temperatures, altered precipitation patterns, and increased CO_2 levels can significantly shift the distribution and behavior of pests and pathogens. For instance, warming climates may allow pests that were previously limited by colder temperatures to expand their ranges into new regions, affecting crops that have not developed resistance to these invaders. Additionally, warmer temperatures can accelerate the life cycles of pests and pathogens, leading to more frequent and severe infestations. Droughts, floods, and extreme weather events can further weaken plant health, making them more vulnerable to biotic stress. This dynamic environmental context requires adaptive and flexible strategies to protect crops. Farmers, researchers, and policymakers must work together to develop resilient crop varieties and integrated approaches that can withstand the shifting landscape of biotic threats in a changing climate.

To tackle these challenges, the adoption of IPM is emerging as a key solution for sustainable agriculture. IPM is an holistic approach that combines biological, chemical, and cultural practices to manage pests and pathogens in an environmentally responsible manner. Instead of relying solely on chemical pesticides, IPM emphasizes the use of biological control agents, such as beneficial insects or microbes, which can naturally suppress pest populations. In addition, cultural practices like crop rotation, intercropping, and the use of cover crops can help reduce pest build-up and break pathogen cycles. Chemical pesticides are still used in IPM but in a more targeted and judicious manner, reducing their overall application and minimizing the risks of resistance development and environmental contamination. The future of IPM also includes the integration of advanced technologies such as precision agriculture, which allows farmers to monitor pest populations and apply treatments more efficiently, reducing waste and limiting the environmental footprint of pesticide use. Genetic engineering and biotechnology also offer promising tools for developing crops with enhanced resistance to a broad range of pathogens and pests, which can be incorporated into IPM strategies to reduce reliance on chemical interventions.

Moving forward, it is essential to recognize that no single solution will suffice in managing biotic stress. Multi-faceted approaches that combine traditional breeding techniques, cutting-edge biotechnology, and sustainable farming practices will be necessary to address the ever-evolving challenges posed by pests and pathogens. Continuous investment in research and development is crucial to discovering new resistance genes, understanding pathogen evolution, and creating adaptive strategies that can help mitigate the impacts of climate change on agricultural systems. By embracing innovative technologies and sustainable practices, agriculture can become more resilient, ensuring food security in the face of a rapidly changing world.

Table 8.1. Key aspects of systemic acquired resistance (SAR) and the Gene-for-Gene Hypothesis in plant defense mechanisms. (Table author's own.)

Aspect	**Systemic Acquired Resistance (SAR)**	**Gene-for-Gene Hypothesis**	**References**
Definition	A plant-wide defense response triggered by a localized infection.	Describes the specific interaction between plant genes and pathogen Avr genes.	Todd (2024)
Activation	Initiated when plants recognize PAMPs or when specific pathogens breach the plant's first line of defense.	Activated when a pathogen's Avr gene corresponds to the plant's resistance gene.	Dodds *et al.* (2024)
Defense mechanism	Leads to a "whole-plant" response, preparing the entire organism for future infections.	Triggers a localized immune response, often resulting in HR (cell death to limit pathogen spread).	Tripathi *et al.* (2024)
Key components	PR proteins, signaling molecules like SA.	Resistance genes in plants, Avr genes in pathogens.	Mishra *et al.* (2024)
Functions of PR proteins	Degrade pathogen cell walls. Produce antimicrobial compounds. Reinforce plant tissues to prevent further invasion.	N/A	Evangelisti and Govers (2024); Marcianò *et al.* (2024)
Role of signaling molecules	SA spreads the defense signal to uninfected tissues, enabling rapid response to subsequent infections.	N/A	Mishra *et al.* (2024)
Response scope	Provides broad-spectrum, long-lasting immunity against a variety of pathogens (bacteria, fungi, etc.).	Highly specific, tailored to the particular pathogen whose Avr gene matches the plant's resistance gene.	Hückelhoven and Schouten (2024)
Example of mechanism efficiency	Plants exhibiting SAR can fend off a wide range of pathogens, showing versatility.	Programmed cell death in the HR prevents the pathogen from spreading by cutting off its nutrient supply	Nadarajah (2024)
Evolutionary aspect	Systemic defense that can be activated repeatedly by different threats.	An evolutionary "arms race" between plants and pathogens; pathogens adapt by altering Avr genes, while plants evolve new resistance genes.	Nigam (2024); von Dahlen (n.d.)
Applications in agriculture	Broad-spectrum resistance is useful for enhancing crop resilience against a wide array of diseases.	Targeted breeding programs for resistance to specific pathogens in crops like rice and wheat.	Cheng *et al.* (2024)

References

Adetuyi, B.O. and Adetunji, C.O. (2024) Application of Plant-Based Nanobiopesticides for Mitigation of Several Biotic Stress. *Handbook of Agricultural Biotechnology.* Volume 1: Nanopesticides, 239.

Agarrwal, R., Ahire, T.P., Manjunatha, N., Pokhare, S., Ojha, R. and Salvi, P. (2024) Imperative Role of R-Genes and Associated Molecular Mechanisms in Plant Disease Resistance. In: Singh, K., Kaur, R. and Deshmukh, R. (eds) *Biotechnological Advances for Disease Tolerance in Plants*. Springer, Singapore, pp. 73–97.

Ahlawat, U., Naruka, A., Changdeo, W.B., Rehsawla, R., Sansanwal, R., Mishra, R. and Singh, V.K. (2024) A review of cutting-edge biotechnological solutions for next-generation farming. *Journal of Experimental Agriculture International* 46, 1187–1204.

Al-Bader, N., Meier, A., Geniza, M., Gongora, Y.S., Oard, J. and Jaiswal, P. (2023) Loss of a premature stop codon in the rice wall-associated kinase 91 (WAK91) gene is a candidate for improving leaf sheath blight disease resistance. *Genes* 14, 1673.

Al-Khayri, J.M., Rashmi, R., Toppo, V., Chole, P.B., Banadka, A. *et al.* (2023) Plant secondary metabolites: The weapons for biotic stress management. *Metabolites* 13, 716.

Ali, S., Tyagi, A., Rajarammohan, S., Mir, Z.A., Bae, H. (2023) Revisiting Alternaria-host interactions: New insights on its pathogenesis, defense mechanisms and control strategies. *Scientia Horticulturae* 322, 112424.

Arora, S., Husain, T. and Prasad, S.M. (2024) Allelochemicals as biocontrol agents: Promising aspects, challenges and opportunities. *South African Journal of Botany* 166, 503–511.

Ayanwale, A.B., Fatunbi, O.A., Kehinde, A.D. and Robin, B. (2024) Emerging technological issues in african agriculture: A systematic review. *Agriculturae Conspectus Scientificus* 89, 1–15.

Ayoub, L., Yaqoob, M., Zahoor, S., Wani, F.F., Irshad, S.S., Gull, A. and Sheikh, M.A. (2024) *Role of Induced Resistance in Insect-Pest Management*. In: Kumar, S. and Furlong, M (eds) *Plant Resistance to Insects in Major Field Crops*. Springer, Singapore, pp. 249–277.

Balasubramani, S., Gayathiri, E., Arya, S.K., Gurusamy, D., Prakash, P., Appu, M. and Moola, A.K. (2024) Role of Heat Shock Proteins in Abiotic and Biotic Stress Response in Plants. In: Kumar Upadhyay, S. (ed.) *Defense-Related Proteins in Plants*. pp. 299–332. Elsevier:

Bhatla, S.C. and Lal, M.A. (2023) *Biotic Interactions, Plant Physiology, Development and Metabolism*. Springer, pp. 717–762.

Bijlwan, A., Ranjan, R., Kumar, M., Gupta, S. and Jha, A. (2024) *Managing Crop Adaptation to Changing Environment*. In: Kumar, N. and Singh, H. *Plant Functional Traits for Improving Productivity*. Springer: Singapore, pp. 257–285.

Biswas, S. and Das, R. (2024) Organic farming to mitigate biotic stresses under climate change scenario. *Bulletin of the National Research Centre* 48, 71.

Castel, B., El Mahboubi, K., Jacquet, C. and Delaux, P.-M. (2024) Immunobiodiversity: Conserved and specific immunity across land plants and beyond. *Molecular Plant*, 17(1), 92–111.

Chaffai, R., Ganesan, M. and Cherif, A. (2024) Abiotic Stress in Plants: Challenges and Strategies for Enhancing Plant Growth and Development. In: *Plant Adaptation to Abiotic Stress: From Signaling Pathways and Microbiomes to Molecular Mechanisms*. Springer, pp. 1–30.

Cheng, X., Zhou, G., Chen, W., Tan, L., Long, Q. *et al.* (2024) Current status of molecular rice breeding for durable and broad-spectrum resistance to major diseases and insect pests. *Theoretical and Applied Genetics* 137, 219.

Chin, S., Johns, S. and Gilroy, S. (2024) Calcium in root growth and development. In: *Plant Roots*. pp. 290–304. CRC Press: Florida, USA.

Choudhury, A. (2023) From plants to animals: A study of phytopathogenic infections. In: *Viral, Parasitic, Bacterial, and Fungal Infections*. pp. 29–36.

DeFalco, T.A. and Zipfel, C. (2021) Molecular mechanisms of early plant pattern-triggered immune signaling. *Molecular Cell* 81, 3449–3467.

Dhiman, V.K., Singh, D., Dhiman, V.K., Pandey, M., Sharma, A., Pandey, H. and Verma, S.K. (2024) *Insects as Biotic Stress in Agriculture*. In: *Plant-Microbe Interaction and Stress Management*. pp. 193–210. Springer: Singapore.

Divekar, P.A., Rani, V., Majumder, S., Karkute, S.G., Molla, K.A. *et al.* (2023) Protease inhibitors: an induced plant defense mechanism against herbivores. *Journal of Plant Growth Regulation* 42, 6057–6073.

Dodds, P.N., Chen, J. and Outram, M.A. (2024) Pathogen perception and signaling in plant immunity. *The Plant Cell* 36, 1465–1481.

Doggett, S.L. and Lee, C.-Y. (2023) Historical and contemporary control options against bed bugs, cimex spp. *Annual Review of Entomology* 68, 169–190.

Evangelisti, E. and Govers, F. (2024) Roadmap to success: how oomycete plant pathogens invade tissues and deliver effectors. *Annual Review of Microbiology* 78, 493–512.

Ghosh, P. and Roychoudhury, A. (2024) Plant peptides involved in abiotic and biotic stress responses and reactive oxygen species (ROS) signaling. *Journal of Plant Growth Regulation* 43, 1410–1427.

Gilbert, G.S. and Parker, I.M. (2023) How to be a plant macroparasite. *The Evolutionary Ecology of Plant Disease*, 71–82.

Hewedy, O.A., Elsheery, N.I., Karkour, A.M., Elhamouly, N., Arafa, R.A. *et al.* (2023) Jasmonic acid regulates plant development and orchestrates stress response during tough times. *Environmental and Experimental Botany* 208, 105260.

Hu, C., Li, Y.-T., Liu, Y.-X., Hao, G.-F. and Yang, X.-Q. (2024) Molecular interaction network of plant-herbivorous insects. *Advanced Agrochem* 3, 74–82.

Hückelhoven, R. and Schouten, A. (2024) Plant immunity and plant defense. In: *Agrios' Plant Pathology*. Elsevier, pp. 161–210.

Islam, A., Elango, S., Saini, P.K., Jan, I. and Sharma, Y. (2023a) Advancing plant pathology and the innovative methods for enhancing disease resistance. *Journal of Plant Biota*.

Islam, M.M., El-Sappah, A.H., Ali, H.M., Zandi, P., Huang, Q. *et al.* (2023b) Pathogenesis-related proteins (PRs) countering environmental stress in plants: A review. *South African Journal of Botany* 160, 414–427.

Jagadeesh Chandra Bose, K., Kaur, S., Sharma, S., Sarwan, J. and Uddin, N. (2024) *Methods from the Field of Synthetic Biology that Aim to Improve Plant Growth and Resistance to Stress through the Use of Genetic Engineering, Metabolomics, Proteomics and Gene Editing Approaches in Biofertilizer Industry: Volume II*. pp. 99–121. Springer Nature: Singapore.

Kamble, A., Michavila, S., Gimenez-Ibanez, S. and Redkar, A. (2024) Shared infection strategy of a fungal pathogen across diverse lineages of land plants, the Fusarium example. *Current Opinion in Plant Biology* 77, 102498.

Khan, S.A., Kojour, M.A.M. and Han, Y.S. (2023) Recent trends in insect gut immunity. *Frontiers in Immunology* 14, 1272143.

Khatri, P., Kumar, P., Shakya, K.S., Kirlas, M.C. and Tiwari, K.K. (2024) Understanding the intertwined nature of rising multiple risks in modern agriculture and food system. *Environment, Development and Sustainability* 26, 24107–24150.

La Vergata, A. (2023) *Struggles for Existence, Images of the Economy of Nature, 1650–1930: From" Nature's War" to Darwin's" Struggle for Life"*. pp. 477–583. Springer: Cham.

Lee, J., Kim, S., Jung, H., Koo, B.-K., Han, J.A. and Lee, H.-S. (2023) Exploiting bacterial genera as biocontrol agents: Mechanisms, interactions and applications in sustainable agriculture. *Journal of Plant Biology* 66, 485–498.

Li, S. (2024) Effect of Modulation of Auxin Response on Clubroot Development Caused by Plasmodiophora brassicae in Arabidopsis.

Liu, C., Liu, Q. and Mou, Z. (2024) Redox signaling and oxidative stress in systemic acquired resistance. *Journal of Experimental Botany*, 75(15), 4535–4548.

Manzoor, S., Nabi, S.U., Rather, T.R., Gani, G., Mir, Z.A. *et al.* (2024) Advancing crop disease resistance through genome editing: A promising approach for enhancing agricultural production. *Frontiers in Genome Editing* 6, 1399051.

Marcianò, D., Kappel, L., Ullah, S.F. and Srivastava, V. (2024) From glycans to green biotechnology: Exploring cell wall dynamics and phytobiota impact in plant glycopathology. *Critical Reviews in Biotechnology*, 45(2), 314–332.

Mathur, V., Sinha, P.G. and Noor, S.A. (2024) *Unraveling the Coevolutionary Arms Race: Insights into the Dynamic Interplay of Plants, Insects and Associated Organisms*. In: *Plant Resistance to Insects in Major Field Crops*. pp. 13–36. Springer: Singapore.

Mishra, S., Roychowdhury, R., Ray, S., Hada, A., Kumar, A. *et al.* (2024) Salicylic acid (SA)-mediated plant immunity against biotic stresses: An insight on molecular components and signaling mechanism. *Plant Stress*, 100427.

Muhammad, M., Basit, A., Wahab, A., Li, W.-J., Shah, S.T. and Mohamed, H.I. (2024) *Response Mechanism of Plant Stresses to Secondary Metabolites Production, Fungal Secondary Metabolites*. Elsevier, pp. 469–492.

Nabi, Z., Manzoor, S., Nabi, S.U., Wani, T.A., Gulzar, H. *et al.* (2024) Pattern-Triggered Immunity and Effector-Triggered Immunity: Crosstalk and cooperation of PRR and NLR-mediated plant defense pathways during host–pathogen interactions. *Physiology and Molecular Biology of Plants*, 4; 30(4), 587–604.

Nadarajah, K.K. (2024) *Defensive Strategies of ROS in Plant–Pathogen Interactions*. In: *Plant Pathogen Interaction*. Springer, pp. 163–183.

Nigam, R. (2024). *Plant Diseases and Food Security in the 21st Century*. Academic Guru Publishing House.

Nikpay, A., Tiwari, A.K., Vejar-Cota, G., Ziaee, M., Wilson, B., Srivastava, S. and Goebel, F.-R. (2023) *Biotic Stresses in Sugarcane Plants and its Management*. In: *Agro-Industrial Perspectives on Sugarcane Production Under Environmental Stress*. Springer, pp. 301–359.

Nitta, A., Natarajan, V., Reddy, A.J. and Rakesh, T. (2024) Impact of climate change on pest biology, behaviour and their distributions. *International Journal of Environment and Climate Change* 14, 46–56.

Onyeaka, H., Ghosh, S., Obileke, K., Miri, T., Odeyemi, O.A., Nwaiwu, O. and Tamasiga, P. (2024) Preventing chemical contaminants in food: Challenges and prospects for safe and sustainable food production. *Food Control* 155, 110040.

Parihar, M., Rakshit, A., Adholeya, A. and Chen, Y. (2024) *Arbuscular Mycorrhizal Fungi in Sustainable Agriculture: Inoculum Production and Application*. Springer Nature: Singapore.

Pittaway, A. (2023) *The Hawkmoths of the Western Palaearctic*. Brill.

Poonia, S.K., Topno, S.E. and Kerketta, A. (2024) Integrated pest and disease management in cucumber and muskmelon. *A Monthly Peer Reviewed Magazine for Agriculture and Allied Sciences*, 1.

Rai, A.K., Sahoo, L.P., Sahoo, M., Seth, T. and Patra, C. (2024) Induced resistance mechanism in plant and its importance in agriculture. *International Journal of Plant and Soil Science* 36, 1–22.

Rathore, D., Naha, N. and Singh, S. (2024) *Phenolic Compounds and Nanotechnology: Application During Biotic Stress Management in Agricultural Sector and Occupational Health Impacts*. In: *Plant Phenolics in Biotic Stress Management*. pp. 503–549. Springer: Singapore.

Ray, R.V. (2024) *Effects of Pathogens and Disease on Plant Physiology*. In: *Agrios' Plant Pathology*. Elsevier, pp. 63–92.

Raza, A., Khandelwal, K., Pandit, S., Singh, M., Kumar, S. *et al.* (2024) Exploring the potential of metallic and metal oxide nanoparticles for reinforced disease management in agricultural systems: A comprehensive review. *Environmental Nanotechnology, Monitoring and Management*, 22, 100998.

Riseh, R.S. and Vazvani, M.G. (2024) Unveiling methods to stimulate plant resistance against pathogens. *Frontiers in Bioscience-Landmark* 29, 188.

Roychowdhury, R., Hada, A., Biswas, S., Mishra, S., Prusty, M.R. (2024). Jasmonic Acid (JA) in plant immune response: Unravelling complex molecular mechanisms and networking of defence signalling against pathogens. *Journal of Plant Growth Regulation*, 44(1), 89–114.

Rufián, J.S., Rueda-Blanco, J., Beuzón, C.R. and Ruiz-Albert, J. (2023) Suppression of NLR-mediated plant immune detection by bacterial pathogens. *Journal of Experimental Botany* 74, 6069–6088.

Saakre, M., Jaiswal, S., Rathinam, M., Raman, K.V., Tilgam, J. *et al.* (2024) Host-delivered RNA interference for durable pest resistance in plants: Advanced methods, challenges, and applications. *Molecular Biotechnology* 66, 1786–1805.

Sahu, B., Choudhary, V.K., Sahu, M., Kumar, K.K., Sujayanand, G. *et al.* (2023) *Biotic Stress Management, Trajectory of 75 years of Indian Agriculture after Independence*. pp. 619–653. Springer Nature: Singapore.

Saraiva, M., Ściślak, M.E., Ascurra, Y.T., Ferrando, T.M., Zic, N. *et al.* (2023) The molecular dialog between oomycete effectors and their plant and animal hosts. *Fungal Biology Reviews* 43, 100289.

Shah, A.A. and Gupta, A. (2023) *Secondary Metabolite Basis of Elicitor-and Effector-Triggered Immunity in Pathogen Elicitation Amid Infections*. In: *Genetic Manipulation of Secondary Metabolites in Medicinal Plant*. Springer, pp. 225–251.

Şimşek, Ö., Isak, M.A., Dönmez, D., Dalda Şekerci, A., İzgü, T. and Kaçar, Y.A. (2024) Advanced biotechnological interventions in mitigating drought stress in plants. *Plants* 13, 717.

Singh, A., Shraogi, N., Verma, R., Saji, J., Kar, A.K. *et al.* (2024) Challenges in current pest management Practices: Navigating problems and a way forward by integrating controlled release system approach. *Chemical Engineering Journal*, 14, 154989.

Singh, M., Pandey, N. and Sharma, O. (2023a) IPM concept and strategies for sustainable agriculture, *Integrated Pest Management in Diverse Cropping Systems*. Apple Academic Press, pp. 31–59.

Singh, P., Pandey, V.K., Sultan, Z., Singh, R. and Dar, A.H. (2023b) Classification, benefits, and applications of various anti-nutritional factors present in edible crops. *Journal of Agriculture and Food Research* 14, 100902.

Somashetty, S., Akbarbasha, R., Krishnamurthy, S., Chowdappa, S. and Konappa, N.M. (2023) A comprehensive review on recent advances in the research on bacterial leaf spot in tomato caused by xanthomonas species. *Plant Archives* 09725210, 23.

Song, W., Shao, H., Zheng, A., Zhao, L. and Xu, Y. (2023) Advances in roles of salicylic acid in plant tolerance responses to biotic and abiotic stresses. *Plants* 12, 3475.

Sun, G., Zhang, X., Liu, Y., Chai, L., Liu, D., Chen, Z. and Lü, S. (2024) Plant chemical defenses against insect herbivores—using the wild tobacco as a model. *Phyton-International Journal of Experimental Botany.*

Sunani, S.K., Rout, A.K. and Kumar, R. (2024) Biotic Stress of Major Pulses and Their Management.

Sutaoney, P., Pandey, D., Joshi, V., Vyas, A., Joshi, N. *et al.* (2023) Use of plant-defense hormones against pathogen diseases. In: *Hormonal Cross-Talk, Plant Defense and Development*. Elsevier, pp. 305–334.

Thiruvengadam, R., Venkidasamy, B., Easwaran, M., Chi, H.Y., Thiruvengadam, M. and Kim, S.-H. (2024) Dynamic interplay of reactive oxygen and nitrogen species (ROS and RNS) in plant resilience: Unveiling the signaling pathways and metabolic responses to biotic and abiotic stresses. *Plant Cell Reports* 43, 1–24.

Todd, T.S. (2024) *HopBA1, a Pathogen Virulence Factor, Reveals Tissue-Specific Immune Responses Within the Pseudomonas syringae Pv. Tomato–Nicotiana Benthamiana Pathosystem*. Colorado State University.

Tripathi, A., Pandey, V.K., Jha, A.K., Srivastava, S., Jakhar, S. *et al.* (2024) Intricacies of plants' innate immune responses and their dynamic relationship with fungi: A review. *Microbiological Research*, 285, 127758.

van Hensbergen, V.P. and Hu, X. (2024) Pattern recognition receptors and the innate immune network. In: *Molecular Medical Microbiology*. Elsevier, pp. 407–441.

Venugopal, E., Krishna, S.B.N., Golla, N. and Patil, S.J. (2024) *Nanofungicides: A Promising Solution for Climate-Resilient Plant Disease Management*. In: *Plant Quarantine Challenges under Climate Change Anxiety*. pp. 513–532. Springer: Singapore.

von Dahlen, J.K. (n.d.) Expression, Regulation and Evolution of the Plant Immune System.

Vukeya, L.R. (2021) Vegetation Cover Dynamics and the Environmental Threats to Biodiversity Conservation in Free State National Botanical Garden, South Africa.

Walker, P. (2023) Investigation of the Brassica Napus-Sclerotinia Sclerotiorum Pathosystem and Development of RNA Interference-Based Technologies to Control Pathogenicity.

Wani, N.R., Rather, R.A., Farooq, A., Padder, S.A., Baba, T.R. *et al.* (2024) New insights in food security and environmental sustainability through waste food management. *Environmental Science and Pollution Research* 31, 17835–17857.

Whitehill, J.G., Bohlmann, J. and Krokene, P. (2023) *Forest Insect – Plant Interactions, Forest Entomology and Pathology*. In: *Volume 1: Entomology*. Springer International Publishing, Cham, pp. 169–204.

Xiang, X., Liu, S., Li, H., Danso Ofori, A., Yi, X. and Zheng, A. (2023) Defense strategies of rice in response to the attack of the herbivorous insect, Chilo suppressalis. *International Journal of Molecular Sciences* 24, 14361.

Xu, X., Ma, X., Dou, J., Chen, W., Chen, J. *et al.* (2024) β-ionone inhibits the grazing of Daphnia sinensis by reducing the activity of acetylcholinesterase. *Journal of Hazardous Materials*, 285, 135690.

Yang, C., Dolatabadian, A. and Fernando, W.D. (2022) The wonderful world of intrinsic and intricate immunity responses in plants against pathogens. *Canadian Journal of Plant Pathology* 44, 1–20.

Yousuf, P., Razzak, S., Parvaiz, S., Rather, Y.A. and Lone, R. (2024) Role of plant phenolics in the resistance mechanism of plants against insects. In: *Plant Phenolics in Biotic Stress Management*. pp. 191–215. Springer: Singapore.

Yu, X.Q., Niu, H.Q., Liu, C., Wang, H.L., Yin, W. and Xia, X. (2024) PTI-ETI synergistic signal mechanisms in plant immunity. *Plant Biotechnology Journal*, (8), 2113–2128.

9 Heavy Metals Stress and Phytoremediation

Muhammad Ahsan Altaf[1,2] and Riaz Ahmad[3*]

[1]Key Laboratory for Quality Regulation of Tropical Horticultural Crops of Hainan Province, School of Breeding and Multiplication (Sanya Institute of Breeding and Multiplication), Hainan University, Sanya 572025, China; [2]Key Laboratory for Quality Regulation of Tropical Horticultural Crops of Hainan Province, School of Tropical Agriculture and Forestry, Hainan University, Haikou 570228, China; [3]Department of Horticulture, The University of Agriculture, Dera Ismail Khan 29111, Pakistan

Abstract

Heavy metals are classified as minimally necessary or non-essential components for healthy crops. They are widely distributed throughout the soil-water system and are easily absorbed by plants, which puts them under oxidative harm. Plants can be defended against heavy metals stress by genetically modifying their genes, applying synthetic chemicals exogenously, or using built-in immune systems. Several intrinsic processes encompass metal discrimination, limited foliar transport, metal retention and differentiation, chelation, and antioxidant enzyme-mediated elimination of harmful free radicals. Phytochelatins combine with heavy metals in the cytoplasm to produce complex systems, which are then stored in vacuoles. Free radicals are scavenged by antioxidants, which in turn transform into non-toxic compounds such as various enzymes. Different stress mitigants such as metallothionein, organic amendments, polyphenols, and tocopherols can help plants to tolerate heavy metals stress. Furthermore, signaling networks aid in the sensing of stress by plants and trigger the stimulation of networks related to the breakdown of hormones, calcium, mitogens, and lipid peroxidation. Plant resistance to heavy metals stress can be increased by using various management techniques. This chapter interprets the precise mechanisms underlying the tolerance of heavy metals stress and focuses on multiple components of genetics, molecular, and cell signaling mechanisms that combine to provide an integrated reaction to heavy metal toxicity. In an attempt to utilize the information regarding endurance in practical applications, several facets of specific model plant species are thoroughly explored.

Keywords: disruption in photosynthesis, metabolic irregularities, poor germination, stunted growth

9.1 Introduction

The health of communities is seriously threatened by contamination in ecosystems, especially when it is caused by dangerous heavy mineral toxins. Toxic metals compounds may pollute ecosystems through urbanization and industrialization (Vardhan *et al.*, 2019). The biological fundamentals of industry, trade, and materialism are all activities that are polluting the environment. Highways, cars, and factories all significantly increase the harmful effects of heavy metals as toxins (Okereafor *et al.*, 2020). Higher heavy metals levels can cause serious harm to plants.

* Email: riazahmadbzu@gmail.com

DOI: 10.1079/9781800628823.0009

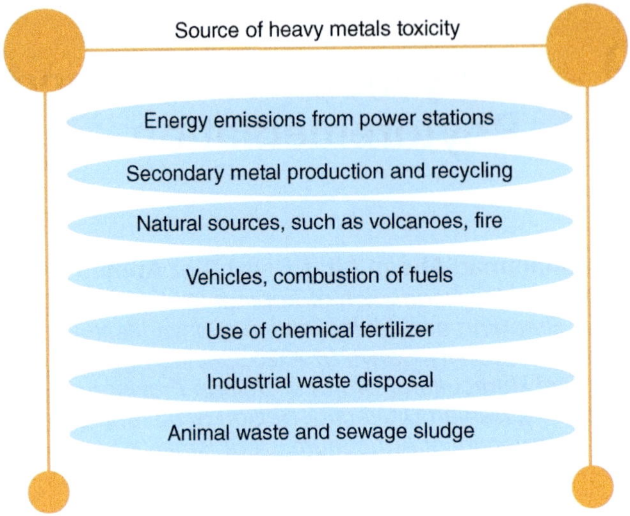

Fig. 9.1. Different sources of heavy metals in surrounding environment. (Figure author's own.)

Higher levels heavy metal toxicity caused leaf oxidative damage by increasing the excessive reactive oxygen species (ROS) accumulation (Fig. 9.1). The majority of heavy metals release ROS continuously into plant cells, which can lead to oxidative harm in plants and unanticipated heavy metals damage (Jaishankar *et al.*, 2014; Bohra *et al.*, 2015).

Plants exposed to heavy metals can suffer rapid impairment of membranes due to lipid oxidation, which serves as one of the main detrimental effects (Parmar *et al.*, 2013; Kumar *et al.*, 2023). Moreover, Shahid *et al.* (2014) revealed that heavy metals contamination was linked to greater amounts of hazardous metals in plant cells, changed mineral nutrient accumulation, reduced seedling growth, and damaged the leaf photosynthetic process. Furthermore, oxidative damage and an uneven antioxidant response were brought on by the heavy metals on musk melon sensitivity (Ghori *et al.*, 2019). Excess of heavy metals disrupts plant growth, and yield of several crops such as potato (Musilova *et al.*, 2011), sweet potato (Antonious *et al.*, 2011; Ragonezi *et al.*, 2022), tomato (Birghila *et al.*, 2023), grapevine (Alagić *et al.*, 2015), peppers (Antonious, 2016), eggplant (Youssef and Abd El-Gawad, 2018), mulberry (Gul *et al.*, 2021), citrus (Abdel-Shafy and El-Khateeb, 2021), guava (Yap *et al.*, 2021), mango (Silva *et al.*, 2012; Širić *et al.*, 2022), okra (Panhwar *et al.*, 2022), spinach (Khaliq *et al.*, 2022) and saffron (Fayssal *et al.*, 2024).

Poor seed germination and hampered seedlings growth was determined in the contaminated soils from heavy metals. Crops growth, productivity, and quality concerns might arise when growing in polluted land and water resources.

Plant tolerance against heavy metal toxicity can be increased by efficient management measures. Phytohormones has a potential as a growth promoter because of its impact on several physiological functions (Saini *et al.*, 2021). Moreover, phytohormones may help as a phytoremediation method to restrict heavy metal toxicities. Previous research (Sytar *et al.*, 2019; Dai *et al.*, 2023) demonstrated that endogenous phytohormones exposure increased plant resilience against heavy metals. Additionally, phytohormones promote the resistance of many crops to climatic toxins in an efficient and cost-effective manner as they are considered to be ecologically sound substances (Nguyen *et al.*, 2021).

Different management strategies are effective for higher seed germination, greater seedlings growth, better biomass, a strong defense system, and better adjustments potential for physiological activities in crops subjected to heavy metals (Goyal *et al.*, 2020; Shah *et al.* 2024). The main objectives of this chapter are to maintain plant production in soils stressed by heavy metals, better understand the melatonin-mediated behavior, and identify indicators in crops that mitigate heavy metal toxicities.

9.2 Production Losses in Horticultural Crops Due to Heavy Metal Toxicities

The severe toxicity of heavy metals results in major loss of productivity. The phytohormonal interplay is the interaction that occurs between different hormones in crops. Recent research indicates that phytohormones play a major role in both plant growth and reaction to stress, and they may even be able to lessen some of the negative effects of metal ions (Gangwar *et al.*, 2014; Tiwari *et al.*, 2024). Phytohormones regulate root development, water loss, and shielding by means of antioxidants, which remove heavy metals by interacting with different other endogenous phytohormones. It is well known that the intrinsic phytohormones in plant tissues are dependent on environmental variables. Several studies found that changes in sunlight, temperatures, toxic substances, and many other climatic factors affected the quantity of hormones that plants stored (Saini *et al.*, 2021; Rahman *et al.*, 2023).

Heavy metal toxicities affect the way phytohormones are stored and produced in various plant tissues (Bücker-Neto *et al.*, 2017). This is a serious concern for plants, as evidenced by reports of some crops, including mulberry (Gul *et al.*, 2021), tomatoes (Saeed *et al.*, 2013), chilies (Lidiková *et al.*, 2021), and hot peppers (Li *et al.*, 2010). Furthermore, antioxidant generation is enhanced in plants under heavy metal toxicity. Studies have shown that cadmium (Cd) pollutants severely damage the structural integrity of cells (Johri *et al.*, 2010; Genchi *et al.* 2020). When cells encounter Cd pollutants, they release serotonin N-acetyltransferase enzymes into the cell interior, which combine with serotonin to generate N-acetylserotonin and ultimately accumulate melatonin (Hoque *et al.*, 2021).

Water hyacinth, a resilient weed detrimental to the environment, can be effectively utilized to extract potentially toxic elements (Du *et al.*, 2020; Galgali *et al.*, 2023). Malar *et al.* (2016) reported a decrease in water hyacinth formation with heavy metals exposure.

The plant increased hormonal levels in different plant tissues appears to be the cause of this resilience, because hormones have potent antioxidant qualities. So, it can strengthen the water hyacinth resilience against heavy metals and greatly increase its potential for phytoremediation (Emamverdian *et al.*, 2020). High hormonal amounts in plants lend confirmation to the hypothesis that the heightened intrinsic content of each of these beneficial antioxidants may be associated with the plant's great tolerance to heavy metals and its use in phytoremediation. One possible elucidation for this event is a rupture of the cell wall. However, plants' chloroplasts were unaffected by aluminum toxicity (Imadi *et al.*, 2016). As a result, plants of all species can have their hormonal formation controlled by the duration and degree of metal contact. The results of Hodzic *et al.* (2021) revealed that both zinc (Zn) and Cd pollutants significantly increased the levels of phytohormones (melatonin) in lemon balm. They found that melatonin, a strong antioxidant, is essential for plants to be able to withstand the detrimental impacts of heavy metals. Therefore, it is possible to maintain that increased hormonal generation results from heavy metal toxicity, which may promote plants to respond to stress brought through heavy metal exposure (Fig. 9.2).

9.3 Morphological and Structural Modifications in Plants Subjected to Heavy Metals

Morphological and anatomical changes in plants were due to exposure of heavy metal toxicities (Gomes *et al.*, 2011; Farzadfar and Zarinkamar, 2012). Batool *et al.* (2015) reported the adverse effects of heavy metal toxicities in plants due to presence of metalloproteins developments because it had potential to restrict enzymatic activities relating with the sulfhydryl side. In the findings of Yadav *et al.* (2021), it has been observed that oxidative injury may occur in plant organelles such as chloroplasts and mitochondria. Singh *et al.* (2015) concluded that an excess of heavy metals in maize resulted in disruption of photosynthesis, water loss and various developmental processes.

The generated heavy metals have detrimental effects on the epidermal tissues of plants, leading to cellular damage, reduced leaf thickness, and disruption of parenchyma cells (Rizvi and Khan, 2017). The rapid decrease in plant

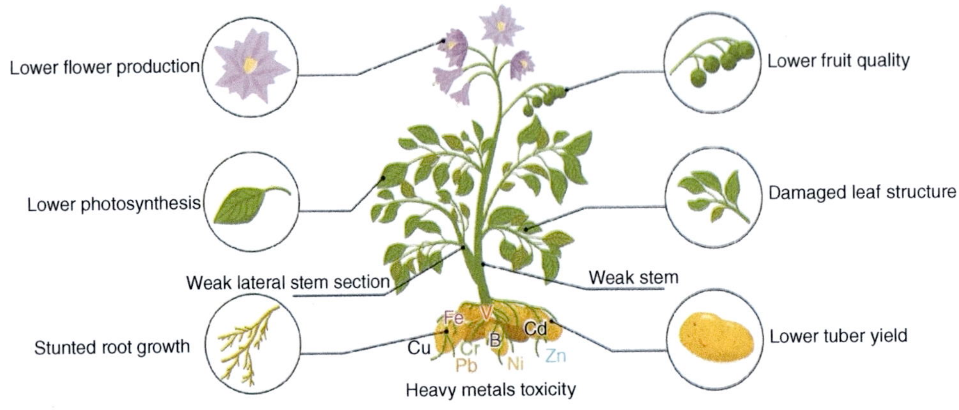

Fig. 9.2. Heavy metal toxicity in plants. (Figure author's own.)

fresh and dried biomass was due to presence of excess of heavy metal toxicity (Azmat et *al.*, 2009). Therefore, a rapid increase in pollution is also a major cause of stunted growth and poor biomass. The limited concentration of heavy metal exposure may increase plant growth due to the presence of higher concentrations of macro- and micronutrients. However, a long-term exposure of heavy metals such as Zn and Cd resulted in the stunted growth of Indian mustard (Sridhar *et al.*, 2005).

The stunted growth, restricted developmental processes and poor yield were due to excess of heavy metals in plants (Madhu and Sadagopan, 2020). Poor leaf thickness was due to excess of pollution, which disturbed the cells and cortex (Yadav *et al.*, 2021). Soil contamination is increasing gradually due to excess of heavy metal toxicity. Different morphological and anatomical changes may occur due to excess of heavy metals resulting in stunted growth and poor yield (Mulenga *et al.*, 2020). Farming in contaminated soils is reducing plant growth and productivity due to translocation of toxic substances from roots towards shoots. Moreover, stomatal pores are influenced by heavy metal toxicity (Bano *et al.*, 2019). Moreover, it has been observed that the restricted growth and poor developmental processes in plants subjected to excess of Cd and Zn levels (Toppi and Gabbrielli, 1999; Asgher *et al.*, 2015). A decrease in plant size, plant biomass and seedlings growth were the drastic effects of excessive concentrations of lead (Pb) and Cd observed by Vassilev and Yordanov (1997). Moreover, a decrease in plant growth and yield were observed due to excess levels of chromium (Cr) in plants by Shanker *et al.* (2005). The alterations in morphology of roots were recorded due to exceeded levels of heavy metal toxicity. Heavy metals are drastically reducing the productivity of crops (Hayat *et al.*, 2012; Kumar *et al.*, 2016).

9.4 Biochemical and Metabolic Adjustments in Plants Subjected to Heavy Metals

When plants are subjected to heavy metal toxicity, it results in the activation of biochemical adjustments within the plant's cells. The accumulation of heavy metals in plant cells may reduce the biomass, chlorophyll pigments, photosynthesis, metabolites, and antioxidants (Maleki *et al.*, 2017). The enhancement of heavy metals in plants reduces the sugars level within the plant cells as reported in many crops (Peralta-Videa *et al.*, 2009; Sytar *et al.*, 2013; Dubey *et al.*, 2018).

Several biological activities are influenced by heavy metal toxicity. For example, the rupturing of cell membranes is due to heavy metal toxicity according to Karlsson *et al.* (2013). However, the plant has the capability to activate different enzymes to repair the cell membrane injury. Plant metabolic processes are influenced by an excess of toxic substances. So, the activation of different metabolites can be effective as a biochemical adjust for repairing the cell membranes under heavy metal toxicity (Jan *et al.*, 2015). Moreover, chlorophyll degradation was recorded due to heavy metal toxicity, and biochemical and metabolic adjustments can effectively reduce the adverse effects of heavy metals focusing on sustainable yield (Berni *et al.*, 2019).

Poor seed germination was recorded in cluster bean due to an excessive concentration of Zn. Moreover, poor growth, low sugars, disruption of chlorophyll pigments, and poor carotenoids content were also recorded in cluster bean due to the toxic effects of elevated Cr concentration (Sangwan *et al.*, 2013).

Phytotoxicity in plants occurs due to an excessive level of Zn within the plants because nutritional imbalances occur within the plants (Ivanov *et al.*, 2021). Excessive Zn levels in the cultivated lands disturbs the metabolic activities that naturally occur in the soil. Moreover, higher translocation level was recorded from roots towards shoots and other plant parts. The elevated reduction in ATPase activity was seen due to excessive Cd concentrations in agronomic crops, such as maize (Tanwir *et al.*, 2015), wheat (Ci *et al.*, 2010), tomato (Nogueirol *et al.*, 2016) and sunflower (Júnior *et al.*, 2015). Several physiological activities such as stomatal conductance, leaf moisture level, and transpiration rate are disturbed due to heavy metal toxicity (Małkowski *et al.*, 2019). Osmotic stress conditions are also causing numerous physiological disorders in plants subjected to excessive levels of heavy metals (Polle and Schützendübel, 2003).

Oxidative injury, chlorosis, and root darkness occurred in crop plants due to an excess of Cd in the root biosphere (Gill *et al.*, 2011). Cd, at a concentration greater than the optimum level, may cause a rapid decline and distress in various enzymes and their structures. An excess of Cd in the soil results in a decrease of friendly microorganisms. However, these microorganisms improve soil fertility through improved uptake and translocation of minerals (Altomare and Tringovska, 2011). Higher concentrations of Cd and Zn greatly influenced the catalytic enzymes in plants (Wyszkowska *et al.*, 2013).

An excess of Pb is toxic for rapeseed as it causes chlorosis, restricted growth, and poor yield (Kanwal *et al.*, 2014). Pb toxicity influences cell division in plants and results in an increase of oxidative injury and decreased growth components (Aslam *et al.*, 2021).

Alterations in biochemical activities such as nutrient restrictions, stunted growth, degradation of enzymes, and a decrease in photosynthesis in plants were observed due to nickel (Ni) toxicity by Hassan *et al.* (2019). The adverse effects of Ni toxicity reduced the crop growth and yield (Yusuf *et al.*, 2011).

Excessive concentrations of mercury are also dangerous for plant health. The adverse effects of mercury are more visible in plants as stunted growth and irregular developmental processes occur. The regulation of stomatal closure is necessary for healthy plants. Zhou *et al.* (2008) observed that an excess of mercury in alfalfa caused disturbances in stomatal conductance and transpiration rate. Irregular transpiration rate and stomatal conductance disturbed the numerous developmental processes of the plants. Moreover, a blockage in water passage was also observed due to excessive levels of mercury in the root biosphere (Chen and Yang, 2012). Raj and Maiti (2019) found that an excess of mercury resulted in higher concentrations of ROS being recorded in mitochondria. This increase in ROS generation resulted in an increase in the rupturing of cell membranes and lipid peroxidation.

Khan and Khan (2010) found that an excess of cobalt (Co) reduced the formation of chlorophyll contents in chickpea plants, where Co was found be higher as compared to the optimal level in plant leaves. Disturbances in biochemical activities were also recorded due to excess of Co content in different parts of plants (Akeel and Jahan, 2020). Higher Co concentration significantly restricted mineral nutrient accumulation from roots towards shoots (Jaleel *et al.*, 2009).

The phytotoxicity of manganese (Mn) has been observed as the cause of necrosis and chlorosis in plant foliage in bean plants (Mahjoubi *et al.*, 2021; Altaf *et al.*, 2023). The synthesis of photosynthetic pigments was disturbed due to an excess of Mn in plants. An excess of Mn increases the generation of ROS and various free radicals, which increase the level of malondialdehyde (MDA) generation (Tavanti *et al.*, 2020). The production of toxic substances within the plants reduces the plant growth and yield. Distressed photosynthesis processes due to excess of Mn were also observed in plants by Gautam and Dubey (2018).

Hence, biochemical and metabolic adjustments contribute to appropriate plant growth and regulate various plant developmental processes under heavy metal toxicity. The activation of the plant defense system was also observed in stress-tolerant genetic resources. Heavy metal toxicity effectively damage the activity of biochemical attributes. The tolerant genetic resources can be further utilized for the development of higher yielding inbred lines. Heavy metal toxicity response to plants are described in Table 9.1.

9.5 Phytoremediation Mechanism Against Heavy Metals

Under heavy metal toxicity, plants have the potential to activate their defense system to reduce the uptake and translocation of heavy metals from the root biosphere. Different phytoremediation mechanisms are also involved in tolerance mechanism of heavy metal toxicity (Yang *et al.*, 2022), among which the application of different phytohormones is effective for mitigation of heavy metal toxicity (Wei *et al.*, 2020).

Various pieces of plant research have revealed that heavy metal toxicity can be mitigated by using phytohormones such as melatonin, salicylic acid (SA), and jasmonic acid (JA). Moreover, the application of silicon can also be effective to reduce the adverse effects of heavy metals (El-Saadony *et al.*, 2021). Several organic amendments, such as biochar application and plant growth-promoting rhizobacteria, can reduce the adverse effects of heavy metals on plants (Liang *et al.*, 2021).

Exclusion is considered an important way to lessen the uptake of heavy metals from root biosphere (Shah and Daverey, 2020). The development of casparian strips around the root's endodermal tissues has the potential to

Table 9.1. Plant responses to heavy metal toxicity. (Table author's own.)

Metals	Crop	Response	Reference
Cd	Wheat	Impaired photosynthesis, increased enzymes activity, oxidative damage	Ci *et al.* (2010)
Co	Maize	Increased Co uptake, reduced mineral nutrient uptake and enhanced ROS production	Jaleel *et al.* (2009)
Cd	Sunflower	Increased metals uptake, ROS and MDA accumulation, lowered mineral uptake, altered enzyme activity	Júnior *et al.* (2015)
Pb	Rapeseed	Reduced ion homeostasis, unbalanced redox homeostasis, declined photosynthesis, and decreased pigment content	Kanwal *et al.* (2014)
Cr	Spinach	Reduced growth, hindered photosynthesis, increased electrolyte leakage (EL), oxidative damage	Khaliq *et al.* (2022)
Ni	Chickpea	Enhanced ROS production, lowered photosynthesis, decreased pigment concentration, altered antioxidant and enzyme activity	Khan and Khan (2010)
Mn	Beans	Reduced leaf area, decreased number of leaves, decreased photosynthesis activity and decreased mineral uptake in leaf	Mahjoubi *et al.* (2021)
Cr	Cluster bean	Lowered mineral nutrient uptake, increased enzyme activity, greater metal accumulation in leaf	Sangwan *et al.* (2013)
Pb	Maize	Reduced leaf area, growth, pigments content, and photosynthesis; oxidative damage	Singh *et al.* (2015)
Zn	Indian mustard	Increased oxidative damage; reduced growth, leaf area, yield, and quality	Sridhar *et al.* (2005)
Cd	Maize	Reduction in leaf area, decreased antioxidant and enzyme activity, decreased proline content and secondary metabolites production	Tanwir *et al.* (2015)

restrict the uptake of heavy metals in excess towards vascular tissues. Moreover, plants may release various organic acids from their roots, which can lessen the uptake of heavy metals by the formation of insoluble complexes in the root biosphere (Pandey *et al.*, 2022).

The active efflux (ATP-dependent transporters) has the potential to eliminate excess heavy metals from plant roots, thereby minimizing their uptake and toxicity. The absorption of heavy metals can be diminished due to competition among various minerals during the absorption process, indicating that the mineral competition mechanism can effectively reduce mineral uptake from roots. Nevertheless, the uptake and translocation of minerals may fluctuate based on the availability of minerals in the root zone, the rate of insoluble complex formation, xylem loading rates, and the translocation of toxic elements to shoots and other plant parts (Priya *et al.*, 2022). Consequently, the exclusion process is effective in mitigating the uptake and translocation of heavy metals. Optimal translocation of heavy metals is beneficial for plant health and yield. Nonetheless, these techniques are efficacious for metallic applications (Bhat *et al.*, 2022). The remediation of contaminants from contaminated soil may be achieved by ecologically benign methods with minimal input costs, such as phyto-accumulation (Fig. 9.3) (Patra *et al.*, 2020).

The removal of pollutants from contaminated soils is very necessary to improve the crop productivity. Phytostabilization can be helpful for the removal of pollutants from contaminated soils affected by the severity of heavy metals (Liu *et al.*, 2022). The process of phytostabilization potentially immobilizes the various contaminants in the root biosphere. Moreover, contaminant accumulation was found to be higher in the root zone of plants and did not allow the pollutants to translocate to other parts of the plant through the roots.

The absorption and assimilation of contaminants is reduced through phytostabilization (Gavrilescu, 2022). Moreover, the bioavailability of pollutants in plants is also reduced through phytostabilization. However, this process helps the cultivation of vegetation on the contaminated lands (Lacalle *et al.*, 2023). The cultivation of tolerant plants can be effective against metals toxicity. Moreover, it will be important to restrict the migrations of pollutants from contaminated lands to other non-contaminated regions. Therefore, cultivation of vegetation can be effective to restrict the pollutant migration (Chen *et al.*, 2021). Hence, phytostabilization is effective to reduce the heavy metals and their adverse effects on plants and soil health.

Phytovolatilization has the potential to mitigate the adversities that occur from heavy metal toxicity in soil and plants (Majeed, 2021).

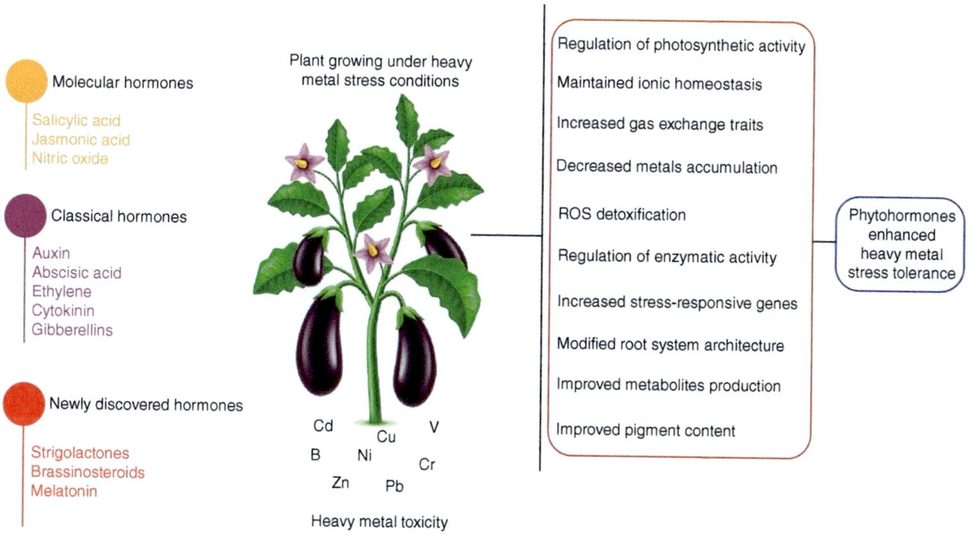

Fig. 9.3. Phytohormones regulate heavy metal stress tolerance in plants. (Figure author's own.)

During this process, pollutants are converted into less toxic form (Kour *et al.*, 2020). The toxic metals are converted into volatile forms, which become less toxic for atmosphere, soils, and plants (Bortoloti and Baron, 2022). However, the removal of toxic metals from soil and the mitigation of adverse effects on plants are necessary to explore for sustainable production to feed huge population with better quality of produce (Timalsina *et al.*, 2022).

9.6 Conclusion

An excess of heavy metals is toxic for soil and plant health. The productivity of crops is decreasing due to a higher uptake and translocation of toxic metals in different plant parts. Plant produce is becoming poisonous and unsuitable for ingestion. It is vital to have food that is healthy and devoid of contaminants. Plants possess the capacity to activate their defensive mechanisms in response to harsh circumstances caused by heavy metals. Management strategies may enhance the plant defense mechanism against excessive heavy metal exposure. Furthermore, tolerant genetic resources have a robust defensive mechanism against elevated levels of heavy metals. Therefore, the discovery, selection, and cultivation of tolerant genotypes, together with the use of appropriate management measures, may enhance agricultural output.

References

Abdel-Shafy, H.I. and El-Khateeb, M. (2021) Heavy metals in citrus fruits as affected by primary treated sewage irrigation. *Egyptian Journal of Chemistry* 64(1), 165–176.

Akeel, A. and Jahan, A. (2020) Role of cobalt in plants: Its stress and alleviation. In: Naeem, M., Ansari, A. and Gill, S. *Contaminants in Agriculture: Sources, Impacts and Management*. Springer, Cham, pp. 339–357.

Alagić, S.Č., Tošić, S.B., Dimitrijević, M.D., Antonijević, M.M. and Nujkić, M.M. (2015) Assessment of the quality of polluted areas based on the content of heavy metals in different organs of the grapevine (*Vitis vinifera*) cv Tamjanika. *Environmental Science and Pollution Research* 22, 7155–7175.

Altaf, M.A., Shahid, R., Lal, P., Ahmad, R., Zulfiqar, F. *et al.* (2023) Current understanding of boosting power of salicylic acid for abiotic stress tolerance in horticultural crops. *South African Journal of Botany* 163, 285–293.

Altomare, C. and Tringovska, I. (2011) Beneficial soil microorganisms, an ecological alternative for soil fertility management. *Genetics, Biofuels and Local Farming Systems*. In: Sustainable Agriculture Reviews, pp. 161–214. Springer: Berlin/Heidelberg, Germany.

Antonious, G.F. (2016) Distribution of seven heavy metals among hot pepper plant parts. *Journal of Environmental Science and Health, Part B* 51(5), 309–315.

Antonious, G.F., Dennis, S.O., Unrine, J.M. and Snyder, J.C. (2011) Heavy metals uptake in plant parts of sweet potato grown in soil fertilized with municipal sewage sludge. *International journal of Geology*, 5(1), 14–20.

Asgher, M., Khan, M.I.R., Anjum, N.A. and Khan, N.A. (2015) Minimising toxicity of cadmium in plants—role of plant growth regulators. *Protoplasma* 252, 399–413.

Aslam, M., Aslam, A., Sheraz, M., Ali, B., Ulhassan, Z. *et al.* (2021) Lead toxicity in cereals: Mechanistic insight into toxicity, mode of action, and management. *Frontiers in Plant Science* 11, 587785.

Azmat, R., Haider, S., Nasreen, H., Aziz, F. and Riaz, M. (2009) A viable alternative mechanism in adapting the plants to heavy metal environment. *Pakistan Journal of Botany* 41(6), 2729–2738.

Bano, C., Amist, N. and Singh, N.B. (2019) Morphological and anatomical modifications of plants for environmental stresses. *Molecular Plant Abiotic Stress: Biology and Biotechnology*, (29), 29–44.

Batool, R., Hameed, M., Ashraf, M., Ahmad, M.S.A. and Fatima, S. (2015) Physio-anatomical responses of plants to heavy metals. *Phytoremediation for Green Energy*, 79–96.

Berni, R., Luyckx, M., Xu, X., Legay, S., Sergeant, K. *et al.* (2019) Reactive oxygen species and heavy metal stress in plants: Impact on the cell wall and secondary metabolism. *Environmental and Experimental Botany* 161, 98–106.

Bhat, S.A., Bashir, O., Haq, S.A.U., Amin, T., Rafiq, A. *et al.* (2022) Phytoremediation of heavy metals in soil and water: An eco-friendly, sustainable and multidisciplinary approach. *Chemosphere* 303, 134788.

Birghila, S., Matei, N., Dobrinas, S., Popescu, V., Soceanu, A. and Niculescu, A. (2023) Assessment of heavy metal content in soil and Lycopersicon esculentum (tomato) and their health implications. *Biological Trace Element Research* 201(3), 1547–1556.

Bohra, A., Sanadhya, D. and Chauhan, R. (2015) Heavy metal toxicity and tolerance in plants with special reference to cadmium: A review. *Journal of Plant Science Research* 31(1), 51.

Bortoloti, G.A. and Baron, D. (2022) Phytoremediation of toxic heavy metals by Brassica plants: A biochemical and physiological approach. *Environmental Advances* 8, 100204.

Bücker-Neto, L., Paiva, A.L.S., Machado, R.D., Arenhart, R.A. and Margis-Pinheiro, M. (2017) Interactions between plant hormones and heavy metals responses. *Genetics and Molecular Biology* 40(1 suppl 1), 373–386.

Chen, J. and Yang, Z.M. (2012) Mercury toxicity, molecular response and tolerance in higher plants. *Biometals* 25, 847–857.

Chen, Y.G., Huang, J.H., Luo, R., Ge, H.Z., Wołowicz, A. (2021) Impacts of heavy metals and medicinal crops on ecological systems, environmental pollution, cultivation, and production processes in China. *Ecotoxicology and Environmental Safety* 219, 112336.

Ci, D., Jiang, D., Wollenweber, B., Dai, T., Jing, Q. and Cao, W. (2010) Cadmium stress in wheat seedlings: growth, cadmium accumulation and photosynthesis. *Acta Physiologiae Plantarum* 32, 365–373.

Dai, Z.H., Guan, D.X., Bundschuh, J. and Ma, L.Q. (2023) Roles of phytohormones in mitigating abiotic stress in plants induced by metal (loid) s As, Cd, Cr, Hg, and Pb. *Critical Reviews in Environmental Science and Technology* 53(13), 1310–1330.

Du, Y., Wu, Q., Kong, D., Shi, Y., Huang, X. *et al.* (2020) Accumulation and translocation of heavy metals in water hyacinth: Maximising the use of green resources to remediate sites impacted by e-waste recycling activities. *Ecological Indicators* 115, 106384.

Dubey, S., Shri, M., Gupta, A., Rani, V. and Chakrabarty, D. (2018) Toxicity and detoxification of heavy metals during plant growth and metabolism. *Environmental Chemistry Letters* 16, 1169–1192.

El-Saadony, M.T., Desoky, E.S.M., Saad, A.M., Eid, R.S., Selem, E. and Elrys, A.S. (2021) Biological silicon nanoparticles improve Phaseolus vulgaris L. yield and minimize its contaminant contents on a heavy metals-contaminated saline soil. *Journal of Environmental Sciences* 106, 1–14.

Emamverdian, A., Ding, Y. and Xie, Y. (2020) The role of new members of phytohormones in plant amelioration under abiotic stress with an emphasis on heavy metals. *Polish Journal of Environmental Studies* 29(2), 1009–1020.

Farzadfar, S. and Zarinkamar, F. (2012) Morphological and anatomical responses of Matricaria chamomilla plants to cadmium and calcium. *Advances in Environmental Biology* 6, 1603–1609.

Fayssal, A.S., Kumar, P., Popescu, S.M., Sardar, H., Ahmad, R. *et al.* (2024) Health risk assessment of heavy metals in saffron (Crocus sativus L.) cultivated in domestic wastewater and lake water irrigated soils. *Heliyon* 10(5), e27138.

Galgali, P., Palimkar, S., Adhikari, A., Patel, R. and Routh, J. (2023) Remediation of potentially toxic elements-containing wastewaters using water hyacinth–a review. *International Journal of Phytoremediation* 25(2), 172–186.

Gangwar, S., Singh, V.P., Tripathi, D.K., Chauhan, D.K., Prasad, S.M. and Maurya, J.N. (2014) Plant responses to metal stress: the emerging role of plant growth hormones in toxicity alleviation. In: *Emerging Technologies and Management of Crop Stress Tolerance*. Academic Press, pp. 215–221. Elsevier Inc.

Gautam, A. and Dubey, R.S. (2018) Metal toxicity in plants: Induction of oxidative stress, antioxidative defense system, metabolic alterations and phytoremediation. In: Hemantaranjan, A. (ed.) *Molecular Physiology of Abiotic Stresses in Plant Productivity*. pp. 256–290. Scientific Publishers: India.

Gavrilescu, M. (2022) Enhancing phytoremediation of soils polluted with heavy metals. *Current Opinion in Biotechnology* 74, 21–31.

Genchi, G., Sinicropi, M.S., Lauria, G., Carocci, A. and Catalano, A. (2020) The effects of cadmium toxicity. *International Journal of Environmental Research and Public Health* 17(11), 3782.

Ghori, N.H., Ghori, T., Hayat, M.Q., Imadi, S.R., Gul, A., Altay, V. and Ozturk, M. (2019) Heavy metal stress and responses in plants. *International Journal of Environmental Science and Technology* 16, 1807–1828.

Gill, S.S., Khan, N.A., Anjum, N.A. and Tuteja, N. (2011) Amelioration of cadmium stress in crop plants by nutrients management: Morphological, physiological and biochemical aspects. *Plant Stress* 5(1), 1–23.

Gomes, M.P., Nogueira, M.D.O.G., Castro, E.M.D. and Soares, Â.M. (2011) Ecophysiological and anatomical changes due to uptake and accumulation of heavy metal in Brachiaria decumbens. *Scientia Agricola* 68, 566–573.

Goyal, D., Yadav, A., Prasad, M., Singh, T.B., Shrivastav, P. *et al.* (2020) Effect of heavy metals on plant growth: An overview. In: *Contaminants in Agriculture: Sources, Impacts and Management*. pp. 79–101. Delve Publishing: Canada.

Gul, A.S., Naveed, F., Ali, M., Ahmad, R. and Saqib, M. (2021) Effect of different wastewater irrigation regimes on growth of mulberry (Morus macroura Miq.). *Erwerbs-obstbau* 63(3), 331–337.

Hassan, M.U., Chattha, M.U., Khan, I., Chattha, M.B., Aamer, M. *et al.* (2019) Nickel toxicity in plants: Reasons, toxic effects, tolerance mechanisms, and remediation possibilities—a review. *Environmental Science and Pollution Research* 26, 12673–12688.

Hayat, S., Khalique, G., Irfan, M., Wani, A.S., Tripathi, B.N. and Ahmad, A. (2012) Physiological changes induced by chromium stress in plants: An overview. *Protoplasma* 249, 599–611.

Hodzic, E., Galijasevic, S., Balaban, M., Rekanovic, S., Makic, H., Kukavica, B. and Mihajlovic, D. (2021) The protective role of melatonin under heavy metal-induced stress in Melissa Officinalis L. *Turkish Journal of Chemistry* 45(3), 737–748.

Hoque, M.N., Tahjib-Ul-Arif, M., Hannan, A., Sultana, N., Akhter, S. *et al.* (2021) Melatonin modulates plant tolerance to heavy metal stress: morphological responses to molecular mechanisms. *International Journal of Molecular Sciences* 22(21), 11445.

Imadi, S.R., Waseem, S., Kazi, A.G., Azooz, M.M. and Ahmad, P. (2016) Aluminum toxicity in plants: an overview. In: *Plant Metal Interaction*, pp. 1–20. Springer Cham: Switzerland.

Ivanov, Y.V., Ivanova, A.I., Kartashov, A.V. and Kuznetsov, V.V. (2021) Phytotoxicity of short-term exposure to excess zinc or copper in Scots pine seedlings in relation to growth, water status, nutrient balance, and antioxidative activity. *Environmental Science and Pollution Research* 28, 14828–14843.

Jaishankar, M., Tseten, T., Anbalagan, N., Mathew, B.B. and Beeregowda, K.N. (2014) Toxicity, mechanism and health effects of some heavy metals. *Interdisciplinary Toxicology* 7(2), 60–72.

Jaleel, C.A., Jayakumar, K., Chang-Xing, Z. and Iqbal, M. (2009) Low concentration of cobalt increases growth, biochemical constituents, mineral status and yield in Zea mays. *Journal of Scientific Research* 1(1), 128–137.

Jan, A.T., Azam, M., Siddiqui, K., Ali, A., Choi, I. and Haq, Q.M.R. (2015) Heavy metals and human health: mechanistic insight into toxicity and counter defense system of antioxidants. *International Journal of Molecular Sciences* 16(12), 29592–29630.

Johri, N., Jacquillet, G. and Unwin, R. (2010) Heavy metal poisoning: The effects of cadmium on the kidney. *Biometals* 23, 783–792.

Júnior, C.A.L., de Sousa Barbosa, H., Galazzi, R.M., Koolen, H.H.F., Gozzo, F.C. and Arruda, M.A.Z. (2015) Evaluation of proteome alterations induced by cadmium stress in sunflower (*Helianthus annuus* L.) cultures. *Ecotoxicology and Environmental Safety* 119, 170–177.

Kanwal, U., Ali, S., Shakoor, M.B., Farid, M., Hussain, S. *et al.* (2014) EDTA ameliorates phytoextraction of lead and plant growth by reducing morphological and biochemical injuries in Brassica napus L. under lead stress. *Environmental Science and Pollution Research* 21, 9899–9910.

Karlsson, H.L., Cronholm, P., Hedberg, Y., Tornberg, M., De Battice, L., Svedhem, S. and Wallinder, I.O. (2013) Cell membrane damage and protein interaction induced by copper containing nanoparticles—Importance of the metal release process. *Toxicology* 313(1), 59–69.

Khaliq, M.A., Javed, M.T., Hussain, S., Imran, M., Mubeen, M. *et al.* (2022) Assessment of heavy metal accumulation and health risks in okra (*Abelmoschus Esculentus* L.) and spinach (*Spinacia Oleracea* L.) fertigated with wastwater. *International Journal of Food Contamination* 9(1), 11.

Khan, M.R. and Khan, M.M. (2010) Effect of varying concentration of nickel and cobalt on the plant growth and yield of chickpea. *Australian Journal of Basic and Applied Sciences* 4(6), 1036–1046.

Kour, J., Khanna, K., Bakshi, P., Sharma, I., Ohri, P. (2020) Role of Beneficial Microbes in the Molecular Phytotoxicity of Heavy Metals. In: *Cellular and Molecular Phytotoxicity of Heavy Metals*, pp. 227–262. Cham: Springer International Publishing.

Kumar, V., Suryakant, P., Kumar, S. and Kumar, N. (2016) Effect of chromium toxicity on plants: A review. *Agriways* 4(1), 107–120. Springer Nature: Singapore.

Kumar, R., Altaf, M.A., Lal, M.K. and Tiwari, R.K. (eds) (2023) *Melatonin in Plants: A Regulator for Plant Growth and Development*. Springer Nature: Singapore.

Lacalle, R.G., Bernal, M.P., Álvarez-Robles, M.J. and Clemente, R. (2023) Phytostabilization of soils contaminated with As, Cd, Cu, Pb and Zn: Physicochemical, toxicological and biological evaluations. *Soil and Environmental Health* 1(2), p.100014.

Li, Q., Cai, S., Mo, C., Chu, B., Peng, L. and Yang, F. (2010) Toxic effects of heavy metals and their accumulation in vegetables grown in a saline soil. *Ecotoxicology and Environmental Safety* 73(1), 84–88.

Liang, M., Lu, L., He, H., Li, J., Zhu, Z. and Zhu, Y. (2021) Applications of biochar and modified biochar in heavy metal contaminated soil: A descriptive review. *Sustainability* 13(24), p.14041.
Lidiková, J., Čeryová, N., Šnirc, M., Musilová, J., Harangozo, Ĺ. *et al.* (2021) Heavy metals presence in the soil and their content in selected varieties of chili peppers in Slovakia. *Foods* 10(8), 1738.
Liu, H., Teng, Y., Zheng, N., Liu, L., Yue, W., Zhai, Y. and Yang, J. (2022) Field experiments of phyto-stabilization, biochar-stabilization, and their coupled stabilization of soil heavy metal contamination around a copper mine tailing site, *Inner Mongolia. Minerals* 12(6), 702.
Madhu, P.M. and Sadagopan, R.S. (2020) Effect of heavy metals on growth and development of cultivated plants with reference to cadmium, chromium and lead – a review. *Journal of Stress Physiology and Biochemistry* 16(3), 84–102.
Mahjoubi, Y., Rzigui, T., Kharbech, O., Mohamed, S.N., Abaza, L. *et al.* (2021) Exogenous nitric oxide alleviates manganese toxicity in bean plants by modulating photosynthesis in relation to leaf lipid composition. *Protoplasma*, 259, 949–964.
Majeed, A. (2021) Heavy metal pollution in agricultural soils consequences and bioremediation approaches. In: *Bioremediation Science*. pp. 227–239. CRC Press: Florida, USA.
Malar, S., Shivendra Vikram, S., Favas, JCP. and Perumal, V. (2016) Lead heavy metal toxicity induced changes on growth and antioxidative enzymes level in water hyacinths [Eichhornia crassipes (Mart.)]. *Botanical Studies* 55, 1–11.
Maleki, M., Ghorbanpour, M. and Kariman, K. (2017) Physiological and antioxidative responses of medicinal plants exposed to heavy metals stress. *Plant Gene* 11, 247–254.
Małkowski, E., Sitko, K., Zieleźnik-Rusinowska, P., Gieroń, Ż. and Szopiński, M. (2019) Heavy metal toxicity: Physiological implications of metal toxicity in plants. In: *Plant Metallomics and Functional Omics: A System-Wide Perspective*, pp. 253–301. Springer: Berlin/Heidelberg, Germany.
Mulenga, C., Clarke, C. and Meincken, M. (2020) Physiological and growth responses to pollutant-induced biochemical changes in plants: A review.
Musilova, J., Bystrická, J., Vollmannová, A. and Melicháčová, S. (2011) Contamination of potato tubers by heavy metals and their influence on the formation of phenolic substances. *Journal of Central European Agriculture* 12(3), 0–0.
Nguyen, T.Q., Sesin, V., Kisiala, A. and Emery, R.N. (2021) Phytohormonal roles in plant responses to heavy metal stress: Implications for using macrophytes in phytoremediation of aquatic ecosystems. *Environmental Toxicology and Chemistry* 40(1), 7–22.
Nogueirol, R.C., Monteiro, F.A., Gratão, P.L., de Alcantara da Silva, B.K. and Azevedo, R.A. (2016) Cadmium application in tomato: Nutritional imbalance and oxidative stress. *Water, Air, and Soil Pollution* 227, 1–20.
Okereafor, U., Makhatha, M., Mekuto, L., Uche-Okereafor, N., Sebola, T. and Mavumengwana, V. (2020) Toxic metal implications on agricultural soils, plants, animals, aquatic life and human health. *International Journal of Environmental Research and Public Health* 17(7), 2204.
Pandey, A.K., Zorić, L., Sun, T., Karanović, D., Fang, P. *et al.* (2022) The anatomical basis of heavy metal responses in legumes and their impact on plant–rhizosphere interactions. *Plants* 11(19), 2554.
Panhwar, A., Faryal, K., Kandhro, A., Bhutto, S., Rashid, U. *et al.* (2022) Utilization of treated industrial wastewater and accumulation of heavy metals in soil and okra vegetable. *Environmental Challenges* 6, 100447.
Parmar, P., Dave, B., Sudhir, A., Panchal, K. and Subramanian, R.B. (2013) Physiological, biochemical and molecular response of plants against heavy metals stress. *International Journal of Current Research* 5(1), 80–89.
Patra, D.K., Pradhan, C. and Patra, H.K. (2020) Toxic metal decontamination by phytoremediation approach: Concept, challenges, opportunities and future perspectives. *Environmental Technology and Innovation* 18, 100672.
Peralta-Videa, J.R., Lopez, M.L., Narayan, M., Saupe, G. and Gardea-Torresdey, J. (2009) The biochemistry of environmental heavy metal uptake by plants: Implications for the food chain. *The International Journal of Biochemistry and Cell Biology* 41(8–9), 1665–1677.
Polle, A. and Schützendübel, A. (2003) Heavy metal signalling in plants: linking cellular and organismic responses. In: *Plant Responses to Abiotic Stress*. Springer Berlin Heidelberg, Berlin, Heidelberg, pp. 187–215.
Priya, A.K., Gnanasekaran, L., Dutta, K., Rajendran, S., Balakrishnan, D. and Soto-Moscoso, M. (2022) Biosorption of heavy metals by microorganisms: Evaluation of different underlying mechanisms. *Chemosphere* 307, 135957.

Ragonezi, C., Nunes, N., Oliveira, M.C.O., de Freitas, J.G., Ganança, J.F.T. and de Carvalho, M.Â.P. (2022) Sewage sludge fertilization—A case study of Sweet Potato yield and heavy metal accumulation. *Agronomy* 12(8), 1902.

Rahman, S.U., Li, Y., Hussain, S., Hussain, B., Riaz, L. *et al.* (2023) Role of phytohormones in heavy metal tolerance in plants: A review. *Ecological Indicators* 146, 109844.

Rai, G.K., Bhat, B.A., Mushtaq, M., Tariq, L., Rai, P.K. *et al.* (2021) Insights into decontamination of soils by phytoremediation: A detailed account on heavy metal toxicity and mitigation strategies. *Physiologia Plantarum* 173(1), 287–304.

Raj, D. and Maiti, S.K. (2019) Sources, toxicity, and remediation of mercury: An essence review. *Environmental Monitoring and Assessment* 191(9), 566.

Rizvi, A. and Khan, M.S. (2017) Biotoxic impact of heavy metals on growth, oxidative stress and morphological changes in root structure of wheat (*Triticum aestivum* L.) and stress alleviation by Pseudomonas aeruginosa strain CPSB1. *Chemosphere* 185, 942–952.

Saeed, A., Sohail, M. and Rashi, N. (2013) Effects of heavy metal toxicity on the biochemical response in tomato plants grown in contaminated silt-soil. *Bangladesh Journal of Scientific and Industrial Research* 48(4), 229–236.

Saini, S., Kaur, N. and Pati, P.K. (2021) Phytohormones: Key players in the modulation of heavy metal stress tolerance in plants. *Ecotoxicology and Environmental Safety* 223, 112578.

Sangwan, P., Kumar, V., Khatri, R.S. and Joshi, U.N. (2013) Chromium (VI) induced biochemical changes and gum content in cluster bean (*Cyamopsis tetragonoloba* L.) at different developmental stages. *Journal of Botany* 2013(1), 578627.

Shah, V. and Daverey, A. (2020) Phytoremediation: A multidisciplinary approach to clean up heavy metal contaminated soil. *Environmental Technology and Innovation* 18, 100774.

Shah, A.A., Altaf, M.A., Tiwari, R.K. and Lal, M.K. (2024) Melatonin interaction with other phytohormones in regulation of abiotic stresses in horticultural plants. *Scientia Horticulturae* 327, p.112869.

Shahid, M., Pourrut, B., Dumat, C., Nadeem, M., Aslam, M. and Pinelli, E. (2014) Heavy-metal-induced reactive oxygen species: Phytotoxicity and physicochemical changes in plants. *Reviews of Environmental Contamination and Toxicology Volume* 232, 1–44.

Shanker, A.K., Cervantes, C., Loza-Tavera, H. and Avudainayagam, S. (2005) Chromium toxicity in plants. *Environment International* 31(5), 739–753.

Sharma, J.K., Kumar, N., Singh, N.P. and Santal, A.R. (2023) Phytoremediation technologies and their mechanism for removal of heavy metal from contaminated soil: An approach for a sustainable environment. *Frontiers in Plant Science* 14, 1076876.

Silva, J., Nascimento, C., Biondi, C.M. and Cunha, K. (2012) Heavy metals in soils and plants in mango orchards in Petrolina, Pernambuco, Brazil. *Revista Brasileira de Ciência do solo* 36, 1343–1354.

Singh, S., Srivastava, P.K., Kumar, D., Tripathi, D.K., Chauhan, D.K. and Prasad, S.M. (2015) Morpho-anatomical and biochemical adapting strategies of maize (Zea mays L.) seedlings against lead and chromium stresses. *Biocatalysis and Agricultural Biotechnology* 4(3), 286–295.

Širić, I., Eid, E.M., El-Morsy, M.H., Osman, H.E., Adelodun, B. *et al.* (2022) Health risk assessment of hazardous heavy metals in two varieties of mango fruit (Mangifera indica L. var. Dasheri and Langra). *Horticulturae* 8(9), 832.

Sridhar, B.M., Diehl, S.V., Han, F.X., Monts, D.L. and Su, Y. (2005) Anatomical changes due to uptake and accumulation of Zn and Cd in Indian mustard (Brassica juncea). *Environmental and Experimental Botany* 54(2), 131–141.

Sytar, O., Kumar, A., Latowski, D., Kuczynska, P., Strzałka, K. and Prasad, M.N.V. (2013) Heavy metal-induced oxidative damage, defense reactions, and detoxification mechanisms in plants. *Acta Physiologiae Plantarum* 35, 985–999.

Sytar, O., Kumari, P., Yadav, S., Brestic, M. and Rastogi, A. (2019) Phytohormone priming: regulator for heavy metal stress in plants. *Journal of Plant Growth Regulation* 38, 739–752.

Tanwir, K., Akram, M.S., Masood, S., Chaudhary, H.J., Lindberg, S. and Javed, M.T. (2015) Cadmium-induced rhizospheric pH dynamics modulated nutrient acquisition and physiological attributes of maize (Zea mays L.). *Environmental Science and Pollution Research* 22, 9193–9203.

Tavanti, R.F.R., David Queiroz, G., Caroline Da Rocha Silva, A., Moya Peres, W., Pereira Paixão, A. *et al.* (2020) Changes in photosynthesis and antioxidant metabolism of cotton (*Gossypium hirsutum* L.) plants in response to manganese stress. *Archives of Agronomy and Soil Science* 66(6), 743–762.

Timalsina, H., Gyawali, T., Ghimire, S. and Paudel, S.R. (2022) Potential application of enhanced phytoremediation for heavy metals treatment in Nepal. *Chemosphere* 306, 135581.

Tiwari, R.K., Altaf, M.A., Kumar, R. and Lal, M.K. (2024) Evolution of melatonin as an oxidative stress mitigator in plant. In *Melatonin in Plants: A Pleiotropic Molecule for Abiotic Stresses and Pathogen Infection*. Springer Nature Singapore, Singapore, pp. 1–20.

Toppi, D.L.S. and Gabbrielli, R. (1999) Response to cadmium in higher plants. *Environmental and Experimental Botany* 41(2), 105–130.

Vardhan, K.H., Kumar, P.S. and Panda, R.C. (2019) A review on heavy metal pollution, toxicity and remedial measures: Current trends and future perspectives. *Journal of Molecular Liquids* 290, 111197.

Vassilev, A. and Yordanov, I. (1997) Reductive analysis of factors limiting growth of cadmium-treated plants: A review. *Bulgarian Journal of Plant Physiology* 23(3–4), 114–133.

Wei, L., Zhang, M., Wei, S., Zhang, J., Wang, C. and Liao, W. (2020) Roles of nitric oxide in heavy metal stress in plants: crosstalk with phytohormones and protein S-nitrosylation. *Environmental Pollution* 259, 113943.

Wyszkowska, J., Borowik, A., Kucharski, M. and Kucharski, J. (2013) Effect of cadmium, copper and zinc on plants, soil microorganisms and soil enzymes. *Journal of Elementology* 18(4), 769–796.

Yadav, V., Arif, N., Kováč, J., Singh, V.P., Tripathi, D.K., Chauhan, D.K. and Vaculík, M. (2021) Structural modifications of plant organs and tissues by metals and metalloids in the environment: A review. *Plant Physiology and Biochemistry* 159, 100–112.

Yang, Z., Yang, F., Liu, J.L., Wu, H.T., Yang, H. (2022) Heavy metal transporters: Functional mechanisms, regulation, and application in phytoremediation. *Science of the Total Environment*, 809, 151099.

Yap, C.K., Razali, A., Nulit, R., Peng, S.H.T., Yap, C.W., Okamura, H. and Ismail, M.S. (2021) Biomonitoring of heavy metals in the guava (Psidium guajava) for their health risk assessment in Kluang, Malaysia. *Food Science and Engineering*, 2, 13–20.

Youssef, M.A. and Abd El-Gawad, A.M. (2018) Accumulation and translocation of heavy metals in eggplant (*Solanum melongena* L.) grown in a contaminated soil. *Journal of Energy, Environmental and Chemical Engineering* 3(1), 9–18.

Yusuf, M., Fariduddin, Q., Hayat, S. and Ahmad, A. (2011) Nickel: An overview of uptake, essentiality and toxicity in plants. *Bulletin of Environmental Contamination and Toxicology* 86, 1–17.

Zhou, Z.S., Wang, S.J. and Yang, Z.M. (2008) Biological detection and analysis of mercury toxicity to alfalfa (Medicago sativa) plants. *Chemosphere* 70(8), 1500–1509.

10 Hormonal Regulation of Stress Responses

Kamran Akbar[1], Javed Iqbal[1], Banzeer Ahsan Abbasi[2], Tabassum Yaseen[1*], Salma Noureen[1], Muhammad Anas[3,4], Sadia Zafar[4] and Farishta Zarshan[1]

[1]Department of Botany, Bacha Khan University, Charsadda 24420, Khyber Pakhtunkhwa, Pakistan; [2]Department of Botany, Rawalpindi Women University, Rawalpindi 46300, Pakistan; [3]Department of Soil Science and Plant Nutrition, University of Selcuk Campus, Konya 42079, Türkiye; [4]Department of Plant Sciences, Faculty of Biological Sciences, Quaid-i-Azam University, Islamabad 45320, Pakistan; [5]Department of Botany, Division of Science and Technology, University of Education, Lahore, Pakistan

Abstract

Plant hormones regulate growth, development, and stress responses, which is essential for plant survival under abiotic stresses like drought, heat, and salinity. By understanding various hormone-mediated mechanisms, we can improve plant tolerance to environmental stresses such as drought and cold etc.

Plants face abiotic stresses and biotic stresses, which impact growth, nutrient uptake, and the microbiome. Phytohormones regulate plant processes at low concentrations, affecting growth and stress adaptation. Under stress, plants activate signaling pathways involving secondary messengers (including Ca^{2+} and reactive oxygen species (ROS)) that regulate stress-responsive genes. Abscisic acid (ABA) is central to abiotic stress, while jasmonic acid (JA), salicylic acid (SA), and ethylene (ET) mediate biotic stress. Hormonal interactions fine-tune stress responses in plants. ABA plays a crucial role in drought adaptation by enhancing ROS scavenging and activating antioxidant enzymes, regulating metabolic pathways for better water balance and promoting the accumulation of secondary metabolites. Cold stress causes oxidative damage, but plants adapt by synthesizing protective proteins and metabolites. Phytohormones, especially ABA, enhance cold-tolerance by regulating stress responses, maintaining membrane integrity, and activating antioxidant systems. Cold-sensitive crops suffer from oxidative stress, while cold-resistant varieties exhibit better antioxidant activity and osmoprotectants. Phytohormones, including ABA, brassinosteroids (BRs), and JA, regulate stress-tolerance mechanisms, improving cold resistance and cellular protection. Climate change exacerbates agricultural challenges, with temperature extremes and erratic rainfall. Understanding phytohormonal regulation can help improve crop resilience to climate-induced stresses, such as drought and cold. Research is needed to explore plant adaptations to these stresses.

Keywords: phytohormone, regulation, mechanism, adaptation, antioxidant

* Email: tabassumyaseen@bkuc.edu.pk

DOI: 10.1079/9781800628823.0010

10.1 Introduction

10.1.1 Importance of stress responses in plants

Signaling molecules, known as plant hormones, control critical facets of growth, development, and response to stressors from the external environment. Abiotic stressors, such as drought, salt, heat, cold, and floods, significantly impact plant development and survival. Adaptation and tolerance to such stressors require sophisticated signaling, sensing, and stress-response mechanisms. Therefore, we must explore the ways in which multiple plant hormones govern plants' responses to abiotic stress and discover hormonal interaction during abiotic stress signaling (Waadt *et al.*, 2022).

10.1.2 Types of stress in plants

Biodiversity and ecosystem services are seriously threatened by plants that encounter various biotic and abiotic stressors, including pollution and climate change. Although a large body of research has shown how plants may adapt to different stressors, little is known about how plants react to several stressors, and how this impacts growth and development. When plants are subjected to biotic and abiotic stressors, such as shortage of water and cadmium at the same time, their aboveground biomass is significantly reduced, their oxidative homeostasis is out of balance, their nutrient uptake is hindered, and their root growth is stunted. This illustrates how multiple stressors can work in concert to produce negative physiological effects. The effects of increased exposure to heavy metals and water shortage go beyond growth and development; they also affect the microenvironment's biodiversity, including the microbiome and nutrient profile of the rhizosphere. These results have important ramifications for how plants interact with stress and how ecosystems work, which calls for quick action to eradicate the effects of pollution and solve global environmental problems in order to encourage plants' long-term tolerance to a variety of stressors.

Here, we examine how plants can withstand different combinations of stresses, emphasizing the necessity of interdisciplinary research and cutting-edge techniques, such as the use of molecular tools and omics, to fully comprehend the underlying mechanisms of stress tolerance. Future studies on tolerance towards plant stress should be approached collaboratively, bringing together experts from other disciplines with a range of resources and experience, in order to hasten the development of tolerance to multiple environmental stressors in plants (Nawaz *et al.*, 2023).

Plant development and production can be restricted by biotic and abiotic environmental stresses, which can lead to significant losses in biodiversity and yield (Rillig *et al.*, 2019). When it comes to crop and other plant productivity, abiotic stress can be brought on by non-living elements including drought, high temperatures, heavy metals, nutrient shortages, and mineral toxicity, whereas biotic stress is brought out by diseases, insect pests, and competition with other plants (Ahmad *et al.*, 2020). Plant physiology and biochemical processes may be negatively impacted by these stresses, leading to disturbed plant development and growth processes (Shaar-Moshe *et al.*, 2017). Knowing how environmental threats (biotic and abiotic) impact plants is essential for sustainable ecosystem development with plants that are more resistant to stress combinations.

10.2 Phytohormones

Other than nutrients, plant growth regulators are organic compounds that, in trace amounts, promote, inhibit, or change a physiological function in plants. Plant hormones acts as signal molecules that are produced inside plants and exist in very small quantities. Plant hormones and growth regulators are chemicals that affect flowering deformation and death of the stem, leaves, roots, aging, and other plant components; suppression or promotion of stem elongation, etc. (Mundiyara *et al.*, 2020).

Plants naturally produce trace amounts of an organic substance called a phytohormone, commonly referred to as a plant growth regulator, which controls development and other processes at a position distant from the plant's point of production (Rastogi, 2011). Despite the fact that bacteria and fungi produce very similar substances that affect plants, plants are the ones that naturally produce them. These chemicals, called plant hormones, in tiny levels promote and affect the development and diversification of tissues and cells in addition to providing sustenance (Opik *et al.*, 2005).

Plant hormone synthesis is often widespread, rather than localized, within plant tissues. Hormones are made of simple substances that are easier for plants to absorb. They are produced in one part of a cell and affect other parts of the same and other cells. Hormone concentrations needed for plant responses must be extremely low (10^{-6} to 10^{-5} mol/l). Since the late 1970s, only scientists have been able to piece together the effects of plant hormones and their linkages to plant physiology, making research on these substances exceedingly difficult.

To transport hormones, the plant makes four distinct motions. The gradual movement of ions and molecules between different cells and cytoplasmic streaming inside cells are the mechanisms that enable localized mobility.

Vascular structures transport hormones to different plant parts. Sieve tubes or phloem deliver sugars to the roots and flowers from leaves, while water and other mineral solutes are carried from roots to the leaves through xylem. A plant growth regulator is another name for a natural or artificial organic molecule that translocates hormones to change or regulate one or more specific physiological processes in a plant (Mundiyara *et al.*, 2020).

The initial definition of hormone refers specifically to mammalian secretory substances that form in distinct organs, including glands, tissues and cells, while traversing venous structures or equivalent biological tissues before inducing metabolic changes in target areas (Kleine *et al.*, 2016).

Seven primary plant growth regulators comprise auxins in combination with gibberellins (GAs), cytokinins (CKs), abscisic acid (ABA), brassinosteroids (BRs), and chemicals linked to jasmonic acid (JA) and ethylene (ET). This list of recognized plant hormones excludes polyamines and phenolic chemicals. These days, the phrase is commonly used with a heading to display the living species and is used to refer to the mobile signals of living organisms, including hormones from plants, animals, insects etc. Hormones are present in all living organisms that are frequently species-specific and represent the characteristics of its life events.

Plant hormones provide explanatory power to understand developmental plasticity as a plant operates through constant creation of new organs throughout its existence. The locations that harbor plant hormones act as the primary sites for new growth development. There are fewer known plant hormones than mammalian hormones. All higher plants appear to contain plant hormones and many of them are also found in fungus, bacteria, and smaller plants. The universality of plant hormones is astounding.

Plant hormones have a variety of diverse actions, with one hormone frequently having several different impacts. The different effects that are seen when a plant hormone is given, however, do not always prove that the hormone is the direct cause of the impact in plants that have not been treated. A plant hormone's site of action may be fairly far from the final effect's location, and its effects might range from direct to indirect. Furthermore, there is proof that many plant hormones interact in quite intricate ways.

Despite lacking a nervous system, plants must be able to sense and utilize environmental cues to control their growth and organ development. Plant response systems use tiny chemical signals that share striking structural similarities with neurotransmitters, although lacking a biological design akin to that of a nervous system. The genome contains the fundamental blueprint for growth and development, which is altered by cues from the environment. Plant hormones are thought to function as mediators, and one of the obvious distinctions from mammalian hormones is the environmental adjustment that results in developmental plasticity, which is manifested in variations throughout the growth and development of plants. The production, catabolism, and transport of plant hormones are influenced by certain environmental cues, such as light, temperature, and moisture. Environmental factors also probably have an impact on sensitivity to plant hormones. In contrast to high light intensity, GA has a higher influence on shoot elongation at low light levels (Mander and Liu, 2010).

10.3 Major Classes of Plant Hormones

10.3.1 Auxin

The stem tip of plants produces auxin, a hormone that promotes cell elongation. The darker side of the plant receives auxin, which causes its cells to enlarge relative to their lighter counterparts. The consequence is a phenomenon known as phototropism, where the tip of the plant curls toward the light. The principal locations of the shoot tip, growing seeds, and buds are where auxin is produced. Auxin has a beneficial effect on root initiation, bud formation, and cell enlargement. Auxin maintains apical dominance by encouraging stem extension and preventing the formation of lateral buds (Mundiyara *et al.*, 2020).

10.3.2 Cytokinins

Plant growth chemicals called cytokinins (CKs) or phytohormones promote cytokinesis, or cell division, in the roots and shoots of plants. In addition to their main function in cell proliferation and differentiation, they influence leaf withering, axillary bud formation, and apical dominance. CKs were initially extracted from yeast cells and were called kinins. In plants, they delay tissue depreciation, manage circulatory auxin amongst plant parts, and influence stem growth and leaf expansion. Together they have very beneficial synergistic effect, and their ratio has a practical effect on growth stages of plant.

Together with ET, CKs promote the formation of lateral buds, which counteracts the effect of auxin-induced apical dominance and encourages the fruit, flower, and leaf abscission (Sipes and Einset, 1983). In plants, the equation A/C = constant, where A is auxins and C is CKS, represents the relationship between them. The two types of CKs are adenine-type (kinetin, zeatin, and 6-benzylaminopurine) and phenylurea-type (diphenylurea and thidiazuron) (Aina *et al.*, 2012).

10.3.3 Gibberellins

Plant hormones called gibberellins (GAs) regulate growth and other developmental processes, such as sex expression, enzyme activation, blooming, stem elongation, germination, dormancy, and senescence of leaves and fruits. On seedless grape cultivars, GAs are used frequently to improve the fruit's size and quality. A pre-bloom spray of 20 parts per million GA lengthens the rachis of the fruit cluster. Consequently, the clusters become less susceptible to illness and looser during the growing season. Additionally, GAs also enhance barley malt production and reduce time consumption for this process. Applying GAs to germinating barley accelerates the synthesis and release of hydrolytic enzymes, including amylases, and raises the endogenous concentration of the hormone (Mander and Liu, 2010).

A crucial regulator of plant development is ABA. The term "abscisic acid" refers to the high concentration of this substance in newly fallen or abscised leaves. In drought-stressed plants, ABA is essential in making their stomata close. When there is a water deficit, a signal reaches the stem or leaves, where it causes the synthesis of ABA precursors, which then send the signal to the roots. This is how water stress affects plants. ABA is subsequently released from the roots and makes its way to the plant's leaf sections via xylem (Ren *et al.*, 2007). This regulates how much potassium and sodium the guard cells take in, which makes them less turgid and causes their stomata to close (Yan *et al.*, 2007). Other plant growth regulators that work under stress include ET, BRs, SA, plant peptide hormones, polyamines, nitric oxide (NO), strigolactones, karrikins, florigen, anthocyanins, vernalin, and morphactins.

10.4 Regulation of Stress Reactions by Plant Hormones

Recognizing biotic and abiotic stressors in plants sets off a series of complex signaling cascades. Many of these early signaling events include generation of secondary signaling molecules such as reactive oxygen species (ROS) and inositol phosphate, kinase cascade activation, and variations in intracellular Ca^{2+} concentration. As Ca^{2+} sensors, calcium-binding proteins sense an increase in intracellular Ca^{2+} levels brought on by unfavorable environmental factors (Kudla *et al.*, 2010). The Ca^{2+} sensors that have been triggered can either bind to cis regions in the promoter of the key stress-responsive genes or interact with DNA-binding proteins that regulate these genes to either activate or inhibit them. Additionally, elevated Ca^{2+} levels can activate calcium/calmodulin-dependent protein kinases (CCaMKs), calcium-dependent protein kinases (CDPKs), or phosphatases, which can phosphorylate or dephosphorylate specific transcription factors (TFs) and control the amount of genes that respond to stress (Reddy *et al.*, 2011).

Ca^{2+} ions collaborate with ROS and second messengers. In response to stressful events they generate a dramatic increase in the formation of ROS, termed the "oxidative burst" (Miller *et al.*, 2008). Some of the ROS molecules that mediate signaling functions are hydrogen peroxide (H_2O_2), superoxide anion (O_2^-) radicals, hydroxyl (OH) radicals, and singlet oxygen ($^1O_2$) (Gadjev *et al.*, 2006). A great example of ROS-mediated stress response is the way H_2O_2 triggers the mitogen-activated protein kinase (MAPK) cascade and causes specific stress-related genes in *Arabidopsis* to be overexpressed (Kovtun *et al.*, 2000).

But the reactions of plants to environmental constraints mediated by ROS and Ca^{2+} are just the start. In reaction to outside stimuli, the intricate stress-response system of plants requires the activation of numerous inter-related pathways. The main endogenous mediators in modulating the plant stress response are plant hormones, which have complex relationships with one another and can influence a wide range of physiological functions. To further facilitate the creation of a strong and effective stress response, the signaling pathways of the nine distinct groups of plant hormones involved in defensive responses are intimately intertwined. As a result, the study gave a succinct summary of how plant hormones help control different stress responses by concentrating on how they interact.

The main hormones that plants produce include strigolactones, GAs, CKs, auxins, ET, SA, GAs, ABA, JA, and BRs. It is well established that JA, ABA, SA and ET influence the plant defense response to abiotic stressors and some pathogens (Nakashima and Yamaguchi-Shinozaki, 2013). Plants' defense against abiotic stressors is often under the control of ABA because these environmental conditions are known to increase ABA levels (Lata and Prasad, 2011). Phytohormones e.g., SA, JA, and ET are crucial in the body's reaction to biotic stressors since their levels increase when diseases are present (Bari, 2009; Jones, 2009). However, there are other components to the stress-response process besides these hormones. According to recent studies, auxins, GAs, and CKs interact with ABA, SA, JA, and ET to control the defensive response in plants (Navarro *et al.*, 2008).

10.5 Plant Hormones' Function in Regulating Immune Responses

Climate change is especially dangerous for agriculture. The creation of new climate-smart agricultural types is essential with enhanced yield and environmental resilience in order to address the challenges presented by global population expansion, climate-related stresses on agricultural output, and changes in dietary preferences. Genomic studies and molecular genetics have demonstrated that alleles in genes, linked to growth regulation mediated by phytohormones, have significantly increased production in key crops.

We now have a much better grasp of the molecular underpinnings of the various phytohormone-mediated processes in plants thanks to plant research conducted by Hirayama and Mochida (2022). The results offer crucial information for raising agricultural output in light of climate change. It was addressed how plant hormonomics, or multiple phytohormone profiling, can be used to improve crops, and new developments in this field were emphasized. Physiological phenotyping was suggested as being a crucial method through plant hormonomics, which is a focused approach on representative plant growth regulators related to increases in agricultural output. Particularly, it surveyed up-to-date methods in plant hormonomics, namely mass spectrometry, and plant hormone mass profiling methods with nanosensors (Hirayama and Mochida, 2022).

In major part, phytohormones control how plants grow and react to their surroundings. The main hormones in plants that control plant immunological responses are ET, JA, and SA. Auxins, GAs, CKs, ABA, and BRs are among the plant hormones that have recently been identified as significant modulators of plant immunity. All these hormones have been previously demonstrated to regulate plant growth and development. Plant hormones function in intricate networks to regulate how the body reacts to environmental and developmental stimuli, lowering the fitness costs related to defense. The knowledge about the molecular mechanisms governing these hormonal networks is limited. Additionally, viruses target hormone signaling networks to upset and evade plant defensive mechanisms. Auxin, ABA, and SA's regulatory roles in plant resistance, along with the complex connections between their signal transduction pathways were studied by Denancé *et al.* (2013).

Numerous attackers (including bacteria, fungus, viruses, oomycetes, and insects) constantly put plants under biotic stress in their natural environments, endangering both the plant and its offspring. Given that most species rely on green plants as their main source of energy, it is no wonder that plants have developed different mechanisms of resistance that may develop naturally or be brought on by illness or insect infestation (Glazebrook, 2005; Panstruga *et al.*, 2009).

To detect pests and diseases and to initiate defensive mechanisms, plants have developed molecular systems. The specialized recognition by plant protein recognition receptors (PRRs) is the intrinsic immune system of plants, the known plant PPRs called pathogen-associated molecular pattern (PAMPs), of relatively conserved pathogen chemicals. This reaction to resistance is known as PAMP-triggered immunity. Effector proteins released by successful pathogens deregulate pattern-triggered immunity (PTI). Plant resistance (R) proteins fight this by identifying effectors and triggering effector-triggered immunity (ETI) (Dodds and Rathjen, 2010).

Kempel *et al.* (2011), along with Walters and Heil (2007), state that metabolites utilized in plant resistance should ideally be detrimental to other physiological processes that have an adverse effect on other plant characteristics such as biomass or seed production and that is why the immune responses should be fine-tuned. The way in which a dynamic and complex network structure has formed is due to these physiological constraints, together with other factors, such as plants co-existing with their natural predators.

The combination of defensive layers made from various physiological processes and cellular components reduces the intrinsic fitness cost of having a strong defense (Chisholm *et al.*, 2006; Panstruga *et al.*, 2009; Schulze-Lefert and Panstruga, 2011). According to Pieterse *et al.* (2009), Santner *et al.* (2009), and Jaillais and Chory (2010), phytohormones are small molecules that can function in a complicated network either constructively or antagonistically to control the resistance response. According to Peng *et al.* (2009), a database with the names of hormone-regulated genes and phenotypic descriptions of mutants linked to hormones has been created due to the significant expansion of our understanding of the intricate hormone network that controls plant immunity. According to Jones and Dangl (2006), along with Dodds and Rathjen (2010), there has been an evolution of sophisticated molecular techniques developing in pathogens that disrupt the signaling mechanisms of hormones and/or deregulate hormone biosynthesis, which facilitates their ability to evade plant defenses. SA and signaling pathways mediated by ET and JA are widely acknowledged to be essential for pathogen resistance (Robert Seilaniantz *et al.*, 2011). According to Glazebrook (2005) and Bari and Jones (2009), SA signaling increases plant resistance towards biotrophic pathogens, for which living tissue is necessary for completion of their life cycle. However, resistance to necrotrophic diseases pathogens that cause plant tissue death after infection and on the presence of herbivorous pests, usually requires ET/JA pathways. Plant resistance to certain necrotrophic infections requires the SA route, yet resistance to certain biotrophic pathogens was found to require the ET/JA system. According to Robert-Seilaniantz *et al.* (2011) and Berrocal-Lobo *et al.* (2002), there are several exceptions to this principle. Other hormones that have recently been identified as key players in plant–pathogen interactions are ABA and auxins. They were initially discovered due to their function in regulating the development processes of plants and their reaction to abiotic stresses (Mauch-Mani and Mauch, 2005; Kazan and Manners, 2009; Ton *et al.*, 2009; Fu and Wang, 2011). The phytohormone routes are all connected by a huge, complex, and as-yet-unidentified network. Auxin, GAs, CKs, ET, ABA, and other hormones are believed to have an impact on the SA–JA signaling backbone (Pieterse *et al.*, 2012).

Developing hormone-based breeding strategies that seek to boost crop resistance to infections requires an understanding of the complex regulation of hormone homeostasis during plant–pathogen interactions and how illnesses modify this regulation. Changing a plant's hormone system can be detrimental to its growth and defense against an alternative pathogen with a different lifestyle, but it can also boost resistance to a particular disease (Holeski *et al.*, 2012).

One of the main obstacles to the productivity of rain-fed agriculture, particularly in the context of climate change, is drought. Many adaptive methods have evolved in plants to recognize stressors and react to them through particular processes. ABA is essential for the growth and development of plants as it acts as a signal that it is essential for plants to react to dryness. ABA triggers a variety of physiological processes, including stomatal closure, root system modulation, soil microbial community structure, transcriptional and post-transcriptional gene expression activation, and metabolic alterations. The processes underlying drought responses mediated by ABA in plants must thus be understood in order to guarantee agricultural production and global food security. Scientists studied how plants adapt their drought response mechanisms by changing the reception of ABA, along with ABA- and drought-related gene expression and metabolic route operations from cellular to whole-plant levels. Knowing how plants interact with drought in the presence of ABA, can help to create crops by using integrated advanced biotechnology approaches. It is crucial to the development of stress biology research that this review clarifies the genetic, biochemical, and molecular processes underlying drought responses mediated by ABA in plants (Aslam *et al.*, 2022).

Because they can't move, plants have to be able to respond quickly to a number of stresses. Both biotic (infections, insects, and herbivores) and abiotic (dryness, salinity, cold, and high temperatures) factors can lead to stress (Lamers *et al.*, 2020). In response to these conditions, plants developed a number of adaptation strategies. Stomatal closure is a typical example to minimize water loss and stop pathogens from entering the leaves (Sussmilch and McAdam, 2017; Agurla *et al.*, 2018). ABA, methyl jasmonate (MJ) and SA are secreted by plants as a response of abiotic stress, and elicitors such as flagellin 22 in response to pathogen attack. Given its involvement in numerous abiotic and biotic stress scenarios, ABA is thought to be an essential and versatile hormone. On the other side, ET, SA, and MJ help in resistance to biotic stress. ABA accumulation in leaves results in stomatal closure to minimize water loss and up-regulation of genes, which supports osmotic adaption when leaves are subjected to drought, salt or cold stress (Lim *et al.*, 2015; Niu *et al.*, 2018). Cross-adaptation against drought and diseases is mediated by plants' elevated ABA levels in addition to insect herbivores (Lee and Luan, 2012; Nguyen *et al.*, 2016).

In order to survive either short-term or long-term responses to drought stress, A number of morpho-physiological, metabolic, and molecular methods have evolved in plants. Rubin *et al.* (2017) assert that phytohormones play a critical role in controlling plant development and mediating environmental stressors such as drought, which negatively impact agricultural output and jeopardize global food security. The most effective breeding and sustainable crop production technique may be phytohormone engineering, and drought stress should be managed with robust and innovative strategies.

The last ten years have seen a rise in interest in comprehending how ABA varies over time to modify plant responses (Müller, 2021). Plants can adapt to environmental stressors such as drought by reorganizing different biochemical and physiological signal transduction pathways when exposed to ABA, which is vital for plant growth (Wani and Kumar, 2015). ABA is also required for senescence, stomatal closure, seed germination, biomolecule synthesis, and root architecture alteration. ABA is an isoprenoid produced in plastids by a mixed ester originating from the methylerythritol phosphate (MEP) pathway from carotenoids (C40), a byproduct of isopentenyl diphosphate (IPP) (Vishwakarma *et al.*, 2017).

Each step of the ABA synthesis process is catalysed by an enzyme. The first step involves the conversion of zeaxanthin to all trans-violaxanthin in the chloroplast production of ABA. Zeaxanthin epoxidase (ZEP) catalyzes the cyclic hydroxylation of epoxycarotenoid to all-xanthin by means of their intermediate antheraxanthin. An unidentified enzymatic activity causes all trans-violaxanthin to undergo cis-isomerization in the subsequent stage, resulting in either violaxanthin or cisneoxanthin. Following the separation of the violaxanthin and neoxanthin cis-isomers by 9-cis-epoxycarotenoid dioxygenase (NCED) enzymes, xanthoxin, a C15 intermediate product, is ultimately transported to the cytoplasm. In turn, short-chain alcohol dehydrogenase/reductase (SDR) catalyzes the initial conversion of xanthoxin into an abscisic aldehyde. Aldehyde oxidase (AO) is the final enzyme that converts abscisic aldehyde into ABA.

Variations in the expression of genes and the activation of ABA production enzymes that correlate to the messenger RNA (mRNA) level are the reasons why plants under drought stress have noticeably increased ABA levels. The TFs ABF, MYC, MYB, NAC, ERF, bZIP, and DREB/CBF have raised the transcript levels of numerous ABA biosynthesis genes, including ZEP/ABA1, AAO3, 9-cis-epoxycarotenoid dioxygenase (NCED3), and molybdenum cofactor sulfurase (MCSU/LOS5/ABA3), by means of both ABA-dependent and ABA-independent pathways. For regulating growth, development, and reaction of plants to different environmental stressors ABA is very essential (Zhu, 2002) and is said to have a major role in mediating drought (Boominathan *et al.*, 2004). Water loss through transpiration is prevented during drought by ABA-mediated stomatal conductance (Wani and Kumar, 2015). ABA trafficking has been shown to include the detoxifying efflux carrier (AtDTX50) and the multidrug and toxic compound extrusion (MATE) transporter family (Saavedra *et al.*, 2010).

According to Zhang *et al.* (2008), the ABA responsive-element binding protein (ABP9), which belongs to the bZIP family, also boosts photosynthetic ability during drought. According to Sridha and Wu (2006), histone acetylation is crucial for ABA-mediated gene regulation, which helps plants adapt to dry conditions. At the transcriptional and proteomic levels, it was found that for some plant species, including *Arabidopsis*, rice, and maize, ABA-mediated drought control depends on the MAPK signaling cascade (Hamel *et al.*, 2012). When everything is said and done, ABA is an essential hormone that regulates how plants respond to drought through complex molecular communication networks. Through breeding programs, studies on ABA regulators may therefore contribute to the development of drought-tolerant crops. Through breeding programs, studies on ABA regulators may therefore aid in the creation of crops that can withstand drought. Due to the extensive investigation of ABA's actions using molecular and genetic approaches, crops that can withstand drought have been grown. Emphasizing the significance of ABA in a diverse array of biochemical and physiological processes, as well as in drought avoidance and adaptation, was the goal of the study. A summary of the molecular underpinnings of ABA's actions and its roles in regulating drought stress responses in conjunction with other hormones was also given (Aslam *et al.*, 2022).

10.6 The Role of ABA in Drought Resistance

Drought adaptation requires the presence of ABA. When drought stress occurs, ABA-mediated stomatal closure slows transpiration, reducing water loss. Additionally, by promoting the elongation of root cells and progressively raising hydraulic conductivity, ABA aids plants' recovery from water-limited environments. Recent developments in plant genomics have sped up the process of identifying and characterizing drought-responsive ABA-dependent candidate genes. In contrast to wild type (WT) plants, mutants of the MATE transporter gene AtDTX50, which is associated with ABA efflux, has reduced stomatal conductance and enhanced resistance to drought (Zhang *et al.*, 2006). It has well been known that in relation to the previously identified protein phosphatases 2C (PP2Cs) and sucrose non-fermenting related kinases 2 (SnRK2s), the initial step of the core ABA signaling pathway is the binding of ABA to the pyrabactin resistance 1/pyrabactin resistance like/regulatory component of ABA (PYR/PYL/RCAR) receptors (Muchhal and Raghothama, 1999). Several species, including rice, tomatoes, and *Arabidopsis*, have been shown to benefit from the PYR/PYL/RCAR proteins' ability to withstand drought. Additionally, by upregulating the expression of CDPK6 in response to drought stress, it has been demonstrated that ABA controls the signaling of CDPKs.

CDPKs interact with a number of fundamental TFs related to ABA, such as ABA-responsive element-binding factors (ABFs/AREBs), to enhance their transcriptional activity. Additionally resistant to environmental stressors such as drought are transgenic plants that overexpress ZEP (Park *et al.*, 2008).

Rice's resistance to drought was improved by OsbZIP72 overexpression, which also increased the expression of the ABA-responsive late embryogenesis abundant (LEA) genes, which might be helpful for creating crops that can withstand drought. *Arabidopsis* plants with the ABCG25 gene overexpressed during drought, exhibited lowered water loss and reduced evapotranspiration. The importance of ABA-related genes in controlling ABA responses to drought by the increased drought sensitivity displayed by AtABCG40 mutants was also emphasized (Kang *et al.*, 2010).

The ABA hormone has been identified as the primary regulator of water shortage in plants. ABA is essential for regulating the production of genes, proteins, and enzymatic processes that help plant cells withstand dehydration, as numerous studies have shown. For instance, ABA levels in *Arabidopsis*, wheat, rice, tomato, soybean, maize, and sesame rose significantly under drought in a study by Zhang *et al.* (2006). These findings demonstrate that for plants to react to drought stress in a suitable manner, ABA-mediated drought resistance is required (Fig. 10.1).

10.7 Drought Responses Mediated by ABA via Physio-biochemical Change

Special mechanisms have developed in plants for adaptation to resist drought stress and mitigate its negative effects. Plant stress response is regulated by signal molecules such as ROS. According to Sreenivasulu *et al.* (2012), plants respond to drought by producing a range of secondary metabolites that help them survive. ABA has the ability to synchronize many processes in plants to help them withstand drought stress. Water limits should therefore be addressed while maintaining plant growth and productivity through the development and engineering of novel and dynamic strategies like ROS and secondary metabolites.

10.7.1 System for scavenging ROS

The complex interplay between ABA and dryness in plants leads to several downstream

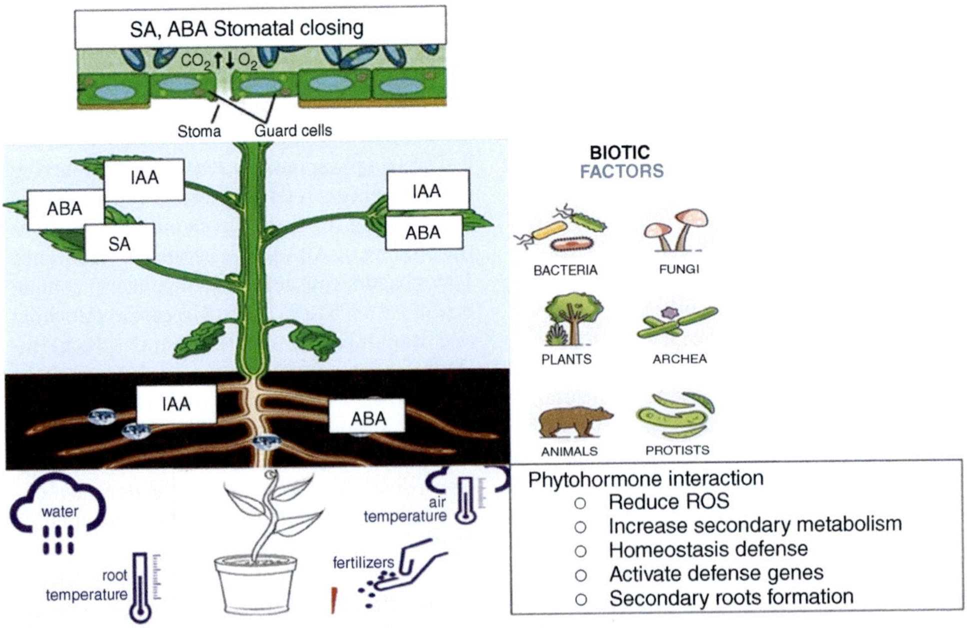

Fig. 10.1. Role of phytohormones in plant responses to biotic and abiotic stresses. (Figure author's own.)

mechanisms that facilitate drought adaptation in an ABA-dependent fashion. The connection between ABA signaling and metabolic status was proposed as a result of the theory that dryness may change the cellular redox and metabolic condition of plants, which in turn affects the cellular susceptibility to ABA accumulation (Aslam *et al.*, 2022). Endogenous ABA quickly rises in response to a water deficiency in the soil, encouraging stomatal closure in plants. Additionally, prior research has shown that ABA is necessary for drought escape brought on by water stress. The antioxidant enzymes superoxide dismutase (SOD), peroxidase (POD), catalase (CAT), ascorbate peroxidase (APX), and glutathione reductase (GR) are triggered by ABA, for example, and may help drought-stricken wheat seedlings scavenge ROS. This would cause the aquaporin to be expressed, controlling osmotic adjustment, decreasing oxidative damage, and increasing root conductivity.

By opening Ca^{2+} channels in the plasma membrane, ABA has been shown to cause stomata to close. This, in turn, causes stomata guard cells to produce H_2O_2 via a membrane-bound NADPH oxidase. According to Aharoni and Galili (2011), the plants in the first metabolism land create a range of primary metabolites that are essential to their survival and have higher nutritional and medicinal value. Primary metabolites are typically involved in the synthesis of polyamines, phosphatidylcholine, redox regulation, protein–disulfide linkage, mRNA capping, and methylation activities, including DNA methylation. Dryness tolerance is believed to be supported by primary metabolites and the metabolic genes associated with them, which are involved in numerous metabolic pathways. An estimated 200,000 metabolites are currently recognized in plants. ABA mostly produces metabolites that contain sulfur, carbohydrates, and nucleosides/nucleotides (Zhu and Assmann, 2017). The tricarboxylic acid (TCA) cycle, pentose phosphate route, shikimate pathway, glycolysis, and the synthesis of secondary metabolites by aliphatic and aromatic amino acids are the primary pathways that result in primary metabolites.

Water availability fluctuations are closely linked to ABA, a signaling molecule that helps plants adapt to drought stress and regulate plant growth (Yoshida *et al.*, 2019). *Camelina sativa* and *A. thaliana* drought tolerance was enhanced by ABA-inducible Wax synthase/acyl-CoA: diacylglycerol acyltransferase (WSD1) by loading wax onto leaves and stems and accumulating wax on the epidermis. Due to its vulnerability to drought, the canola crop experiences significant losses in yield. It would be possible to create crops with more drought resistance, nevertheless, if it were known what genes underline ABA-mediated drought tolerance.

Recent omics approaches have revealed that many proteins in metabolism, photosynthesis, protein synthesis, membrane transport activities, protein folding/transport and degradation, and stress/defense response have an ABA-induced or ABA-repressed condition. In order to learn more about the function of ABA signaling in drought tolerance, a variety of repressed and ABA-induced metabolites have been employed as indicators.

Environmental stresses can affect plants in many different ways, and their reactions are all comparable. There has been an analysis of plant metabolic profiles by researchers to anticipate their responses to drought. Osmotic adjustment and the build-up of certain secondary metabolites are two different ways that plants have adapted to lessen their reactions to drought. During stress reactions, the plant stores metabolites such as proline, glycine betaine, phenolic compounds, soluble sugars, and any other appropriate solutes. By regulating water potential, cell turgor, osmotic adjustment, stabilizing proteins and membrane lipid bilayer structures during drought, and guaranteeing life, these metabolites support proper physiological functioning. Conversely, in drought-stressed plants, secondary metabolites reduce oxidative stress by scavenging free radicals.

According to Urano *et al.* (2009), metabolic profiling in response to drought identified ABA-inducible metabolic networks that promote the accumulation of important dehydration-inducible genes, including lysine ketoglutarate reductase/saccharopine de hydrogenase (AtLKR/SDH), delta 1-pyrroline-5-carboxylase (P5CS), arginine decarboxylase, branch-chain aminotransferase (AtBCAT2Tryptophan), glutamine, alanine, proline, aspartate, leucine, isoleucine, ornithine, valine, and citric acid cycle precursors, including cis-aconitate, succinate, and 2-oxoglutarate, as well as flavonoids, like cyanidin and quercetin, and lipids, like acylated sterylglycosides and glycosyl inositol phosphoceramides, were all more prevalent in *Arabidopsis* and a few crop plants, like maize, barley, and rice, when drought was experienced. ZmPIS, a phosphatidylinositol synthase, effectively increased drought tolerance in maize by changing the lipid content of the membrane and ABA production (Liu *et al.*, 2013).

Glycine max's overexpression of ABF3 caused notable changes in primary metabolites and secondary metabolites, including glycerophospholipids, glycolipids, fatty acyls, prenollipids, and their derivatives in the root system during a drought. According to Hong *et al.* (2013), in order to help the plant develop and adapt, its root system clings to the soil and absorbs water and minerals from the surroundings. ABA is primarily used to modify the growth restriction, development pattern, and structure of roots. It controls the growth of lateral roots and is closely linked to disturbances in the environment. ABA-deficient mutants (aba2-1, aba3-1) have a greater incidence of lateral root emergence in genetic testing. Additionally, it was shown that the lateral roots must be inhibited for peanuts to produce ABA endogenously. One interesting finding is that plants that are ABA-insensitive (abi4) have more lateral roots. Thus, the development of lateral roots is influenced by ABA signaling, in addition to ABA levels. Conversely, water stress regulates the elongation and cellular integrity of roots by monitoring ROS emissions, which indicate the expression of the respiratory burst oxidase homolog (RBOH) gene. But according to genetic studies conducted on *Medicago truncatula* mutants, ABA is essential for maintaining the root meristem (Liang *et al.*, 2007).

Research has shown that ABA-deficient mutants require ABA to maintain root development during drought. Auxin-dependent pathways may have a role in the ABA-mediated root growth of tomatoes in soil drying conditions, according to Zhang *et al.* (2006). While tomato mutant *notabilis* (not) in drought uses exogenous IAA to lengthen primary roots, the main roots of maize ABA-deficient mutants vp5 and vp14 are stunted. ABA offers a modest way to modify the root system to cope with drought stress, but its effects on root architecture can vary depending on environmental variables and the genotypic background. Root exudates are crucial components that support responses to environmental disturbances, accounting for 11% to 40% of photosynthetically derived carbon (Williams and de Vries, 2020). In addition, they can be utilized as chemoattractants or as a source of food. The exudates of wild oats, grassland species, *A. thaliana* and *Brassica napus* are rich in organic acids, carbohydrates, fatty acids, amino acids, and secondary metabolites but the kinds and rates of exudation depend on species (Rudrappa *et al.*, 2008). Plant-secreted compounds contribute to the formation of diverse plant–microbial interactions, hence enhancing a plant's ability to endure environmental stressors (Naylor and Coleman-Derr, 2018). ABA mediates dryness by enhanced osmolyte synthesis, which includes proline, organic acids, and protective proteins. Increased carbon remobilization was shown to be positively correlated with both ABA concentration and sucrose phosphate synthase activity by Yang *et al.* (2002). As expected, during water stress, increased endogenous root ABA concentrations could improve root exudates. During drought stress, plant metabolites, particularly ABA, were significantly elevated in root exudation, according to Gargallo-Garriga *et al.* (2018). According to these findings, carbon remobilization and root exudation may involve ABA to lower plant stress tolerance. Settling in the root underground along with symbolizing the dark portion of the soil, the underground environment also shows variations in a number of abiotic stresses, such as temperature, water content, and mineral availability. One of the most difficult issues in ecology is predicting the responses of different species to environmental upsets. The root-hormonal state, root growth, and rhizobacteria count are all impacted by ABA. For instance, rhizobacteria are able to produce, break down, and get their nourishment from phytohormones or their precursors (Dodd *et al.*, 2010). The ABA concentration in the roots remained unchanged when rice seeds were inoculated with P6W and P1Y bacterial strains, however the ABA concentration in the shoot dropped to 14% by P6W and 22% by P1Y, according to the earlier study (Zhao, 2025). When treated with P6W strains, the ABA-deficient tomato mutant *Flacca* (flc) *notabilis* (not) markedly increased the biomass of WT plants while preventing primary root extension. CRDT-EB-1 (*B. marisflavi*), which was isolated from the mustard seedlings' rhizosphere, was found to greatly increase drought tolerance by secreting ABA analog/xanthoxin (Gowtham *et al.*, 2021). Additionally, the flacca and sitiens of ABA-deficient mutants inoculated with *B. megaterium* showed decreased growth. The ABA content, which encourages the development of wild-type plant growth, is essential to *B. megaterium* development (Fig. 10.2). The ability of ABA-metabolizing rhizobacteria to change the root-phytohormonal state and stimulate plant development may therefore be implied.

10.8 Responses of Biphasic Root Growth Depending on ABA Dosage

ABA has been demonstrated to either promote or prevent the formation of roots. In areas with enough watering, it also prevents the growth of plant roots and shoots. More severe drying inhibits root growth, whereas mild drying promotes it. Similarly, ABA prevents plants from growing when there is a lack of water. Although there were slight biphasic effects on root growth from dry soil, severe water deficit slowed root growth while water deficit accelerated it. Low amounts of ABA encouraged root growth, whereas high levels hindered it, according to Luo *et al.* (2014), who demonstrated the complicated biphasic effects of exogenous ABA on root growth under well-watered settings. By promoting auxin influx and ET generation, ABA also inhibits the growth of *Arabidopsis* roots. Similarly, the ET-insensitive mutant's etr1-1 and ein2-1; ein3-1, as well as auxin-insensitive mutants (axr2-1/iaa7) and auxin influx mutants (aux 1-7; aux 1-T) cannot react to ABA's enhanced inhibitory impact. Auxin efflux inhibitors also reduced the stimulatory impact of low ABA concentrations (pin2/eir1-1), while auxin efflux mutants (iaa7/axr2-1) displayed a lesser extent of this effect. ABA may be necessary for the biphasic root growth of plants when coupled.

10.9 ABA-Mediated Drought Regulation Molecular Mechanism

Under drought, ABA regulates a number of physiological and molecular processes. Stress

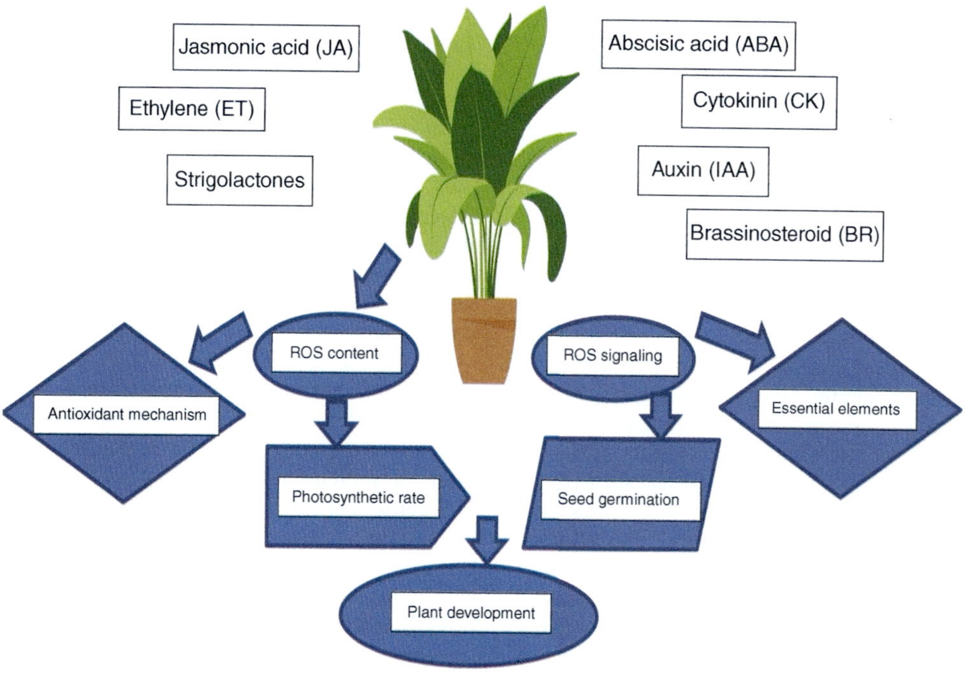

Fig. 10.2. Phytohormones and their interaction. (Figure author's own.)

stimuli are received to plants via their roots and leaves, which then relay this information to their shoots so they can produce ABA. In response to drought, protein kinases act as positive regulators of metabolism, transport, and ABA signaling (Fujita *et al.*, 2009). An ABA-responsive G protein receptor (GCR2), for example, inhibited the synthesis of ABA-inducible genes and induced insensitivity during germination. There have also been demonstrated interactions between ABA and the flowering-time control protein (FCA) and the Mg-chelatase H subunit (del Carmen Rodríguez-Gacio *et al.*, 2009).

When sensors perceive drought stimuli on the membrane, they initiate an activation of drought-responsive genes. After that, these signals are transmitted via a number of signal transduction channels. AAO3, ABA2, NCED1, NCED3, and other classes of ABA biosynthesis, along with transporter proteins AtABCG25 and AtABCG40 were recently found in the plasma membrane, showed that plants have sophisticated systems for monitoring and reacting to environmental changes. ABA absorption into cells is mediated by nitrate transporter 1/peptide transporter protein (NPF), formerly known as AIT1, which may also play a role in intercellular ABA transfer.

Many studies have examined how the root system's genetic control is affected by drought stress. Zhang *et al.* (2021a) conducted a transcriptome analysis of pearl millet during drought at one, three, and seven-hour time periods. They determined that 12 genes were unregulated at all-time points. However, at the one hour, three hour, and seven-hour time points, respectively, a total of 2004, 1538, and 605 genes had differential expression. Several highly-expressed genes were connected to plant hormone signaling, metabolic activities, ABA signal transduction pathway, and the MAPK signaling system. This could give researchers a genetic foundation for comprehending how other plants withstand drought. Similarly, by acting as a positive regulator, OsMAD23 modifies ABA biosynthesis and increases rice's resistance to drought, as shown by Luo *et al.* (2014). Through the activation of several genes involved in ABA and proline biosynthesis, such as OsNCED2/3/4 and OsP5CR, OsMAD23 facilitates the generation of endogenous ABA and the accumulation of proline. The findings suggest a novel approach of developing rice that is more resistant to salt and drought. ABFs, DRE, ABRE, AREBs, and many more TFs have their expression controlled by ABA. According to a piece of transactivation research, under drought and other stressors, the ABA sensitive TF OsBZIP72 does mediates the transportation and distribution of sucrose in rice by binding with the promoters of OsSWEET13 and OsSWEET15 (Mathan *et al.*, 2021). And as a drought-adapted strategy, *Arabidopsis* ZmbZIP33 was found to dramatically improve root length and chlorophyll content in maize (Zhang *et al.*, 2021b).

Climate change is currently having an adverse influence on agricultural production worldwide. To ensure food safety in light of climate change and the world's rapidly growing population, agricultural productivity must significantly rise (Farooq *et al.*, 2022). Significantly, climate estimates show that with global carbon dioxide emissions, unpredictable rainfall, and rising temperatures, the agricultural productivity of field crops that are commercially valued is negatively impacted (Bouabdelli *et al.*, 2022). A deeper comprehension of biodiversity expansion, natural events, and natural plant systems, along with their adaptation to various climatic circumstances, requires an awareness of the interplay between temperature and crop phenology. The various environmental conditions (biotic and abiotic stresses) that plants face in field settings can significantly impair a variety of physiological, morphological, molecular, biochemical and cellular processes (Shahzad *et al.*, 2021). Food security is seriously threatened by weather extremes, such as heat and cold stress, which affects agricultural production. The combination of temperature stress and other stressors may cause the plant to suffer oxidative damage (Considine and Foyer, 2021). According to Zahra *et al.* (2021), cold stress slows down plant development via altering phytohormonal signaling, lowering antioxidant defense enzyme activity and membrane integrity, and reducing the synthesis of proteins and secondary metabolites. This causes cellular damage or even cell death. Plant tissues/organisms (leaves, shoots, or roots), lant species, cold stress interval, growth phase (vegetative or reproductive), and other environmental factors all influence how cold stress impacts crop plants.

In temperate and polar regions, plants may be able to freeze without damage by transporting their nutrients from regions in the leaves and shoots to storage organs (Thorsen and Höglind, 2010). According to Smallwood and Bowles (2002), freezing stress results in dehydrative stress by blocking the absorption and supply of water to aerial plant components. Because too much cold stress causes cellular water to freeze and expand, it damages plant tissues. According to Song *et al.* (2021), "cold hardening" is the physiological process by which plants adapt to avoid cellular damage at extremely cold temperatures. Research indicates that plants that have been acclimated can withstand temperatures of up to −30°C, whereas those that have not can only tolerate minimum temperature of −5°C. (Song *et al.*, 2021b) Tropical crops, such *Zea mays* and *Solanum lycopersicum*, cannot tolerate freezing temperatures or cold harden. The three steps that plants use to adapt to cold stress are to detect low temperatures and convert the signal inhabiting appropriate genes expression to stimulate the plant. To counteract the stresses caused by extremely cold temperatures, which affect cell activity, plants use the genes. According to Thorsen and Höglind (2010) and Song *et al.* (2021b), the vast majority of genes and plant responses to cold stress are similar with those to drought or salinity.

Small chemical molecules called phytohormones influence plant development and stress resistance (Sabagh *et al.*, 2021). Phytohormones are crucial for balancing various signal transduction pathways and acting as chemical messengers that control stress responses to synchronize cellular activity and altering both internal and exterior reactions to stress (Ku *et al.*, 2018; Sabagh *et al.*, 2021).

10.10 Current Understanding Regarding Crop Plant Responses to Cold Stress

There are three categories of plants in response to cold temperatures: (i) frost-resistant (capable of withstanding very low temperatures); (ii) chilling-sensitive (severely harmed at temperatures >0 °C and <15 °C); and (iii) chilling-resistant (capable of withstanding low temperatures) (Ding *et al.*, 2016). In response to freezing stress, plants primarily alter the fluidity of their cell membranes, disrupt the stability and structure of proteins, decrease the activity of antioxidant enzymes, and increase the buildup of ROS. These modifications cause ion leakage and membrane degradation, which affect photosynthesis. Cold stress ultimately affects the production of proteins and gene expression (Ding *et al.*, 2016). However, cold stress affects plants more and can lead to death of the plant. This freezing stress dries cells and forms ice nuclei within the apoplastic space that could cause irreparable damage (Ding *et al.*, 2016). One of the cold stress phenomena described by Noctor and Foyer (2016) is the excessive production of ROS and lipid peroxidation that harms plant organelles, particularly in the spring. Moreover, chilling stress prevents roots from absorbing water, resulting in drought stress and water scarcity. It also prevents nutrients from being absorbed by roots and transferred to shoots (Hassan *et al.*, 2021).

Complex systems that reduce cold stress harm have been established by plants (Juurakko and Walker, 2021). Activating ROS scavenging systems and increasing the production of stress proteins and osmoprotectants such as proline, glycine betaine, sucrose, and trehalose, makes plants more resilient to frost stress in the future (Masouleh *et al.*, 2020). Furthermore, cold stress dramatically raises the amount of LEA protein, which restores unfolded proteins, serves as a hydration buffer, shields molecules and membranes, and prevents cellular dehydration (Kosová *et al.*, 2021). The genomes of sorghum subjected to a variety of stressors, including cold, revealed 68 LEA genes from eight gene families. According to Nagaraju *et al.* (2019), cis-regulatory elements found in LEA genes suggest that these genes work in a particularly conservative manner in stressed plants.

Additionally, cold stress dramatically changes the number of plant physio-biochemical functions. For example, cold stress (4°C for 7 days) reduced photosynthetic efficiency, stomatal conduction, and transpiration in wheat (*Triticum aestivum* L.) while raising levels of ROS and MDA, which in turn inhibited roots and shoots development (Repkina *et al.*, 2021). MJ treatment enhanced proline content and enhanced SOD and POD activities to stop ROS production in wheat, which decreased the negative effects of cold stress (Repkina *et al.*, 2021). In a different study, in cold-sensitive chickpeas (*Cicer arietinum* L.)

genotypes (ICC 15,567, and GPF 2), cold stress increased stomatal conductance, cell membrane damage, and leaf hydration status while decreasing the rate of photosynthetic activity during the reproductive phase (Rani *et al.*, 2021). In the reproductive period, cold stress usually results in chickpea flower and pod abortion, which drastically reduces seed production. Despite having higher antioxidant enzyme activity (SOD, CAT, AXP, and GR) in cold-tolerant chickpea genotypes (ICC 17,258, and ICC 16,349), non-enzymatic antioxidant substances, such as glutathione (GSH) and ascorbic acid (Asc), and osmoprotectants (proline and sucrose), they did not respond similarly to cold sensitive genotypes in which an increase in ROS and MDA that caused damages to anthers and ovules as well (Rani *et al.*, 2021). The extent of plant damage varies with cold stress length. In turnip rape (*Brassica rapa* L.) seeds exposed to cold stress for 6–12 days the biomass, plumule and radicle development, and germination index significantly decreased; however, in turnip rape seedlings exposed to cold stress for 72 hours, germination and phenolic compound synthesis increased (Ilyas *et al.*, 2022).

10.11 The Role of Phytohormones in Enhancing Resistance to Cold Stress

The diverse mechanisms of phytohormones, including physiological and biochemical control, are crucial for increasing cold stress resistance and promoting cold stress-induced plant growth and development. A number of hormones contribute to the cold stress-tolerance of different agricultural plants, as discussed in the next section.

10.11.1 Abscisic acid

One phytohormone that is essential for controlling how plants react physiologically to stress is abscisic acid (ABA). The isoprenoid molecule ABA is produced from β-carotene (C40) via the carotenoid route (Chen *et al.*, 2022). According to Rai *et al.* (2011) and Chen *et al.* (2022), there are numerous processes at play, such as the synthesis of antioxidant enzymes, osmoprotectants, osmotic stress tolerance, biotic and abiotic stress responses, cell division, elongation, stomatal closure, germination, cell turgor maintenance, embryo maturation, seed dormancy, germination, stomatal closure, leaf senescence, and floral induction.

Osmotic stress caused by low temperatures causes plant cells to have less water available, which leads to the production and biosynthesis of more ABA (Heidarvand and Maali Amiri, 2010; Chen *et al.*, 2022). Carried from roots to leaves, endogenous ABA builds up in guard cells, causing stomatal closure, a reduction in transpiration rate, and a delayed pace of cell growth (Llanes *et al.*, 2015). Plants normally require very little ABA to grow, and excess ABA is removed by controlling its endogenous production, according to Li *et al.* (2022).

In cold-tolerant plants, elevated ABA levels function as a chemical signal within cells that starts the synthesis of new proteins when acclimated to cold temperatures. Additionally, exogenous ABA reduces cold stress damage symptoms in a variety of plants (Shi and Yang, 2014) and increases cold stress resistance (Qin *et al.*, 2019) by promoting the transcription of antioxidant enzymes for ROS scavenging through ABA signaling. For example, when exogenous ABA, at a concentration of 1×10^{-5} mol/l, was applied to wheat plants cultivated under cold stress (0°C to –24°C) there was a resulting decrease in H_2O_2 generation via increasing CAT, SOD, POD, APX, and GR activities (Yu *et al.*, 2020). According to Lee and Luan (2012), ABA also increased putrescine production, which made rice more resistant to cold, indicating that ABA may control the levels of polyamine in plant (Li *et al.*, 2022).

In plants exposed to cold stress, ABA preserves the photosynthetic system by changing the chloroplasts' ultrastructure (Venzhik *et al.*, 2016). While appressed thylakoid membrane length and grana number increased at 4°C, wheat plants treated with 0.1 mM ABA at 22°C showed an increase in photosynthetic membrane lengths, both appressed and non-appressed thylakoid membranes (Venzhik *et al.*, 2016). Exogenous ABA (100 μM) at cold stress (4°C) decreased H_2O_2 and malondialdehyde (MDA) concentrations in Bermuda grass (*Cynodon dactylon* L.) by strengthening antioxidant mechanisms, reducing cell membrane damage and electrolyte leakage (EL), and raising photosystem II

(PSII) activity and other photosynthetic parameters (Huang *et al.*, 2015). Field grown chickpeas exposed to cold stress (average 10–12/2–4°C; day/night), had decreased EL, Chl content and relative water content (RWC) when subjected to external application of ABA (10 μmol/l) at flowering stage, improved pollen viability and germination, flower preservation as well as seed yield and decreased the oxidative damage caused by increasing SOD, CAT and APX activities and by increasing ABA, GSH, and proline contents (Kumar *et al.*, 2008).

According to Huang *et al.* (2015), two types of sugarcane (*Saccharum officinarum* L.) showed enhanced electrical conductivity, cell membrane damage, and elevated levels of GA3, proline, MDA, and ABA following 14 days of exposure to cold stress (0°C). ABA is essential for cold stress tolerance, as evidenced by the fact that exogenous ABA (100 μmol/l) reduced MDA and GA3 levels and cell membrane damage while raising proline and ABA levels (Huang *et al.*, 2015).

Plants also use the induction of leaf senescence as a defense against the cold winter temperatures. In yellow flag (*Iris pseudacorus* L.), for example, ABA enhanced cold stress-tolerance by controlling leaf senescence (Caselles *et al.*, 2021). Yellow flag leaves under chilling stress showed greater levels of ABA and lower levels of violaxanthin, which are essential for ABA synthesis, but decreased levels of CK, Chl, and PSII efficiency (Fv/Fm) (Caselles *et al.*, 2021). Nonetheless, rhizomes have greater levels of ABA and carotenoid (Caselles *et al.*, 2021).

10.11.2 Brassinosteroids

Plant physiological processes such as cell division, elongation, photo morphogenesis, germination and xylem differentiation can be regulated by brassinosteroids (BRs), a family of polyhydroxylated phytohormones with a steroidal polyhydroxylated group that includes 24-epibrassinolide (EBR) and brassinazole, an inhibitor of BR production (Hafeez *et al.*, 2021). Additionally, BRs react to a range of biotic and abiotic stimuli, which aids plants in adapting to their surroundings (Planas-Riverola *et al.*, 2019; Hafeez *et al.*, 2021). Leaves, roots, flowers, and seeds are among the many plant parts and organs that contain BRs.

BRs also have a significant impact on cold stress modulation and plant life by controlling a number of physio-chemical processes, such as growth, biomass, and antioxidant system activity. To boost plant development, for instance, foliar application of exogenous BRs (0.1 μmol/l) to tomato increased the quantum yield of PSII (Fv/Fm) and decreased the negative impacts of cold stress (4°C) while lowering oxidative stress and EL (Xia and Møller, 2018). Giving plants exposed to cold stress a BR can reduce ROS and H_2O_2 levels, which are signs of oxidative damage. In the chloroplasts, for example, cold stress speeds up the oxygenation reaction, which is linked to higher H_2O_2 and glyoxylate production in peroxisomes (Devireddy *et al.*, 2021). Cold stress may also have a negative impact on chloroplast structure and other photosynthetic processes, including Rubisco and other Calvin cycle enzymes. When BR-deficient mutant plants are exposed to cold stress, their reduced photosynthetic efficiency may be the result of chlorophyll degradation (Devireddy *et al.*, 2021). Through CK function, BRs indirectly affect Chl biosynthesis, as evidenced by the lower CK levels in BR-mutant plants, which promotes Chl production and chloroplast growth. EBR treatment enhanced the Fv/Fm ratio, stabilized Chl molecules and thylakoid membranes, and enhanced photosynthetic efficiency in plants exposed to cold stress.

Cold stress-exposed *Cucumus sativus* L. seedlings (12/8°C; day/night) showed decreased levels of GA, JA, and inositol tetrakisphosphate, but zeatin riboside, ABA, and IAA contents increased when BRs (0.1 μM exogenous BR) were applied externally (Anwar *et al.*, 2018). According to Anwar *et al.* (2018), EBR decreased the generation of ROS and MDA, while enhancing seedling development, chlorophyll content, photosynthetic efficiency, antioxidant mechanisms, and membrane stability. Li *et al.* (2022) demonstrated that pre-treatment with 1 nmol/l EBR enhanced stomatal conductance and photosynthetic activity, induced enzymatic antioxidant activities (SOD, CAT, GR, APX), and reduced the production of ROS, prevented lipid peroxidation, and thus made *B. rapa* more cold stress resilient. According to Pociecha *et al.* (2017), exogenous BR therapy increased the antioxidant enzyme activities (SOD, POD, and CAT), total phenol content, and phenylalanine-ammonia-lyase (PAL) activity in plants exposed to cold stress, including tomato and ryegrass (*Secale cereale* L.). Therefore, BRs can prevent peroxidation and preserve and stabilize cell membranes.

10.11.3 Auxin

The production, distribution, and signaling of auxin and indole-3-acetic acid (IAA), especially in stressful situations, are poorly understood. According to studies, cold stress has an impact on how auxins that are given to plants exogenously are transported (Morris, 1979). Additionally, according to Morris (1979), cold stress suppresses fruit set, pollen grain viability, and plant IAA concentration. Auxin transport was restored after being blocked by chilling stress at 4°C in a wild-type *Arabidopsis* when the temperature was brought to normal (Wyatt *et al.*, 2002). To enhance the growth of plants in both ordinary and cold stress conditions, auxin's cellular and molecular reactions to cold stress have been assessed in a number of research studies on *Arabidopsis* roots (Shibasaki *et al.*, 2009). Consequently, auxin mutants displayed decreased gravity responses and root lengths. A direct auxin transport assay and an expression study of the auxin responsive marker IAA2GUS confirmed that intracellular auxin transport is the initial target of cold stress (Shibasaki *et al.*, 2009). In another study, by decreasing auxin accumulation, CS limited meristem growth and cell quantity, which in turn decreased meristematic cells' ability to divide. This could have occurred because gene expression was decreased by CS, related to PIN1/3/7 and auxin synthesis. The inflorescence stems also responded less to CS in terms of gravity, a stimulus that monitors the direction of plant growth, as a result of the asymmetric auxin distribution in *Arabidopsis* (Makila *et al.*, 2023).

By controlling the expression of TFs like WRKYs and interacting with other phytohormones like ABA and JA, auxin can enhance cold stress tolerance in rice seedlings (Yang *et al.*, 2020). The combination of ABA and auxin greatly altered the stress-responsive gene expression in rice cultivars under cold stress indicating that endogenous ABA and auxin may be able to reduce the negative effects from cold stress (Du *et al.*, 2012). Abiotic stress and auxin function, particularly cold stress, are extremely difficult to study. Auxin-mediated cold stress control will be better understood if upcoming molecular-level research in plant–soil systems sheds insight on the mechanisms underlying auxin-facilitated plant responses to cold stress.

10.11.4 Salicylic acid

A unique phenolic molecule called salicylic acid (SA) modifies the expression of proteins linked to illness. Along with increasing abiotic stress resistance, including cold stress, SA also speeds up plant growth, ripening, and expansion. The ammonia-lyase pathways for phenylalanine and isochorismate are used to generate SA. Exogenous SA has been demonstrated in research to improve some plant species' cold stress-resistance mechanisms (Kazemi-Shahandashti *et al.*, 2014). In *Magnolia wufengensis*, grown at cold stress (0, −5, −10, −15, −20, −25, and −30°C), SA supplementation altered the metabolism of total soluble sugar by promoting the accumulation of total soluble sugar. Furthermore, soluble carbohydrates like fructose and glucose preserved the liquid-crystalline state of membrane phospholipids, maintaining structural damage management and boosting freezing tolerance (Duan *et al.*, 2022). In mountain rye (*Secale montanum* L.), hormonal crosstalk, including exogenous SA (0, 25, 50, 75, and 100 ppm), dramatically reduced cold stress harm via controlling defensive mechanisms, which in turn accelerated seed germination. According to Kaur and Gupta (2017), administering SA (0.1 mM) to musk melon (*Cucumis melo* L.) under cold stress conditions enhanced seed germination and a variety of antioxidant enzymes.

10.11.5 Gibberellic acids

Gibberellic acid (GA) is another name for gibberellin, a type of tetracyclic diterpenoid carboxylic acid. Some of these GAs—mainly GA1 and GA4—have growth hormone-like effects in higher plants. During germination, hydrolytic enzyme activity may be inhibited or diminished by chilling (Ghannad *et al.*, 2022). GA is a growth hormone that aids seed germination.

While GA3 treatment markedly enhanced emergence, low-temperature stress inhibited the emergence of chickpea seedlings. According to Aziz and Pekşen (2020), GA seed priming reduced EL, preserved relative water content, and enhanced crop development and stand establishment in chickpeas. While the germination index, germination rate, and weight and length of radicals and coleoptiles increased, the mean germination time decreased along with the seed weight, when wheat seeds were presoaked in GA3 (5 μmol/l) and sodium nitroprusside (SNP, 0.1 mmol/l) at low temperatures (Li *et al.*, 2013). Exogenous GA3 and SNP also enhanced amylase activity, promoted starch breakdown, and enhanced seed respiration. Lei *et al.* (2019) investigated the association of endogenous hormones to cold tolerance in two cultivars of rapeseed (*Brassica napus* L.): freezing (–2 and –4°C) and chilling (4 and 2°C). While freezing stress reduced the GA3 content in 1801, chilling stress raised the ABA and GA3 levels in 1801, but not in C20. This suggests that 1801 had a greater survival rate (by 100 and 75% under –2 and –4°C) than C20. Lei *et al.* (2019) found that the 1801 cultivar exhibited greater antioxidant enzyme, hormone, and osmolyte activity under cold stress compared to C20. During cold storage, the effects of exogenous SA injection on chilling damage (CI) in post-harvest green tomatoes were investigated. All of these factors play a part in preserving cellular redox homeostasis and adapting to cold during cold stress. Exogenous SA promoted the breakdown of DELLA protein and upregulated the expression of the GA3ox1 gene, which raised the amount of bioactive GA. This implies that the responses of tomato fruit to oxidative and chilling damage during extended cold storage depend on SA–GA signaling. After cooling, tomatoes had an exogenous GA3 treatment, which increased proline content, decreased MDA and EL, and promoted antioxidant enzyme activities, the CI index fell. According to Ding *et al.* (2016), GA seems to have a role in controlling the cold stress response in tomato fruit.

10.11.6 Strigolactones (SLs)

Strigolactones (SLs), a class of newly-discovered hormones, are derived from carotenoids. Carotenoids activate a variety of plastid-based and cytosolic enzymes, resulting in the production of SLs as apo-carotenoid complexes (Cooper *et al.*, 2018).

SLs, which are released by 79% of terrestrial plant roots, have been linked to the symbiotic interactions of arbuscular mycorrhizae fungi (AMF) (Cooper *et al.*, 2018). SLs promote stress tolerance via controlling metabolic, physiological, and molecular processes. Exogenous GR24 reduces the effects of cold stress via increasing proline concentration, increasing cell survival, increasing soluble protein, increasing antioxidant enzyme activities, increasing photosynthesis, and reducing the production of ROS. Zhang *et al.* (2020) investigated how GR24 affected *Brassica rapa* growth and development at cold stress (4°C). By altering several physiological and biochemical traits, they found that cold stress inhibited *Brassica rapa* seedlings. GR24 boosted the expression of genes such as NADPH, cold-regulated genes, and MAPK, as shown by qRT-PCR. GR24 enhanced cold stress tolerance in *Brassica rapa* seedlings by reducing H_2O_2 production, according to additional research (Zhang *et al.*, 2020).

Huang *et al.* (2021) applied exogenous SLs (0, 0.5, 1, 2, and 4 μmol/l) to mature strawberry fruits after precooling them at 0°C for a whole night. They then examined the effects of cold storage on the generation of nitric oxide (NO), hydrogen sulfide (H_2S), phenylpropanoid metabolism, and antioxidant system in post-harvest strawberries. The ability to scavenge DPPH, OH, and O^{-2} was enhanced, whereas polyphenol oxidase and POD activities were notably reduced by 1 μmol/l SL. Incorporation of exogenous SLs increased the total phenols, lignin and flavonoids levels by increasing activities of phenylalanine aminelyase and 4 coumaric acid coenzyme A ligase. According to these findings, SLs preserve the quality of strawberry fruit by enhancing the antioxidant system and the metabolism of NO, H_2S, and phenylpropanoid during cold storage (Huang *et al.*, 2021).

In a different investigation by Ma *et al.* (2022), the impact of exogenous GR24 on the antioxidant capacity and quality deterioration of post-harvest sweet oranges (*Citrus sinensis* L.) stored in cold storage was considered. Fruit hardness (9.2%), titratable acidity (11.7%), citrus color index (22.34%), soluble sugar content (8.17%), and total soluble solids (13.13%) all increased with the GR24 treatment, whereas MDA (27.59%) and H_2O_2 (30.08%) buildup decreased with improved enzyme and antioxidant activities. According to Ma *et al.* (2022), the antioxidant system was strengthened by GR24, which prolonged fruit senescence when it was being stored in a cold environment.

Few studies have looked into the function of SLs in cold stress to date. The processes through which SLs reduce cold stress require more investigation. Under chilling stress, Omoarelojie *et al.* (2021) biochemically and physiologically examined the effect of different rac-GR24 (0, 1, 10 μmol/l) dosages application on mung bean (*Vigna radiata* L.) seedlings. According to the findings, GR24 improved the hydration status by promoting proline accumulation and lowering ROS and MDA levels. GR24 resulted in promotion of the activity of phenylalanine ammonia lyase and tyrosine ammonia lyase, and mung bean chilling tolerance by influencing the phenylpropanoid pathway enzymes, lipoxygenase, and SOD levels through these enzymes (Omoarelojie *et al.*, 2021).

In a different study, SL biosynthesis was stimulated in tomatoes under both cold and heat stress. Conversely, SL signaling or biosynthesis impairment had the opposite impact from the synthetic SL analog GR24, which increased resistance to heat and cold. ABA, CBF1 transcription, heat shock protein (HSP) 70 (HSP70), and antioxidant enzyme activity were all associated with SL-mediated cold and heat tolerance. SL-induced resistance to heat and cold stimuli is linked to ABA-dependent HSP synthesis, antioxidant metabolism, and CBF synthesis, according to the findings (Chi *et al.*, 2021).

A. thaliana and pea cold stress tolerance depend on SLs. In *Arabidopsis* max (max2–1 and max4–1), pea rms4 and wild type, photosynthesis was significantly reduced by cold stress in the mutants but was resistant to dark chilling. This implies that SLs protect photosynthesis from dark chilling damage and thereby improve shoot biomass production. Yang *et al.* (2020) reported that cold stress improved the lifespan of MdD14-OE, the SL receptor gene "MdD14", which was cloned from apples and discovered to be present in the cytoplasm and nucleus. Additionally, transgenic apple calli, which overexpresses the protein, provides evidence that the SL signal transduction system and a variety of abiotic stress tolerance, including as drought, salt, and cold stress, depend on apple MdD14 (Yang *et al.*, 2020).

Under normal and cold stress circumstances, the levels of both D17 and D27 transcriptions in non-mycorrhizal rice roots were comparable, but Os01g0701500 and D10 gene expression was downregulated in these roots. In mycorrhizal rice roots exposed to cold stress, the aforementioned genes were remarkably elevated, increasing the production of SL (Liu *et al.*, 2015). The detrimental effects of cold stress on plants must be stopped by SLs. Ongoing research using high output technology is studying the mechanisms behind how SL can stimulate plant growth in the presence of cold stress.

10.11.7 Jasmonic acid

According to Raza *et al.* (2021), jasmonate, jasmonate isoleucine conjugate (JA-Ile), and MeJA are examples of fatty acid derivatives known as jasmonic acid or jasmonates (JA). α-linolenic acid in plants is the source of JA, a well-studied phytohormone that naturally controls plant growth. Jasmonate, a product of JA, regulates how plants react to cold stress and other abiotic stresses. According to recent research, JA controls root formation to enhance plant growth and development. Internal browning in peach fruit kept at 4°C was reduced by exogenous JA (100 μl/l) through enhanced expression of ET-induced genes related to lipid metabolism via increasing ET generation. According to Zhao *et al.* (2013), ET synthesis helped stabilize the membrane by raising the amounts of phospholipid and glycerolipid in cold-stored peach fruit. Tomatoes kept in liquid nitrogen at –80°C showed improved cold stress tolerance when JA and MeJA application (0.05 mmol/l) raised the arginine decarboxylase, arginase, and ornithine decarboxylase activities as well as the expression level of SlMYC2 genes, thereby increasing the polyamine content (Zhou *et al.*, 2019). In the same way, JA enhances cold stress tolerance by upregulating cold stress-responsive genes. For example, MeJA stimulant (0.1 mmol/l) activation of MYC2 TFs was observed in bananas stored at –80°C. According to Niu *et al.* (2014), Additionally, MeJA reduced CI in peach fruit and caused less damage to the membranes when exposed to −80°C by upregulating the JA-mediated CBF pathway and decreasing relative EL.

10.12 Conclusion

For regulating growth and development, and developing stress responses, plant hormones are the key elements. They govern processes that allow plants to adapt to both biotic and abiotic stresses, ensuring their survival under challenging environmental conditions. Key hormones such as auxins, CKs, ABA, SA, JA, ET, and GAs interact within complex signaling networks to manage these adaptations. In drought stress responses, ABA is essential because it controls mechanisms like stomatal closure, water conservation, and root development. It enhances drought tolerance by modulating gene expression and supporting osmotic adjustment, and its interaction with other hormones ensures a balanced stress response. Similarly, other hormones like SA, JA, and ET are involved in plant immune responses, helping plants cope with both abiotic and biotic challenges.

The impact of climate change is exacerbating environmental stress on agriculture, making it more crucial to understand these hormonal pathways. Phytohormones influence not only stress tolerance but also improve metabolic pathways, root architecture, and antioxidant defenses. Advances in genomics and molecular tools are enabling the identification of stress-responsive genes, offering potential for developing crops with enhanced resilience to drought, salinity, temperature extremes, and pollution.

Understanding the crosstalk between hormones like ABA, JA, SA, and ET can uncover new ways to fine-tune stress responses in plants, improving their resilience. By focusing on hormonal interactions, we can better address issues such as water scarcity, salinity, and heat stress, thereby safeguarding food production in a changing climate. Further research into plant hormonomics—using technologies like mass spectrometry—holds great promise for enhancing our ability to breed crops that can thrive under harsh conditions. These advancements can contribute significantly to agricultural sustainability and food security. Plant hormones are integral to optimizing plant growth and stress resistance, and a deeper understanding of these mechanisms is essential for the development of crops that can withstand future environmental challenges. By leveraging hormonal pathways, we can ensure food security and support sustainable agriculture practices.

References

Agurla, S., Gahir, S., Munemasa, S., Murata, Y. and Raghavendra, A.S. (2018) Mechanism of stomatal closure in plants exposed to drought and cold stress. *Advances in Experimental Medicine and Biology* 1081, 215–232. DOI: 10.1007/978-981-13-1244-1_12.

Aharoni, A. and Galili, G. (2011) Metabolic engineering of the plant primary–secondary metabolism interface. *Current Opinion in Biotechnology* 22, 239–244.

Ahmad, Imtiaz, María del Mar Jiménez-Gasco and Mary E. Barbercheck (2020) The role of endophytic insect-pathogenic fungi in biotic stress management. *Plant Stress Biology: Strategies and Trends*, 379–400.

Aina, O., Quesenberry, K. and Gallo, M. (2012) Thidiazuron-induced tissue culture regeneration from quartered-seed explants of Arachis paraguariensis. *Crop Science* 52(3), 1076–1083.

Ali, S., Hayat, K., Iqbal, A. and Xie, L. (2020) Implications of abscisic acid in the drought stress tolerance of plants. *Agronomy* 10, 1323.

Anwar, A., Xianchang, Y.U. and Yansu, L.I. (2020) Seed priming as a promising technique to improve growth, chlorophyll, photosynthesis and nutrient contents in cucumber seedlings. *Notulae Botanicae Horti Agrobotanici Cluj-Napoca*, 48(1), pp.116–127.

Aslam, M., Waseem, M., Jakada, B.H., Okal, E.J., Lei, Z., Saqid, H.S.A. *et al.* (2022) Mechanisms of abscisic acid-mediated drought stress responses in plants. *International Journal of Molecular Sciences* 23(3), 1084.

Aziz, T. and Pekşen, E. (2020) Seed priming with gibberellic acid rescues chickpea (Cicer arietinum L.) from chilling stress. *Acta Physiologiae Plantarum* 42, 1–10.

Bari, R. and Jones, JD. (2009) Role of plant hormones in plant defence responses. *Plant Molecular Biology* 69(4), 473–88.

Berrocal-Lobo, M., Molina, A. and Solano, R. (2002) Constitutive expression of ethylene-response-factor1 in Arabidopsis confers resistance to several necrotrophic fungi. *The Plant Journal* 29, 23–32.

Boominathan, P., Shukla, R., Kumar, A., Manna, D., Negi, D., Verma, P.K. and Chattopadhyay, D. (2004) Long term transcript accumulation during the development of dehydration adaptation in Cicer arietinum. *Plant Physiology* 135, 1608–1620.

Bouabdelli, S., Zeroual, A., Meddi, M. and Assani, A. (2022) Impact of temperature on agricultural drought occurrence under the effects of climate change. *Theoretical and Applied Climatology* 148(1), 191–209.

Caselles, V., Casadesús, A. and Munné-Bosch, S. (2021) A dual role for abscisic acid integrating the cold stress response at the whole-plant level in Iris pseudacorus L. growing in a natural wetland. *Frontiers in Plant Science*, 12, p.722525.

Chen, G., Zheng, D., Feng, N., Zhou, H., Mu, D., Zhao, L., Shen, X., Rao, G., Meng, F. and Huang, A. (2022) Physiological mechanisms of ABA-induced salinity tolerance in leaves and roots of rice. *Scientific Reports*, 12(1), p.8228.

Chi, C., Xu, X., Wang, M., Zhang, H., Fang, P., Zhou, J. and Yu, J. (2021) Strigolactones positively regulate abscisic acid-dependent heat and cold tolerance in tomato. *Horticulture Research* 8.

Chisholm, S.T., Coaker, G., Day, B. and Staskawicz, B.J. (2006) Host-microbe interactions: shaping the evolution of the plant immune response. *Cell* 124, 803–814.

Considine, M.J. and Foyer, C.H. (2021) Stress effects on the reactive oxygen species-dependent regulation of plant growth and development. *Journal of Experimental Botany* 72(16), 5795–5806.

Cooper, J.W., Hu, Y., Beyyoudh, L., Yildiz Dasgan, H., Kunert, K., Beveridge, C.A. and Foyer, C.H. (2018) Strigolactones positively regulate chilling tolerance in pea and in Arabidopsis. *Plant, Cell and Environment* 41(6), 1298–1310.

del Carmen Rodríguez-Gacio, M., Matilla-Vázquez, M.A. and Matilla, A.J. (2009) Seed dormancy and ABA signaling: The breakthrough goes on. *Plant signaling and behavior*, 4(11), 1035.

Denancé, N., Sánchez-Vallet, A., Goffner, D. and Molina, A. (2013) Disease resistance or growth: the role of plant hormones in balancing immune responses and fitness costs. *Frontiers in plant science*, 4, 155.

Devireddy, A.R., Tschaplinski, T.J., Tuskan, G.A., Muchero, W. and Chen, J.G. (2021) Role of reactive oxygen species and hormones in plant responses to temperature changes. *International Journal of Molecular Sciences* 22(16), 8843.

Ding, Y., Zhao, J., Nie, Y., Fan, B., Wu, S. *et al.* (2016) Salicylic-acid-induced chilling-and oxidative-stress tolerance in relation to gibberellin homeostasis, C-repeat/dehydration-responsive element binding factor pathway, and antioxidant enzyme systems in cold-stored tomato fruit. *Journal of Agricultural and Food Chemistry* 64(43), 8200–8206.

Dodd, I.C., Zinovkina, N.Y., Safronova, V.I., Belimov, A.A. (2010) Rhizobacterial mediation of plant hormone status. *Annals of Applied Biology* 157, 361–379.

Dodds, P.N. and Rathjen, J.P. (2010) Plant immunity: Towards an integrated view of plant–pathogen interactions. *Nature Reviews Genetics* 11, 539–548.

Du, Y.L., Wang, Z.Y., Fan, J.W., Turner, N.C., Wang, T. and Li, F.M. (2012) β-Aminobutyric acid increases abscisic acid accumulation and desiccation tolerance and decreases water use but fails to improve grain yield in two spring wheat cultivars under soil drying. *Journal of Experimental Botany*, 63(13), pp.4849–4860.

Duan, X., Zhu, Z., Yang, Y., Duan, J., Jia, Z. *et al.* (2022) Salicylic acid regulates sugar metabolism that confers freezing tolerance in Magnolia wufengensis during natural cold acclimation. *Journal of Plant Growth Regulation*, 1–9.

Farooq, M.S., Uzair, M., Raza, A., Habib, M., Xu, Y. *et al.* (2022) Uncovering the research gaps to alleviate the negative impacts of climate change on food security: A review. *Frontiers in Plant Science* 13, 927535.

Fu, J. and Wang, S. (2011). Insights into auxin signaling in plant–pathogen interactions. *Frontiers in Plant Science* 2, 74. DOI: 10.3389/fpls.2011.00074.

Fujita, Y., Nakashima, K., Yoshida, T., Katagiri, T., Kidokoro, S. *et al.* (2009) Three SnRK2 protein kinases are the main positive regulators of abscisic acid signaling in response to water stress in Arabidopsis. *Plant Cell Physiology* 50, 2123–2132.

Gadjev, I., Vanderauwera, S., Gechev, T.S., Laloi, C., Minkov, I.N. *et al.* (2006) Transcriptomic footprints disclose specificity of reactive oxygen species signaling in Arabidopsis. *Plant Physiology* 141(2), 436–45

Gargallo-Garriga, A., Preece, C., Sardans, J., Oravec, M., Urban, O., Peñuelas, J. (2018) Root exudate metabolomes change under drought and show limited capacity for recovery. *Scientific Reports* 8, 12696.

Ghannad, R., Haghjou, M.M., Raza, A. and Hasanuzzaman, M. (2022) Induction of hydrolytic enzyme activities in dormant seeds of Dracocephalum kotschyi Boiss. causes improvement of germination and seedling vigor indices. *Acta Physiologiae Plantarum* 44(4), 48.

Glazebrook, J. (2005) Contrasting mechanisms of defense against biotrophic and necrotrophicpathogens. *Annual Review of Phytopathology* 43, 205–227.

Gowtham, H.G., Duraivadivel, P., Ayusman, S., Sayani, D., Gholap, S.L., Niranjana, S.R. and Hariprasad, P. ABA analogue produced by Bacillus marisflavi modulates the physiological response of Brassica juncea L. under drought stress. *Applied Soil Ecology.* 2021, 159,

Hafeez, M.B., Zahra, N., Zahra, K., Raza, A., Batool, A., Shaukat, K. and Khan, S. (2021) Brassinosteroids: Molecular and physiological responses in plant growth and abiotic stresses. *Plant Stress*, 2, p.100029.

Hamel, L.-P., Nicole, M.-C., Duplessis, S. and Ellis, B.E. (2012) Mitogen-activated protein kinase signaling in plant-interacting fungi: Distinct messages from conserved messengers. *Plant Cell* 24, 1327–1351.

Hassan, M.A., Xiang, C., Farooq, M., Muhammad, N., Yan, Z. *et al.* (2021) Cold stress in wheat: plant acclimation responses and management strategies. *Frontiers in Plant Science* 12, 676884.

Heidarvand, L. and Maali Amiri, R. (2010) What happens in plant molecular responses to cold stress? *Acta Physiologiae Plantarum* 32, 419–431.

Hirayama, T. and Mochida, K. (2022) Plant hormonomics: a key tool for deep physiological phenotyping to improve crop productivity. *Plant and Cell Physiology* 63(12), 1826–1839.

Holeski, L.M., Jander, G. and Agrawal, A.A. (2012) Transgenerational defense induction and epigenetic inheritance in plants. *Trends in Ecology and Evolution* 27, 618–626.

Hong, J.H., Seah, S.W., Xu, J. (2013) The root of ABA action in environmental stress response. *Plant Cell Reports* 32, 971–983.

Huang, X., Chen, M.H., Yang, L.T., Li, Y.R. and Wu, J.M. (2015) Effects of exogenous abscisic acid on cell membrane and endogenous hormone contents in leaves of sugarcane seedlings under cold stress. *Sugar Tech* 17, 59–64.

Huang, D., Wang, Y., Zhang, D., Dong, Y., Meng, Q., Zhu, S. and Zhang, L. (2021) Strigolactone maintains strawberry quality by regulating phenylpropanoid, NO, and H2S metabolism during storage. *Postharvest Biology and Technology* 178, 111546.

Ilyas, M., Khan, W.A., Ali, T., Ahmad, N., Khan, Z. *et al.* (2022) Cold stress-induced seed germination and biosynthesis of polyphenolics content in medicinally important Brassica rapa. *Phytomedicine Plus* 2(1), 100185.

Jaillais, Y. and Chory, J. (2010) Unraveling the paradoxes of plant hormone signaling integration. *Nature Structural and Molecular Biology* 17, 642–645.

Jones, J. D. and Dangl, J. L. (2006) The plant immune system. *Nature* 444, 323–329

Juurakko, C.L. and Walker, V.K. (2021) Cold acclimation and prospects for cold-resilient crops. *Plant Stress* 2, 100028.

Kang, J., Hwang, J.-U., Lee, M., Kim, Y.-Y., Assmann, S.M., Martinoia, E. and Lee, Y. (2010) PDR-type ABC transporter mediates cellular uptake of the phytohormone abscisic acid. *Proceedings of the National Academy of Sciences of the United States of America* 107, 2355–2360.

Kaur, S. and Gupta, N. (2017) Effect of proline and salicylic acid on germination and antioxidant enzymes at different temperatures in Muskmelon (Cucumis melo L.) seeds. *Journal of Applied and Natural Science* 9(4), 2165–2169.

Kazan, K. and Manners, J. M. (2009) Linking development to defense: Auxin in plant–pathogen interactions. *Trends in Plant Science* 14, 373–382.

Kazemi-Shahandashti, S.S., Maali-Amiri, R., Zeinali, H., Khazaei, M., Talei, A. and Ramezanpour, S.S. (2014) Effect of short-term cold stress on oxidative damage and transcript accumulation of defense-related genes in chickpea seedlings. *Journal of Plant Physiology* 171(13), 1106–1116.

Kempel, A., Schädler, M., Chrobock, T., Fischer, M. and van Kleunen, M. (2011) Tradeoffs associated with constitutive and induced plant resistance against herbivory. *Proceedings of the National Academy of Sciences of the United States of America* 108, 5685–5689.

Kleine, B. and Rossmanith, W.G. (2016). Hormones: Some Definitions. In *Hormones and the Endocrine System: Textbook of Endocrinology*, pp. 11–17. Springer International Publishing: Cham.

Kosová, K., Klíma, M., Prášil, I.T. and Vítámvás, P. (2021) COR/LEA proteins as indicators of frost tolerance in Triticeae: A comparison of controlled versus field conditions. *Plants* 10(4), 789.

Kovtun Y, Chiu WL, Tena G, Sheen J. (2000) Functional analysis of oxidative stress-activated mitogen-activated protein kinase cascade in plants. *Proceedings of the National Academy of Sciences of the United States of America* 97(6), 2940–5.

Ku, Y.S., Sintaha, M., Cheung, M.Y. and Lam, H.M. (2018) Plant hormone signaling crosstalks between biotic and abiotic stress responses. *International Journal of Molecular Sciences* 19(10), 3206.

Kudla J, Batistic O and Hashimoto K. (2010) Calcium signals: The lead currency of plant information processing. *Plant Cell* 22(3), 541–63.

Kumar, S., Kaur, G. and Nayyar, H. (2008) Exogenous application of abscisic acid improves cold tolerance in chickpea (Cicer arietinum L.). *Journal of Agronomy and Crop Science* 194(6), 449–456.

Lamers, J., van der Meer, T. and Testerink, C. (2020) How plants sense and respond to stressful environments. *Plant Physiol.* 182, 1624–1635. doi: 10.1104/ pp.19.01464

Lata, C. and Prasad M. (2011) Role of DREBs in regulation of abiotic stress responses in plants. *Journal of Experimental Botany* 62(14), 4731–48.

Lee, S.C. and Luan, S. (2012) ABA signal transduction at the crossroad of biotic and abiotic stress responses. *Plant, Cell and Environment* 35, 53–60. DOI: 10.1111/j.1365-3040.2011.02426.x.

Lei, Y.A.N., Shah, T., Cheng, Y., Yan, L.Ü., Zhang, X.K. and Zou, X.L. (2019) Physiological and molecular responses to cold stress in rapeseed (Brassica napus L.). *Journal of Integrative Agriculture* 18(12), 2742–2752.

Li, X., Jiang, H., Liu, F., Cai, J., Dai, T., Cao, W. and Jiang, D. (2013) Induction of chilling tolerance in wheat during germination by pre-soaking seed with nitric oxide and gibberellin. *Plant Growth Regulation* 71, 31–40.

Li, Z., Qiu, Z., Ge, H. and Du, C. (2022) Long-term dynamic of cold stress during heading and flowering stage and its effects on rice growth in China. *Atmosphere* 13(1), 103.

Liang, Y., Mitchell, D.M. and Harris, J.M. (2007) Abscisic acid rescues the root meristem defects of the Medicago truncatula latd mutant. *Developmental Biology* 304, 297–307.

Lim, C.W., Baek, W., Jung, J., Kim, J.H. and Lee, S.C. (2015) Function of ABA in stomatal defense against biotic and drought stresses. *The International Journal of Molecular Sciences* 16, 15251–15270. DOI: 10.3390/ijms160715251.

Liu, X., Zhai, S., Zhao, Y., Sun, B., Liu, C., Yang, A. and Zhang, J. (2013) Overexpression of the phosphatidylinositol synthase gene (ZmPIS) conferring drought stress tolerance by altering membrane lipid composition and increasing ABA synthesis in maize. *Plant, Cell and Environment* 36, 1037–1055.

Liu, Z., Li, Y., Wang, J., He, X. and Tian, C. (2015) Different respiration metabolism between mycorrhizal and non-mycorrhizal rice under low-temperature stress: A cry for help from the host. *The Journal of Agricultural Science* 153(4), 602–614.

Llanes, A., Devinar, G. and Luna, V. (2015) Role of abscisic acid in legumes under abiotic stress. *Legumes under Environmental Stress: Yield, Improvement and Adaptations*, 145–160.

Luo, X., Chen, Z., Gao, J. and Gong, Z. (2014) Abscisic acid inhibits root growth in Arabidopsis through ethylene biosynthesis. *The Plant Journal: Cell and Molecular Biology* 79, 44–55.

Ma, Q., Lin, X., Zhan, M., Chen, Z., Wang, H., Yao, F. and Chen, J. (2022) Effect of an exogenous strigolactone GR24 on the antioxidant capacity and quality deterioration in postharvest sweet orange fruit stored at ambient temperature. *International Journal of Food Science and Technology* 57(1), 619–630.

Mäkilä, R., Wybouw, B., Smetana, O., Vainio, L., Solé-Gil, A., Lyu, *et al.* (2023) Gibberellins promote polar auxin transport to regulate stem cell fate decisions in cambium. *Nature plants* 9(4), 631–644.

Mander, L. and Liu, H.W. (2010) *Comprehensive Natural Products II: Chemistry and Biology* (Vol. 1). Elsevier: Amsterdam, Netherlands.

Mathan, J., Singh, A. and Ranjan, A. (2021) Sucrose transport in response to drought and salt stress involves ABA-mediated induction of OsSWEET13 and OsSWEET15 in rice. *Physiologia Plantarum*, 171(4), 620–637.

Masouleh, S.S.S., Aldine, N.J. and Sassine, Y.N. (2020) The role of organic solutes in the osmotic adjustment of chilling-stressed plants (vegetable, ornamental and crop plants). *Ornamental Horticulture* 25, 434–442.

Mauch-Mani, B. and Mauch, F. (2005) The role of abscisic acid in plant– pathogen interactions. *Current Opinion in Plant Biology* 8, 409–414.

Miller G, Shulaev V and Mittler R. (2008) Reactive oxygen signaling and abiotic stress. *Plant Physiology* 133(3), 481–9.

Morris, D.A. (1979) The effect of temperature on the velocity of exogenous auxin transport in intact chilling-sensitive and chilling-resistant plants. *Planta* 146, 603–605. https://doi.org/10.1007/BF00388839

Muchhal, U.S. and Raghothama, K.G. (1999) Transcriptional regulation of plant phosphate transporters. *Proceedings of the National Academy of Sciences of the United States of America* 96, 5868–5872.

Muhammad Aslam, M., Waseem, M., Jakada, B.H., Okal, E.J., Lei, Z., Saqib, H.S.A. and Zhang, Q. (2022) Mechanisms of abscisic acid-mediated drought stress responses in plants. *International Journal of Molecular Sciences* 23(3), 1084.

Müller, M. (2021) Foes or friends: ABA and ethylene interaction under abiotic stress. *Plants* 10, 448.

Mundiyara, R., Sodani, R. and Bhati, S.S. (2020) Role of plant growth regulators in crop production. *Agriculture and Food* 2(6), 822–825.

Nagaraju, M., Kumar, S.A., Reddy, P.S., Kumar, A., Rao, D.M. and Kavi Kishor, P.B. (2019) Genome-scale identification, classification, and tissue specific expression analysis of late embryogenesis abundant (LEA) genes under abiotic stress conditions in Sorghum bicolor L. *PloS one* 14(1), e0209980.

Nakashima, K. and Yamaguchi-Shinozaki, K. (2013) ABA signaling in stress-response and seed development. *Plant Cell Reports* 32(7), 959–70.

Navarro, L., Bari, R., Achard, P., Lison, P., Nemri, A., Harberd, N.P. *et al.* (2008) DELLAs control plant immune responses by modulating the balance of jasmonic acid and salicylic acid signaling. *Current Biology* 18(9):650–5.

Nawaz, H., Irum, A., Nasim, W., Hussain, N., Usman, M. and Alam, J. (2023) Hormonal cross-talk mechanisms and plant immunity or defense: An overview. Hormonal Cross-Talk, *Plant Defense and Development*, 1–12.

Naylor, D. and Coleman-Derr, D. (2018) Drought stress and root-associated bacterial communities. *Frontiers in Plant Science* 8, 2223.

Nguyen, D., D'Agostino, N., Tytgat, T.O., Sun, P., Lortzing, T. *et al.* (2016) Drought and flooding have distinct effects on herbivore-induced responses and resistance in Solanum dulcamara. *Plant, Cell and Environment* 39, 1485–1499. DOI: 10.1111/pce.12708.

Niu, S., Gao, Q., Li, Z., Chen, X. and Li, W. (2014) The role of gibberellin in the CBF1-mediated stress-response pathway. *Plant Molecular Biology Reporter* 32, 852–863.

Niu, M., Xie, J., Chen, C., Cao, H., Sun, J. *et al.* (2018) An early ABA-induced stomatal closure, Na+ sequestration in leaf vein and K+ retention in mesophyll confer salt tissue tolerance in Cucurbita species. *Journal of Experimental Botany* 69, 4945–4960. DOI: 10.1093/jxb/ery251.

Noctor, G. and Foyer, C.H. (2016) Intracellular redox compartmentation and ROS-related communication in regulation and signaling. *Plant Physiology* 171(3), 1581–1592.

Omoarelojie, L.O., Kulkarni, M.G., Finnie, J.F. and Van Staden, J. (2021) Strigolactone analog (rac-GR24) enhances chilling tolerance in mung bean seedlings. *South African Journal of Botany* 140, 173–181.

Opik, H., Stephen, R.A., Arthur, W.J. and Herbert, E.S. (2005) *The physiology of flowering plants (4th ed.)*. Cambridge University Press: Cambridge.

Panstruga, R., Parker, J.E. and SchulzeLefert, P. (2009) SnapShot: Plant immune response pathways. *Cell* 136, 978.e1–978.e3.

Park, H.-Y., Seok, H.-Y., Park, B.-K., Kim, S.-H., Goh, C.-H. *et al.* (2008) Overexpression of Arabidopsis ZEP enhances tolerance to osmotic stress. *Biochemical and Biophysical Research Communications* 375, 80–85.

Peng, Z.Y., Zhou, X., Li, L., Yu, X., Li, H. *et al.* (2009) Arabidopsis hormone database: A comprehensive genetic and phenotypic information database for plant hormone research in Arabidopsis. *Nucleic Acids Research* 37, D975–D982.

Pieterse, C.M.J., Leon-Reyes, A., Van der Ent, S. and Van Wees, S.C.M. (2009) Networking by small molecule hormones in plant immunity. *Nature Chemical Biology* 5, 308–316.

Pieterse, C.M.J., van der Does, D., Zamioudis, C., Leon-Reyes, A. and van Wees, S.C.M. (2012) Hormonal modulation of plant immunity. *The Annual Review of Cell and Developmental Biology* 28, 489–521.

Planas-Riverola, A., Gupta, A., Betegón-Putze, I., Bosch, N., Ibañes, M. and Caño-Delgado, A.I. (2019) Brassinosteroid signaling in plant development and adaptation to stress. *Development* 146(5), dev151894.

Pociecha, E., Dziurka, M., Waligórski, P. *et al.* (2017) 24-Epibrassinolide Pre-Treatment Modifies Cold-Induced Photosynthetic Acclimation Mechanisms and Phytohormone Response of Perennial Ryegrass in Cultivar-Dependent Manner. *Journal of Plant Growth Regulation* 36, 618–628. https://doi.org/10.1007/s00344-016-9662-6

Qin, T., Tian, Q., Wang, G. and Xiong, L. (2019) LOWER TEMPERATURE 1 enhances ABA responses and plant drought tolerance by modulating the stability and localization of C2-domain ABA-related proteins in Arabidopsis. *Molecular Plant* 12(9), 1243–1258.

Rai, M.K., Shekhawat, N.S., Harish, Gupta, A.K., Phulwaria, M., Ram, K. and Jaiswal, U. (2011) The role of abscisic acid in plant tissue culture: a review of recent progress. *Plant Cell, Tissue and Organ Culture (PCTOC)* 106, 179–190.

Rani, A., Kiran, A., Sharma, K.D., Prasad, P.V., Jha, U.C., Siddique, K.H. and Nayyar, H. (2021) Cold tolerance during the reproductive phase in chickpea (Cicer arietinum L.) is associated with superior cold acclimation ability involving antioxidants and cryoprotective solutes in anthers and ovules. *Antioxidants* 10(11), 1693.

Rastogi, V.B. (2011) A Complete Course in ISC Biology-Vol. II.

Raza, A., Charagh, S., Zahid, Z., Mubarik, M.S., Javed, R., Siddiqui, M.H. and Hasanuzzaman, M. (2021) Jasmonic acid: A key frontier in conferring abiotic stress tolerance in plants. *Plant Cell Reports* 40(8), 1513–1541.

Reddy, A.S., Ali, G.S., Celesnik, H. and Day, I.S. (2011) Coping with stresses: Roles of calcium and calcium/calmodulin-regulated gene expression. *Plant Cell* 23(6), 2010–32.

Ren, H., Gao, Z., Chen, L., Wei, K., Liu, J. *et al.* (2007) Dynamic analysis of ABA accumulation in relation to the rate of ABA catabolism in maize tissues under water deficit. *Journal of Experimental Botany* 58(2), 211–219.

Repkina, N., Ignatenko, A., Holoptseva, E., Miszalskl, Z., Kaszycki, P. and Talanova, V. (2021) Exogenous methyl jasmonate improves cold tolerance with parallel induction of two cold-regulated (COR) genes expression in Triticum aestivum L. *Plants* 10(7), 1421.

Rillig, M.C., Lehmann, A., de Souza Machado, A.A. and Yang, G. (2019) Microplastic effects on plants. *New Phytologist* 223(3), 1066–1070.

Robert-Seilaniantz, A., Grant, M. and Jones, J.D. (2011) Hormone crosstalk in plant disease and defense: more than just jasmonate-salicylate antagonism. *Annual Review of Phytopathology* 49, 317–343.

Rubin, R.L., van Groenigen, K.J. and Hungate, B.A. (2017) Plant growth promoting rhizobacteria are more effective under drought: A meta-analysis. *Plant Soil* 416, 309–323.

Rudrappa, T., Czymmek, K.J., Paré, P.W. and Bais, H.P. (2008) Root-secreted malic acid recruits beneficial soil bacteria. *Plant Physiology* 148, 1547–1556.

Saavedra, X., Modrego, A., Rodriguez, D., Gonzalez-Garc, M.P., Sanz, L., Nicolas, G. and Lorenzo, O. (2010) The nuclear interactor PYL8/RCAR3 of Fagus sylvatica FsPP2C1 is a positive regulator of abscisic acid signaling in seeds and stress. *Plant Physiology* 152, 133–150.

Sabagh, A.E., Mbarki, S., Hossain, A., Iqbal, M.A., Islam, M.S. *et al.* (2021) Potential role of plant growth regulators in administering crucial processes against abiotic stresses. *Frontiers in Agronomy* 3, 648694.

Santner, A., Calderon-Villalobos, L.I.A. and Estelle, M. (2009) Plant hormones are versatile chemical regulators of plant growth. *Nature Chemical Biology* 5, 301–307.

Schulze-Lefert, P. and Panstruga, R. (2011) A molecular evolutionary concept connecting nonhost resistance, pathogen host range, and pathogen speciation. *Trends in Plant Science* 16, 117–125.

Shaar-Moshe, L., Blumwald, E. and Peleg, Z. (2017) Unique physiological and transcriptional shifts under combinations of salinity, drought, and heat. *Plant Physiology* 174(1), 421–434.

Shahzad, A., Ullah, S., Dar, A.A., Sardar, M.F., Mehmood, T. *et al.* (2021) Nexus on climate change: Agriculture and possible solution to cope with future climate change stresses. *Environmental Science and Pollution Research* 28, 14211–14232.

Shi, Y. and Yang, S. (2014) ABA regulation of the cold stress response in plants. In: *Abscisic Acid: Metabolism, Transport and Signaling*. 337–363.

Shibasaki, K., Uemura, M., Tsurumi, S. and Rahman, A. (2009) Auxin response in Arabidopsis under cold stress: underlying molecular mechanisms. *The Plant Cell*, 21(12), 3823–3838.

Sipes, D.L. and Einset, J.W. (1983) Cytokinin stimulation of abscission in lemon pistil explants. *Journal of Plant Growth Regulation* 2, 73–80.

Smallwood, M. and Bowles, D. J. (2002). Plants in a cold climate. *Philosophical Transactions of the Royal Society of London. Series B: Biological Sciences* 357(1423), 831–847.

Song, Y., Zhang, X., Li, M., Yang, H., Fu, D. *et al.* (2021) The direct targets of CBFs: In cold stress response and beyond. *Journal of Integrative Plant Biology* 63(11), 1874–1887.

Sreenivasulu, N., Harshavardhan, V.T., Govind, G., Seiler, C. and Kohli, A. (2012) Contrapuntal role of ABA: Does it mediate stress tolerance or plant growth retardation under long-term drought stress? *Gene* 506, 265–273.

Sridha, S. and Wu, K. (2006) Identification of AtHD2C as a novel regulator of abscisic acid responses in Arabidopsis. *The Plant Journal* 46, 124–133.

Sussmilch, F.C. and McAdam, S. (2017) Surviving a dry future: abscisic acid (ABA)-mediated plant mechanisms for conserving water under low humidity. *Plants* 6, 54. DOI: 10.3390/plants6040054.

Thorsen, S. M. and Höglind, M. (2010) Modelling cold hardening and dehardening in timothy. Sensitivity analysis and Bayesian model comparison. *Agricultural and Forest Meteorology* 150(12), 1529–1542.

Ton, J., Flors, V. and Mauch-Mani, B. (2009) The multifaceted role of ABA in disease resistance. *Trends in Plant Science* 14, 310–317

Urano, K., Maruyama, K., Ogata, Y., Morishita, Y., Takeda, M. *et al.* (2009) Characterization of the ABA-regulated global responses to dehydration in Arabidopsis by metabolomics. *The Plant Journal* 57, 1065–1078.

Venzhik, Y., Talanova, V. and Titov, A. (2016) The effect of abscisic acid on cold tolerance and chloroplasts ultrastructure in wheat under optimal and cold stress conditions. *Acta Physiologiae Plantarum* 38, 1–10.

Vishwakarma, K., Upadhyay, N., Kumar, N., Yadav, G., Singh, J. *et al.* (2017) Abscisic acid signaling and abiotic stress tolerance in plants: A review on current knowledge and future prospects. *Frontiers in Plant Science* 8, 161.

Waadt, R., Seller, C.A., Hsu, P.K., Takahashi, Y., Munemasa, S. and Schroeder, J.I. (2022) Plant hormone regulation of abiotic stress responses. *Nature Reviews Molecular Cell Biology* 23(10), 680–694.

Walters, D. and Heil, M. (2007) Costs and trade-off associated with induced resistance. *Physiological and Molecular Plant Pathology* 71, 3–17.

Wani, S.H. and Kumar, V. (2015) Plant stress tolerance: Engineering ABA: A potent phytohormone. *Transcriptomics* 3, 1000113.

Williams, A. and de Vries, F.T. (2020) Plant root exudation under drought: Implications for ecosystem functioning. *New Phytologist* 225, 1899–1905.

Wyatt, S.E., Rashotte, A.M., Shipp, M.J., Robertson, D. and Muday, G.K. (2002) Mutations in the gravity persistence signal loci in Arabidopsis disrupt the perception and/or signal transduction of gravitropic stimuli. *Plant Physiology* 130(3), 1426–1435.

Xia, C. and Møller, A.P. (2018) Long-lived birds suffer less from oxidative stress. *Avian Research*, 9(1), p.41.

Yan, J., Tsuichihara, N., Etoh, T. and Iwai, S. (2007) Reactive oxygen species and nitric oxide are involved in ABA inhibition of stomatal opening. *Plant, Cell and Environment* 30(10), 1320–1325.

Yang, J., Zhang, J., Wang, Z., Zhu, Q. and Liu, L. (2002) Abscisic acid and cytokinins in the root exudates and leaves and their relationship to senescence and remobilization of carbon reserves in rice subjected to water stress during grain filling. *Planta* 215, 645–652.

Yang, Y.Y., Ren, Y.R., Zheng, P.F., Zhao, L.L., You, C.X., Wang, X.F. and Hao, Y.J. (2020) Cloning and functional identification of a strigolactone receptor gene MdD14 in apple. *Plant Cell, Tissue and Organ Culture (PCTOC)* 140, 197–208.

Yoshida, T., Obata, T., Feil, R., Lunn, J.E., Fujita, Y., Yamaguchi-Shinozaki, K. and Fernie, A.R. (2019) The role of abscisic acid signaling in maintaining the metabolic balance required for Arabidopsis growth under nonstress conditions. *Plant Cell* 31, 84–105.

Yu, J., Cang, J., Lu, Q., Fan, B., Xu, Q., Li, W. and Wang, X. (2020) ABA enhanced cold tolerance of wheat 'dn1'via increasing ROS scavenging system. *Plant Signaling and Behavior* 15(8), 1780403.

Zhang, J., Jia, W., Yang, J. and Ismail, A.M. (2006) Role of ABA in integrating plant responses to drought and salt stresses. *Field Crops Research* 97(1),111–119.

Zhang, X., Wollenweber, B., Jiang, D., Liu, F. and Zhao, J. (2008) Water deficits and heat shock effects on photosynthesis of a transgenic Arabidopsis thaliana constitutively expressing ABP9, a bZIP transcription factor. *Journal of Experimental Botany* 59, 839–848.

Zhang, X., Zhang, L., Sun, Y., Zheng, S., Wang, J. and Zhang, T. (2020) Hydrogen peroxide is involved in strigolactone induced low temperature stress tolerance in rape seedlings (Brassica rapa L.). *Plant Physiology and Biochemistry* 157, 402–415.

Zhang, A., Ji, Y., Sun, M., Lin, C., Zhou, P. *et al.* (2021a) Research on the drought tolerance mechanism of Pennisetum glaucum (L.) in the root during the seedling stage. *BMC Genomics* 22, 568.

Zhang, Q., Yuan, W., Wang, Q., Cao, Y., Xu, F., Dodd, I.C. and Xu, W. (2021b) ABA regulation of root growth during soil drying and recovery can involve auxin response. *Plant, Cell and Environment*, 45, 871–883.

Zhao, J. (2025). *Rice growth promotion and cold stress alleviation by an endophytic bacterium Microbacterium testaceum M15 isolated from rice seed.* PhD thesis.

Zhao, M.L., Wang, J.N., Shan, W.E.I., Fan, J.G., Kuang, J.F. *et al.* (2013) Induction of jasmonate signalling regulators MaMYC2s and their physical interactions with MaICE1 in methyl jasmonate-induced chilling tolerance in banana fruit. *Plant, Cell and Environment* 36(1), 30–51.

Zahra, N., Shaukat, K., Hafeez, M.B., Raza, A., Hussain, S., Chaudhary, M.T., Akram, M.Z., Kakavand, S.N., Saddiq, M.S. and Wahid, A. (2021) Physiological and molecular responses to high, chilling, and freezing temperature in plant growth and production: consequences and mitigation possibilities. In: *Harsh environment and plant resilience: molecular and functional aspects* (pp. 235–290). Springer International Publishing: Cham.

Zhou, Y., He, R., Guo, Y., Liu, K., Huang, G. *et al.* (2019) A novel ABA functional analogue B2 enhances drought tolerance in wheat. *Scientific Reports* 9, 2887.

Zhu, J.-K. (2002) Salt and drought stress signal transduction in plants. *The Annual Review of Plant Biology* 53, 247–273.

Zhu, M. and Assmann, S.M. (2017) Metabolic signatures in response to abscisic acid (ABA) treatment in Brassica napus guard cells revealed by metabolomics. *Scientific Reports* 7, 12875.

11 Genetic Engineering for Stress Tolerance

Muhammad Anwar[1]*, Saquib Waheed[2], Muhammad Mohsin Altaf[1] and Hafiz Muhammad Rizwan[3]

[1]School of Tropical Agriculture and Forestry (School of Agricultural and Rural Affairs, School of Rural Revitalization), Hainan University, Haikou 570228, China; [2]Department of Neurology, Yongchuan Hospital of Chongqing Medical University, Chongqing, China; [3]Shenzhen Key Laboratory of Food Nutrition and Health, College of Chemistry and Environmental Engineering, Shenzhen University, Shenzhen 518060, China

Abstract

The chapter discusses biotechnological solutions to the dismal agricultural productivity due to environmental stressors such as drought, salinity and high temperatures. It argues that though genetic engineering of genes to deal with direct protective mechanisms will provide one approach it will be even more effective to manipulate the regulatory mechanisms that coordinate the overall response of a plant to stress. The focus of this is transcription factors which are the master proteins that regulate the expression of many downstream genes. Specific families have been discussed: WRKY proteins are described as central integrators that coordinate abiotic and biotic stress signals via complex hormonal crosstalk; NAC factors are emphasized due to dual activation of development and the induction of adaptive responses to such conditions as water scarcity; and MYB regulators are mentioned because of their impact on protective metabolic responses. The conclusion of the synthesis of the research is that alteration of these genetic regulators presents a strong platform on which to produce resilient crops. The key to the future of this area lies in understanding the complex interaction of these regulation networks and applying these results to specific solutions to sustainable agriculture in the climate change scenario.

Keywords: biotic and abiotic stress, metabolic and genetic engineering, stress tolerance, transcriptional factors, phytohormones

11.1 Introduction

Unfavorable climatic conditions, such as high salt, water drought, extremes in temperature, flooding, and nutrient deficiency, significantly impact plant growth and survival (Goel and Madan, 2014; Sehgal *et al.*, 2018; Kobata, 2024; Kumar *et al.*, 2024). These abiotic stressors have an impact on plant development, lowering productivity, and decreasing seed quality, which lowers the potential yield (Goel and Madan, 2014; Sehgal *et al.*, 2018). Climate change is projected to reduce agricultural productivity by 10–25%, with expectations of further declines over the next 50 years. At present, insect pests are responsible for 38% of agricultural losses, whereas weeds account for 34% of the losses. Abiotic stressors are thought to be the cause of a greater than 50% decrease in yield (Junaid and Gokce, 2024).

* Email: anwar_uaar@yahoo.com

DOI: 10.1079/9781800628823.0011

Plants have developed a diverse array of mechanisms for detecting, responding to, and adjusting to alterations in their environment. The existing knowledge regarding the mechanisms that enable crops to sustain yield in the face of abiotic stress is limited. Recent developments in 'omic' technologies are creating opportunities for the discovery of transcriptional, translational, and post-translational mechanisms, as well as signaling pathways that govern plant responses to various stresses (Ahuja *et al.*, 2010). The employment of model plants, including *Arabidopsis thaliana*, *Brachypodium distachyon*, and *Medicago truncatula*, has offered essential resources for comprehending the genetic and biochemical foundations of adaptations to abiotic stress (Bohnert *et al.*, 2006; Hirayama and Shinozaki, 2010). Nonetheless, there has been some degree of success in developing cultivars that are tolerant to abiotic stress via genetic engineering, although it is important to note the relatively small number of transgenic crops that have been introduced to the market to date (Peleg *et al.*, 2011).

A crucial factor to evaluate in the breeding of crops that are tolerant to abiotic stress is the assessment of the success of transgenic plants. From a physiological standpoint, the primary characteristic indicative of plant stress tolerance is survival or recovery. However, from an agronomic perspective, the yield of the crop serves as the principal measure of the effectiveness of stress-tolerant varieties. Consequently, although numerous studies have documented resistance to abiotic stress, most of this research has been conducted using model plants subjected to artificially extreme conditions, such as elevated salinity, significant dehydration, and osmotic shock (Chaudhary *et al.*, 2024; Kumar *et al.*, 2024; Lallawmkimi *et al.*, 2024). The recovery of plants following a stress event has been utilized as a measure of their tolerance. Nevertheless, in natural field conditions, crops must contend with various environmental stresses that differ in their timing, duration, and intensity (Mittler and Blumwald, 2010).

One significant method of producing plants with greater resistance to abiotic stress is through genetic engineering techniques. Plants may withstand stressful environments by upregulating the expression of many genes in response to abiotic stress. These genes that are induced by abiotic stress are divided into two categories: (1) genes that directly protect against stress through their products, such as chaperones, antifreeze proteins, osmoprotectants, and detoxification enzymes; and (2) genes that control the expression of genes related to stress and signal transduction in the stress response, such as transcription factors (TFs), protein kinases, etc. (Yamaguchi-Shinozaki and Shinozaki, 2007). Through the use of an Agrobacterium-mediated transformation system, the stress-inducible genes and their regulatory genes can be changed, transferred, and expressed in different species. Alternatively, the genes can be expressed ectopically, inducing biochemical, molecular, and physiological changes that enhance plant growth, development, and yield under stressful circumstances (Bhatnagar-Mathur *et al.*, 2008). Strong constitutive promoters such as CaMV 35S can regulate the overexpression of stress-inducible genes, hence enhancing stress tolerance. However, this uncontrollably high level of expression depletes cells' reserves of energy, leading to growth retardation. In recent times, the employment of stress-inducible promoters to express stress-inducible genes has guaranteed optimal gene expression at a time-specific level (Kasuga *et al.*, 2004).

This chapter covers several methods of enhancing abiotic stress tolerance through ion transport channels, TFs, and osmoprotectant genetic engineering, overexpression of stress signaling genes, and genetic engineering in phytohormones.

11.2 Overexpression of Genes for Transcriptional Regulation

The strategy of altering individual genes that encode specific metabolic pathways to enhance crop tolerance to abiotic stress has yielded only modest results. The various mechanisms that contribute to plant adaptation under stress, along with the intricate nature of their interactions, partially elucidate the reasons why this approach may be ineffective in practical applications. Furthermore, plants consistently strive to re-establish metabolic homeostasis, which allows them to function in opposition to the altered enzyme. Conversely, focusing on essential regulatory genes that influence various pathways related to metabolic fluxes may aid in re-establishing metabolic homeostasis in plants during periods of stress, thereby enhancing the likelihood of successful outcomes (Reguera *et al.*, 2012). Proteins known as TFs play a crucial role in nearly all biological processes, making them promising candidates for targeting in the development of crops that can withstand stress (Yamaguchi-Shinozaki and Shinozaki, 2006; Reguera *et al.*, 2012). TFs bind to the cis-elements of promoter regions to increase the production of genes that indicate stress, hence improving resistance to various stressors (Bhatnagar-Mathur *et al.*, 2008). The TFs that are involved in stress signalling, include MYB, NAC transcriptional factors, zinc-finger domains, dehydration response elements (DREs), and basic leucine zippers (Xiong *et al.*, 2014). The overexpression of TFs in transgenic plants has demonstrated enhanced stress tolerance, making it an effective technique for the production of crops resistant to drought (Xiong *et al.*, 2014).

11.3 WRKY Transcriptional Factor

Plants are constantly exposed to a range of potential biotic and abiotic stress. These include pathogens, pests, the direct influence of animals and humans, and extreme temperature variations due to global warming and other unfavorable environmental conditions (Singh *et al.*, 2023). Abiotic stress is one of the major constraints on plant growth, resulting in a significant decline in agricultural productivity worldwide (Kopecká *et al.*, 2023). Drought and heat are essential stresses that significantly impact agricultural productivity by causing cellular damage and limiting resources (Oshunsanya *et al.*, 2019).

Over the past decade, extreme temperature variations have been observed worldwide due to global warming, substantially impacting plant functions such as seed germination, photosynthesis, reproduction, and yield (Ainsworth and Ort, 2010). Plants have developed intricate molecular signaling mechanisms that recognize environmental stress or pathogen attack and initiate several signaling cascades to ensure survival and adapt to these varying environmental factors. Gene expression regulation is a crucial regulatory factor of the defense system and is primarily controlled by various TFs present in the promoter region of the target gene (He *et al.*, 2023). TFs are known to play vital roles in regulating various cellular mechanisms. They are critical regulators of plant growth and development by turning gene expression on and off to ensure they are upregulated or downregulated in the desired cell at the right time and location (Akmakjian and Bailey-Serres, 2022). During the last decade, a great effort has been dedicated to breeding initiatives by employing diverse approaches to achieve more stress- tolerant plants.

TFs could be pivotal in plants' resilience to withstand adverse environmental conditions. Recently, a significant focus of plant molecular biologists has been on the WRKY TFs, which contribute to plant growth and development, play a vital role in various biotic and abiotic responses, and regulate physiological processes (Li *et al.*, 2024a). The WRKY family is one of the largest and most functionally diverse TF families in plants. It was originally identified in sweet potatoes SPF1 (Sweet Potato Factor1) containing one or two WRKY domains (Liu *et al.*, 2022). The recent research on the regulatory effects of the WRKY TF has shown critical responses to external stresses, particularly effects on plant secondary metabolism and its role in plants' signalling networks that control various biological processes.

11.3.1 Structure and classification of WRKY transcription factors

WRKY proteins comprise a large family of TFs characterized by one or two highly conserved WRKY domains, including the hallmark N-terminal core motif WRKYGQK and a zinc-finger structure distinct from other known zinc-finger motifs (Xu *et al.*, 2020). WRKY TFs are categorized as members of the novel Zn-chelating DNA-binding domain (DBD), which share four standard folds and are classified as a WRKY-GCM1 family (Xu *et al.*, 2020). The WRKY and GCM1 are metal-chelating DBDs of mutant and mutant-like element (MULE) transposases (Babu *et al.*, 2006). The core conserved feature of GCM1 consists of one or two N-terminal WRKY sequences and a C-terminal Zn-finger motif, and both of these are necessary for the binding to the promoter W-box (C/T) TGAC (T/C) and regulate the transcription (Babu *et al.*, 2006). According to a study by Yamasaki *et al.* (2005), in *A. thaliana*, WRKY4 protein's WRKY domain comprises a four-stranded β-sheet with Cys/His residues, revealing a novel zinc-binding pocket and enabling extensive hydrophobic interactions (Yamasaki *et al.*, 2005). The hydrophobic interactions are made possible with the help of the Gly residue located in the middle of the N-terminal β-strand, further enhancing the structural stability of the β-sheet. A structural model predicts the β-strand containing the WRKYGQK motif connects with a 6 bp long region, compatible with the W-box's (TTGACY) length (Yamasaki *et al.*, 2005). These different binding capabilities are critical for plant transcriptional reprogramming during various stress responses.

WRKY TFs are classified into three main groups based on their structure:

- **Group I**: possesses two WRKY domains, one at the N-terminus and one at the C-terminus, and have a C2H2-type zinc-finger motif.
- **Group II**: possesses one WRKY domain and is subdivided into numerous subgroups, including IIa, IIb, IIc, IId, and IIe, based on additional structural features. They also contain a C2H2-type zinc-finger motif.
- **Group III**: possesses one WRKY domain with a C2HC-type zinc-finger motif, distinguishing them from other known zinc-finger motifs.

Numerous past studies have shown that overexpression lines, or knock-down or knock-out mutants, of WRKY genes can positively or negatively regulate different aspects of plants' defense responses through WRKY TFs. In higher plants, WRKY proteins are encoded by large families, with more than 70 members in *A. thaliana* alone and over 100 in rice. This suggests that the WRKY TF has evolved to perform a wide range of functions, especially during various stress responses.

11.3.2 Activation of defense genes by WRKY TFs

WRKY TFs play a crucial role in helping plants cope with various biotic and abiotic stress responses by modulating the transcription (Li *et al.*, 2024a). Studies have shown that most WRKY genes are involved in the response pathways of significant abiotic stresses, including drought, cold, heat, oxidative stress, and other defense mechanisms (Li *et al.*, 2020; Wang *et al.*, 2023). Under abiotic stress conditions, various WRKY TFs activate or downregulate the transcription of downstream genes involved in stress adaptation. Several studies have demonstrated that WRKY significantly modulates stress-responsive hormones and the expression of genes related to reactive oxygen species (ROS) (Khoso *et al.*, 2022; Wang *et al.*, 2023; Chen *et al.*, 2024a). The overexpression of TaWRKY133 negatively regulates the expression of stress-related genes by binding to the W-box elements of the promoter region of the target genes, ultimately leading to a decreased tolerance and disturbing the water balance in *A. thaliana* (Lv *et al.*, 2022). Similarly, another recent study highlighted the role of OsWRKY54 in *Oryza sativa*, enhancing salt tolerance through ROS scavenging and ion homeostasis (Huang *et al.*, 2022). The upregulation of rice mutant *OsWRKY54* led to increased expression of *OsNHX4* and *OsHKT1;5* genes. It significantly impacted ROS and sodium ion transport detoxification, leading to salt tolerance and enhanced plant growth and development during saline conditions (Huang *et al.*, 2022).

The role of the WRKY TF in regulating various hormones, specifically abscisic acid (ABA), under stress response has also been well-documented. ABA acts as an endogenous messenger and a premier signal for plants to regulate the water status during drought and salinity (Tuteja, 2007). The WRKY TF governs signaling pathways and modulates the expression of ABA-responsive genes by interacting with elements in the promoters of ABA genes, thereby conferring stress tolerance upon plants. For instance, WRKY40 is a central negative regulator of ABA-responsive genes and represses the expression of AB15 in *A. thaliana*, a key TF in the ABA signalling pathway (Shang *et al.*, 2010). Similarly, the downregulation of WRKY40 results in the activation of ABA-responsive genes, leading to an enhanced plant stress tolerance.

WRKY TFs are also crucial in assisting plants with abiotic challenges and enhancing plant defense. For example, WRKY70 in *A. thaliana* is a positive regulator of defense responses by regulating the pathogenesis-related (PR) genes, including PR1, PR2, and PR5 (Li *et al.*, 2004). It also plays a vital role in maintaining the balance between jasmonic acid (JA) and salicylic acid

(SA)-mediated defense and fine-tuning the plant's resistance against biotrophic pathogens (Li *et al.*, 2004). Another study revealed that WRKY33 in *A. thaliana* directly regulates the biosynthesis of camalexin and increases its accumulation during a pathogenic attack by modulating the expression of CYP71A13 and PAD3 genes (Qiu *et al.*, 2008; Mao *et al.*, 2011; Birkenbihl *et al.*, 2017). Other than PR genes, WRKY TFs also strengthen the plant cell wall, which is a physical barrier against attack pathogens. WRKY15 plays a crucial role in regulating the expression level of genes involved in the biosynthesis of callose and lignin (Ge *et al.*, 2020), which are essential constituents of the cell wall of plants, limiting pathogen entry and restricting the spread of infection.

The advancements in transcriptomic analyses and geolocation studies have further clarified the role of WRKY TFs in plant defense. For instance, a study by Li *et al.* (2021) used CRISPR/Cas9 to knockout WRKY3 and WRKY4 to generate *A. thaliana* mutants, which led to a compromised resistance against pathogens and emphasized the importance of TFs in defense regulation (Li *et al.*, 2021). Similarly, a comprehensive transcriptome analysis by Liu *et al.* (2022) revealed that WRKY TFs are part of a more extensive regulatory network that includes other TF families, such as MYB, NAC, and bZIP, which coordinate a robust defense response. Future research will likely focus on unravelling the complex interactions between WRKY TFs and other signalling molecules and their roles in different plant species and under various environmental conditions.

11.3.3 Crosstalk between signaling pathways

Plants, being sessile organisms, must cope with various stress conditions, and it is the most critical hurdle in plant growth and development. Salinity and/or elevated temperature often accompany drought. These conditions passively affect plant growth and productivity by causing cellular damage, disrupting metabolic processes, and limiting resources. During stress, plants may also have a nutrient deficiency, resulting in stunted growth and decreased plant productivity and/or fertility (Francis *et al.*, 2023). To tackle these challenges, plants have evolved complex, intricate signalling networks that allow them to perceive and integrate multiple stress conditions. Combined abiotic and biotic stress events may result in synergistic or antagonistic interactions. Under stress conditions, a plant's defense mechanisms respond by triggering complex signalling networks that relay extracellular cues and crosstalk with other signalling pathways to make the response more robust and ensure a coordinated and efficient adaptation process (Trotta *et al.*, 2014; Kouhen *et al.*, 2023). The intricate interactions and crosstalk between key signalling pathways during stress conditions highlight their importance in plant survival and resilience.

Several well-characterized signalling pathways primarily mediate different stress responses in plants. Hormonal signalling pathways are the most extensively studied, as they play a central role during stress conditions. Among the hormonal signaling pathways, the ABA signalling pathway plays a significant role, as it is synthesized during stress response and activates other downstream TFs and signalling pathways (Kumar *et al.*, 2019; Rai *et al.*, 2024). In *A. thaliana* TF WRKY46, increased drought resistance was demonstrated by regulating the ABA signaling pathway and positively regulating the genes responsive to abiotic stress (Ding *et al.*, 2014). Similarly, WRKY TFs, along with WRKY46, integrate ABA signaling and help in crosstalk with other signaling pathways to fine-tune drought stress. Another study suggested that in *A. thaliana*, WRKY25 and WRKY26 are associated with heat stress by regulating the heat shock proteins (HSPs) and are involved in crosstalk with other stress responses, providing enhanced plant thermotolerance (Li *et al.*, 2011). The phytohormones ethylene (ET), JA, and SA regulate key defense signaling pathways, in part by modulating the production of ROS.

ROS possess a dual function: they can be a damaging agent at high levels but also critical signaling molecules. Their accumulation is perceived as a stress signal, triggering a downstream antioxidant defense mechanism that is mediated by WRKY TFs. Similarly, calcium ions (Ca^{2+}) are secondary messengers mediating numerous protein and stress-related genes. A study by Rushton *et al.* (2010) demonstrated that WRKY70 plays a dual role in *A. thaliana* by upregulating the SA and downregulating JA because JA and SA act antagonistically; JA is primarily involved in defense against necrotic pathogens and SA against biotrophic pathogens. These studies indicate that WRKY TFs are pivotal in crosstalk between different signaling pathways in balancing plant defense against pathogens (Fig. 11.1).

11.4 NAC Transcription Factor

NAC is a class of TFs that plants express in response to stress and has been demonstrated to be crucial for development. The NAC domain, a highly-conserved area, distinguishes the NAC protein. The first NAC gene, NAM, was discovered in petunia (Souer *et al.*, 1996). The highly varied transcriptional regulatory region at the C-terminal and the highly homologous DBD at the N-terminal define a standard NAC protein (Puranik *et al.*, 2012). The NAC domain is located at the N-terminus, contains approximately 150–160 amino acids, and is divided into five subdomains, designated A–E (Ooka *et al.*, 2003). It is possible that subdomain A contributes to the creation of functional heterodimers or homodimers. As DNA-binding sites,

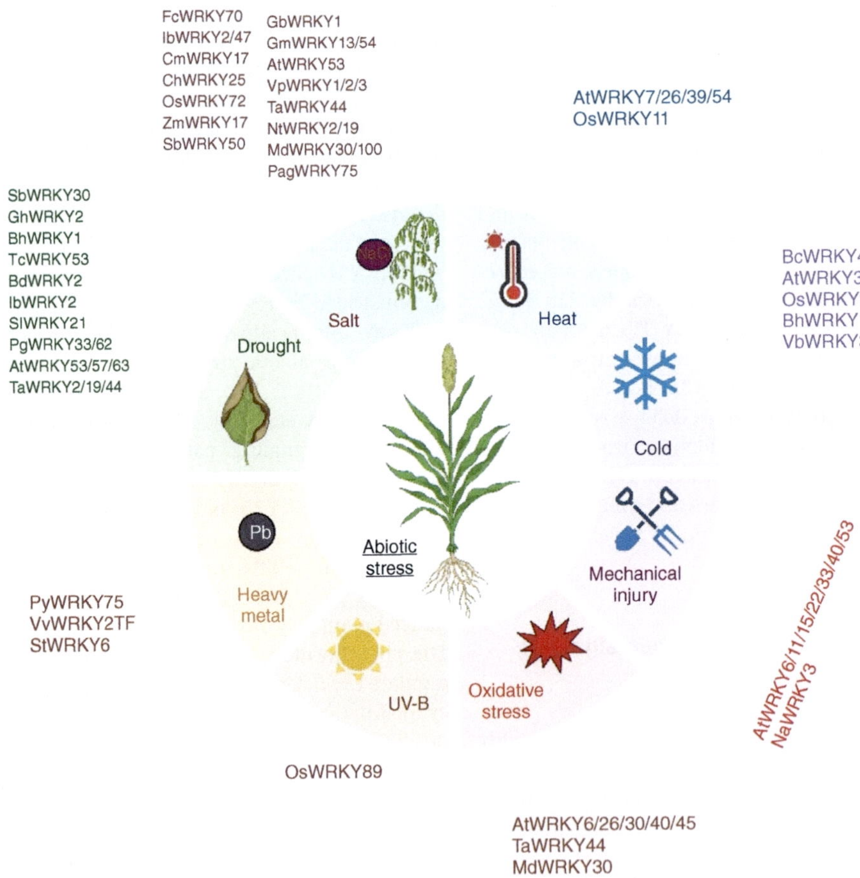

Fig. 11.1. The involvement of WRKY TFs in regulating the plant's defense mechanism against various abiotic stress conditions. (Figure author's own.)

subdomains C and D are positively charged and have been conserved throughout evolution. Subdomains B and E most likely have different functions within NAC proteins. The NAC protein's C-terminal varies greatly between species. The C-terminal controls the transcription of genes that are downstream since it is a transcriptional regulatory region. Transmembrane motifs found in the C-terminal region of certain NAC proteins may localize the proteins to the endoplasmic reticulum (ER) or plasma membrane (Seo *et al.*, 2010). So far, 148 NAC genes in *Zea mays* (Peng *et al.*, 2015), 488 in *Triticum aestivum* (Guerin *et al.*, 2019), 283 in *Gossypium hirsutum* (Sun *et al.*, 2018), and 139 in *Glycine max* (Hussain *et al.*, 2017) have been discovered. 75 NAC genes have been predicted for rice (Ooka *et al.*, 2003).

GhNAC2, which is a drought-tolerance germplasm line and have been isolated from Jinmian 19. When compared to normal watering, the 15% PEG treatment dramatically increased the expression level of GhNAC2 in leaves at 8 hours (Shang *et al.*, 2020). Compared to wildtype controls, *Arabidopsis* ectopically expressing *GhNAC3* has produced significantly greater rates of seed germination and longer root lengths on a medium containing 500 mmol/l mannitol (Xia *et al.*, 2023).

Abiotic stressors are also known to activate GhNAC4, and overexpression of this gene in tobacco significantly boosts drought tolerance by upregulating the expression levels of several genes that respond to stress and stomatal movement (Trishla and Kirti, 2021). The overexpression of CaNAC064 in pepper was strongly induced by ABA and SA and increased cold tolerance (Hou *et al.*, 2020). Overexpression of NtNAC053 in tobacco significantly increased tolerances to drought and salt stresses (Li *et al.*, 2022). The productivity of horticultural crops is significantly impacted by stress caused by low temperatures. The active regulation of cold-responsive genes is essential for enhancing cold tolerance in plants. Overexpression of SlNAC3 in tomatoes increased the cold tolerance, while reducing cold tolerance upon silencing (Wang *et al.*, 2024c). The list of NAC transcriptional factors involved in biotic and abiotic stress is listed in Table 11.1.

Table 11.1. Role of NAC transcriptional factors in plants against abiotic and biotic stress. (Table author's own.)

Gene	Species	Functions	Method	Reference
GhJUB1L1 J	Cotton	Drought tolerance	VIGS	Chen *et al.* (2021)
StNACb4	Potato	Positive regulation bacterial wilt	Overexpression	Chang *et al.* (2020)
SlNAC1	Tomato	Positive regulation cold, heat, drought and salt	Overexpression/RNA	Niu *et al.* (2022)
GhATAF1	Cotton	Salt tolerance, Verticillium susceptibility	Overexpression	He *et al.* (2016)
StNAC2	Tomato	Salt tolerance	Overexpression	Xu *et al.* (2014)
CaNAC064	Pepper	Positive regulation cold stress	Overexpression/VIGS	Hou *et al.* (2020)
GhNAC12	Cotton	Leaf senescence	Ectopic expression in *Arabidopsis thaliana*	Zhao *et al.* (2016)
StNAC103	Potato	Positive regulation of antibacterial response	RNA interference (RNAi)	Dastmalchi *et al.* (2021)
NOR-like1	Tomato	Positively regulates fruit softening	Overexpression	Peng *et al.* (2024)
NtNAC053	Tobacco	Positive regulation of salt and drought	Overexpression	Li *et al.* (2022)
GhNAC3	Cotton	Drought tolerance	Ectopic expression in *Arabidopsis thaliana*/ VIGS	Xia *et al.* (2023)
SlNAC3	Tomato	Cold tolerance	Overexpression/silencing	Wang *et al.* (2024c)

11.5 MYB Transcriptional Factor Mediates Stress Tolerance in Plants

The MYB gene family contained the first TFs discovered in plants, which were linked to a bright phenotypic (Albert and Allan, 2021). Because the protein v-MYB promotes oncogenes, a modified version of the cellular c-MYB, it is referred to as MYB TFs. The MYB TF was discovered for the first time using the avian leucosis virus, also known as the 'v-MYB' virus. One of the most famous classes of TFs, MYBs were first found in animals, but they are more common in plants (Wang *et al.*, 2024d). The ZmMYBC1 gene, which was initially identified in higher plants, regulates the generation of anthocyanins in maize (Li *et al.*, 2019). Following the original discovery, a large number of MYB genes from model organisms and crops were identified. Subsequently, studies were conducted to ascertain the activity of these genes and to identify the TF partners of these genes. Many MYB TFs have been identified in different species, the majority of which function as activators. The activator MYBs have cis-elements that are found in gene promoters, which helps to induce the transcriptional machinery and improves the expression of genes. Even though MYB TFs with repressive functions have been known about for a while, people have just recently become aware of their frequency and importance (Anwar *et al.*, 2019; Anwar *et al.*, 2021).

The conserved DBD located at the N-terminal region of MYB TFs is used to categorize them. This domain consists of one to four 50–53 amino acid long incomplete MYB repeats (Pu *et al.*, 2020). The C-terminus of MYB proteins contains a number of activation domains that contribute to the wide range of regulatory roles of these TFs (Anwar *et al.*, 2021). These groups consist of MYB-related proteins, R2R3-MYB, 4R-MYB, and R1R2R3-MYB (Dubos *et al.*, 2010). R2R3-MYB proteins are mostly found within the plant kingdom. R2R3 MYB genes have been extensively studied and shown to be crucial for controlling phenylpropanoid pathways in plants, as well as additional pathways like terpenoid, secondary metabolism and secondary walls (Anwar *et al.*, 2023; Anwar *et al.*, 2024; Tang *et al.*, 2024).

The ability of a plant to tolerate abiotic stress is directly correlated with the expression of stress-related genes that are functionally regulated by TFs. The ability of a plant to tolerate abiotic stress is directly correlated with the expression of stress-related genes that are functionally regulated by TFs. For plants to survive, therefore, abiotic stress responses and adaptations are crucial (Li *et al.*, 2015; Wang *et al.*, 2024d). Abiotic stress regulation of genes and proteins depends critically on TFs (Su *et al.*, 2024).

Cellular patterning and morphology can be controlled by MYB TFs. Therefore, in response to external stressors, plants can regulate the expression of pertinent functional genes within their systems by signalling pathways, which initiate a series of physiological reactions (Li *et al.*, 2019). An efficient and well-structured signal conditioning network is created by biochemical processes to reduce or eliminate plant harm. A large number of genes, signaling pathways, and gene products are involved in the plant's response to these circumstances. Transcriptional control is therefore essential for plants to react to environmental stressors. For example, MaMYBR30 from mulberry has great potential to positively regulate drought resistance (Liu *et al.*, 2024b). Under salinity and drought conditions, VhMYB15 was highly expressed in roots and mature leaves, and higher survival rates were observed (Han *et al.*, 2024). Overexpression of LpMYB1 in *L. pumilum* roots displayed increased salt tolerance, and chlorophyll, proline, superoxide dismutase (SOD), peroxidase (POD) and catalase (CAT) enzymes were significantly higher (So *et al.*, 2024). The expression level of FvMYB108 was higher in the roots and leaves at the young stage. The higher expression levels of AtCOR47, AtCBF1, AtDREB1A, and AtERD10 genes were observed in overexpression lines of *Arabidopsis*. These genes are involved in low-temperature stress. The genes, including tCCA1, AtRD29a, AtP5CS1, and AtSnRK2.4, which are associated with salt stress tolerance, also showed higher expression. So, FvMYB108 may be involved in combating the serious threats of various stress (Song *et al.*, 2024). Ectopic expression of IbMYB330 in tobacco was subjected to enhanced expression levels of genes including SOD, POD, ascorbate peroxidase (APX), and pyrroline-5-carboxylate synthase (P5CS) (Wang *et al.*, 2024a). PagMYB147, isolated from poplar, provides defense against *Melampsora magnusiana* through the involvement in the regulation of ROS homeostasis, JA signaling, and SA biosynthetic pathway (Wang *et al.*, 2024b). MYB8 controlled the transcript of iron-regulated transporter 1 (IRT1) and positively regulated in response to iron deficiency stress (Gong *et al.*, 2024). By controlling the expression of genes linked to the anthocyanin synthesis and cold-responsive gene pathways, PbMYB1L improves the anthocyanin accumulation and cold tolerance of transgenic *Arabidopsis* (Zhou *et al.*, 2024). Overexpression of GhTT2 in *Arabidopsis* increases the production of proanthocyanidin and improves the osmotic tolerance (Mei *et al.*, 2024). GbMYB5 from cotton improves the drought tolerance in transgenic tobacco and cotton (Chen *et al.*, 2015). The list of MYB transcriptional factors involved in biotic and abiotic stress is listed in Table 11.2.

The salt overly sensitive (SOS) biosynthesis pathway during salt stress in roots and shoots is regulated by MYB transcriptional factors. Figure 11.2 shows a potential model for salinity adaptation mediated by MYB TFs.

11.6 Zinc-finger Domains

One of the biggest groups of TFs, the zinc-finger proteins (ZFPs), get their name from the "finger-like" zinc finger and are widely found throughout the kingdom of plants (Davletova *et al.*, 2005). DNA binding motifs known as the traditional or TFIIIA-type fingers, or Cys-2/His-2-type zinc fingers, are present in eukaryotic TFs. With a consensus sequence of Cx2-4Cx9-12Cx2Cx-4Cx2Hx5HxC, where x indicates any amino acid, they have roughly 30 amino acids. Two pairs of conserved Cys and His are coupled tetrahedrally to a zinc ion (Zn^{2+}) (Pabo *et al.*, 2001). The length and sequence of the long spacer that divides adjacent fingers in multiple-fingered proteins varies greatly from protein to protein. The zinc-figure motifs of plants' TFIIIA-type ZFPs, which comprise the EPF family, are distinguished by the extremely conserved sequence QALGGH (Frugier *et al.*, 2000). This family includes SCOF-1 from soybean, involved in cold tolerance, STZ/ZAT10 from *Arabidopsis*, involved in salt tolerance

Table 11.2. Role of MYB transcriptional factors in plant tolerance. (Table author's own.)

Gene	Species	Functions	Method	References
LpMYB1	*L. pumilum*	Salt tolerance increased	Overexpression	So *et al.* (2024)
MYB8	*Arabidopsis*	Positively regulates Fe deficiency stress	Expression	Gong *et al.* (2024)
MaMYBR30	Mulberry	Salt and drought stress	VIGS/ Overexpression	Liu *et al.* (2024b)
AsMYB1R	*Avena sativa*	Regulates ROS homeostasis	Expression	Chen *et al.* (2024b)
VhMYB15	*Arabidopsis*	Increases salinity and drought tolerance	Overexpression	Han *et al.* (2024)
FvMYB108	*Fragaria vesca*	Salt and cold tolerance of *Arabidopsis*	Ectopic overexpression	Song *et al.* (2024)
IbMYB330	Sweet potato	Drought and salt stress tolerance	Ectopic overexpression	Wang *et al.* (2024b)
PbMYB1L	*Pyrus bretschneideri*	Cold tolerance	Ectopic overexpression	Zhou *et al.* (2024)
GhTT2	*Gossypium hirsutum*	Osmotic tolerance and proanthocyanin production	Ectopic overexpression	Mei *et al.* (2024)
GbMYB5	*Gossypium hirsutum*	Drought stress	Ectopic overexpression	Chen *et al.* (2015)
OsMYB91	Rice	Salt stress	Expression	Zhu *et al.* (2015)
PagMYB147	Poplar	Defense against *Melampsora magnusiana*	Overexpression	Wang *et al.* (2024a)
OsMYB6	Rice	Salt stress	Overexpression	Tang *et al.* (2019)

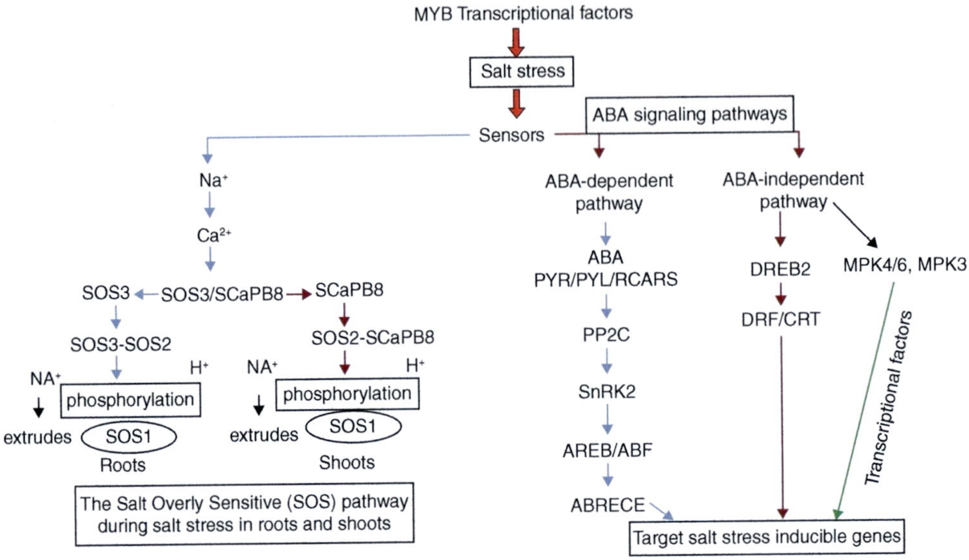

Fig. 11.2. A potential model for salinity adaptation mediated by MYB transcriptional factors. (Figure author's own.)

(Lippuner *et al.*, 1996, Kim *et al.*, 2001). ZFP (Zat12) responds to various biotic and abiotic stresses. It is believed that Zat12 is involved in the signalling of *Arabidopsis* in response to oxidative stress and cold (Davletova *et al.*, 2005). ZmAN13, from the maize, is a putative member of the ZnF-AN1 gene family. In comparison to the wild-type, *Arabidopsis* with an overexpression of ZmAN13 exhibited enhanced sensitivity to salt and drought during the germination and seedling stages but also acquired tolerance to cold (Ben Saad *et al.*, 2019).

11.7 Nuclear Factor Y (NF-Y) and Stress Response Regulation

The TFs known as nuclear factor Y (NF-Y) are crucial for various aspects of plant development and their physiological responses and stress-response regulation (Wang *et al.*, 2021). Nevertheless, the connection between NF-Y, plant hormones, and the resilience of tropical crops to stress is still not well understood. In recent research, the MeNF-YC15 gene was identified in the NF-Y family, which has a great perspective to respond to *Xanthomonas axonopodis* pv. *manihotis* (Xam) treatment. It was observed that overexpression of MeNF-YC15 positively regulated against cassava bacterial blight (CBB) diseases. In conclusion, the interaction between MePP2C1 and MeNF-YC15 plays a crucial role in regulating ET biosynthesis and enhancing disease resistance in cassava, thereby contributing to the genetic improvement of this crop through a defined gene network (Zheng *et al.*, 2024).

The overexpression of OsNF-YB9 in rice results in abnormalities in seed development and the production of smaller grains (Niu *et al.*, 2021). In *Arabidopsis*, the NF-Y TF COTYLEDON1 (LEC1) plays a crucial role in seed development through its interactions with various TFs (Song *et al.*, 2021). Furthermore, NF-Y proteins play a crucial role in regulating plant development and stress response by influencing the transcriptional activity of downstream genes. The NF-Y TF complex, comprising NF-YC, NF-YB, and NF-YA in conjunction with PYRABACTIN RESISTANCE1 (PYR) and PYR1-LIKE (PYL), enhances the tolerance of soybean plants to abiotic stress through the modulation of the ABA signaling pathway (Yu *et al.*, 2021). The NF-Y subunit in plants plays a crucial role in enhancing disease resistance, particularly in rice (Tan *et al.*, 2022).

The analysis of qRT-PCR expression indicates that the six EcNF-Y genes, specifically EcNF-YA1, EcNF-YA5, EcNF-YA16, EcNF-YB6, EcNF-YB10, and EcNF-YC2, may be linked to tolerance against both dehydration and salinity stress during early stress conditions. This suggests that these genes play a role in the multiple stress responses observed in the tolerant genotype. Further functional characterization of these genes is required (Rani *et al.*, 2024a).

It is important to highlight that the role of the NF-Y gene is frequently associated with the hormonal signaling pathway. For example, in rice, the OsNF-YAs TFs play a detrimental role in the plant's antiviral capacity by engaging with the core components of JA, specifically myelocytomatosis (OsMYC2/3), thereby suppressing the JA signaling pathway (Tan *et al.*, 2022).

JA is a recognized phytohormone that plays a crucial role in the plant's response to biotic stress. The trimeric complex composed of NF-YA1, NF-YB2, and NF-YC9 plays a crucial role in enhancing the expression of genes responsive to salinity. In contrast, the JASMONATE-ZIM DOMAIN protein 8 (JAZ8) engages directly with all three subunits, serving as a primary repressor that inhibits the formation of the NF-YA1-YB2-YC9 trimeric complex and diminishes the transcriptional activation capabilities of this complex (Li *et al.*, 2024b). The TF nuclear factor Y (NF-Y) and the receptor proteins for pyrabactin-resistance 1 (PYL) demonstrate the regulation of abiotic stress in finger millet. The PYL-NF-Y module of finger millet plays a crucial role in the ABA signal transduction pathway by regulating genes that respond to abiotic stress (Rani *et al.*, 2024b). In *Arabidopsis*, the three homologues of NF-YC, namely NF-YC3, NF-YC4, and NF-YC9, play a redundant role in the regulation of seed germination influenced by gibberellin (GA) and ABA through their direct interaction with REPRESSOR OF GA1-3 (RGA)-LIKE 2 (Liu *et al.*, 2016).

11.8 Phytohormones and Biotic and Abiotic Stresses

Plant hormones regulate growth throughout the plant life cycle and play a role in modulating adaptive responses to biotic and abiotic stresses (Shu *et al.*, 2016). In seeds, hormone homeostasis keeps the mature seed dormant until the right conditions are met for germination. The main phytohormones that, in response to environmental stimuli, either hinder or stimulate germination are GAs and ABA (Alonso *et al.*, 2024). Other phytohormones have been discovered to be involved in germination in recent years, although it is most likely that the primary phytohormones, ABA and GA, are responsible for their biosynthesis and signaling (Pan *et al.*, 2023).

Phytohormones are crucial yet opposing agents in how plants respond to salt stress. The salinity of soil represents a significant challenge to the sustainable advancement of food and feed production. The growing prevalence of salinity poses a significant threat to arable land, primarily as a result of inadequate irrigation and drainage practices, the intrusion of seawater, and the excessive use of chemical fertilizers, among other factors. Such saline conditions adversely affect plant growth, diminish crop yields, and can ultimately result in plant mortality (Tao *et al.*, 2015, Ulas *et al.*, 2020).

ABA is recognized as a beneficial modulator of the plant defensive response, whereas the influence of ET on salt tolerance varies among different species. In *Cucurbita pepo*, ET has a potential response to early plant growth and salt stress during seed germination (Alonso *et al.*, 2024). In *Rosa hybrida*, salt stress stimulates the degradation of the ET receptor ethylene response 3 (RhETR3). Overexpression of RhETR3 improved the salt sensitivity of *Rosa hybrida* (Zhao *et al.*, 2024). In the roots of sweet potatoes, overexpression of IbPSS1 increases salt tolerance by stimulating ET-signaling-dependent lignin synthesis (Liu *et al.*, 2024a). In *Arabidopsis thaliana*, ET improves arsenic stress. ethylene-insensitive 3 (EIN3) regulated the expression of genes, including cad1–3, pad2–1, abcc1 and abcc2, which are involved in arsenic detoxication (Zou *et al.*, 2024).

Under abiotic stresses, ET regulates plasticity in rice root (Qin *et al.*, 2024). The occurrence of monsoon and irregular rainfall patterns is common, leading to waterlogging that adversely affects wheat production in hypoxic environments. Utilizing wheat genome data, the physicochemical characteristics of the SAMS gene family were examined through bioinformatics techniques, leading to the selection of TaSAMS10 due to its favorable response to waterlogging conditions. In *Arabidopsis* lines with TaSAMS10 overexpression, older leaves and roots exhibited ET accumulation, which was associated with early flowering and enhanced survival rates following a 20-day waterlogging treatment (Cheng *et al.*, 2024). SlNAC3 regulates the transcription of genes involved in ET biosynthesis, resulting in an increased release of ET when exposed to cold stress. In fact, the suppression of these genes has been associated with enhanced cold tolerance (Wang *et al.*, 2024c).

In several plant species, ET triggers the emergence of dormancy and encourages germination (Corbineau *et al.*, 2014). Additionally, it has been observed that ET receptors are involved in the interaction between ET and ABA (Iglesias- Moya *et al.*, 2023).

Auxin, by itself, is not regarded as a primary regulator of seed germination; however, it has been demonstrated to interact with ABA during the germination process (Wang *et al.*, 2011). Brassinosteroids (BRs) facilitate seed germination and have been shown to interact with ABA via BRASSINOSTEROID INSENSITIVE 2, a crucial repressor of BR signaling. This interaction leads to the phosphorylation and stabilization of ABI5, thereby influencing ABA signaling pathways (Hu and Yu, 2014). Recent studies have demonstrated that exogenous auxin acts synergistically with JA to amplify the ABA-induced delay in the germination of *Arabidopsis*. It has been suggested that JA functions, in conjunction with ABA, to postpone seed germination in *Arabidopsis*. In this process, ABA facilitates the biosynthesis of JA via the SAPK10-bZIP72-AOC pathway, while JA enhances ABA signaling through the direct interaction between JAZ and ABI3/ABI5 (Ju *et al.*, 2019, Pan *et al.*, 2023). Nevertheless, mutants in JA signaling genes such as the receptor CORONATINE INSENSITIVE1 (COI1), JAZ, and MYC2 (a bHLH TF) do not show any deficiencies in germination (Pan *et al.*, 2023). Under typical circumstances, SA prevents germination by downregulating the expression of α-amylase genes that are induced by GAs (Xie *et al.*, 2007).

11.9 Conclusion and Future Perspective

Abiotic stress poses a substantial challenge to global agriculture, affecting both farmers and researchers alike. Abiotic factors such as salinity, temperature extremes, and drought conditions can adversely affect the quality and yield of crops, leading to considerable economic losses for agricultural producers and potential food scarcity for local populations. The presence of these stress factors may lead to a decline in both the quality and productivity of crops. Advances in the understanding of the fundamental mechanisms that govern abiotic stress tolerance in plants have led researchers to develop effective strategies aimed at enhancing plant resilience to these environmental challenges. These methods include traditional breeding techniques, genetic modification, and the use of biostimulants, along with other agents that promote plant development. Nonetheless, considerable endeavors remain necessary to develop sustainable and effective solutions that can be implemented on a global scale. The impact of abiotic stress on agricultural crops is intensifying due to climate change and it is anticipated that the occurrence of extreme weather events will increase in both frequency and intensity in the future. Furthermore, the impact of abiotic stressors on agricultural systems is often contingent upon specific contexts, varying according to factors such as regional agricultural methods, water resources, and soil characteristics. Therefore it is crucial to create region-specific solutions that consider the special difficulties faced by farmers around the world.

The WRKY TFs are essential regulatory proteins that regulate stress responses and mediate development and physiological processes. Their transcriptional regulation is crucial in establishing plant defense by establishing crosstalk between various signaling pathways and regulating gene expression through epigenetic mechanisms, making them versatile and critical components of the plant defense system. Future research can further improve our understanding of the diverse functions of WRKY TFs during myriad environmental threats. By understanding the mechanisms through which these TFs can improve the knowledge about plant resistance, scientists can develop new ways to improve plant resistance against environmental stresses, leading to enhanced food security in the face of increasing global challenges.

The function of JA in the process of germination is not yet fully understood. Further investigation is required to find the potential role of JA in seed gemination under various stress conditions. Consequently, the hormonal network that regulates germination is significantly more intricate than merely involving the interactions of phytohormones such as ABA or GAs, and comprehension of this network is still incomplete. The connections among NF-Y, plant hormones, and stress resistance in tropical crops are yet to be elucidated.

References

Ahuja, I., de Vos, R.C., Bones, A.M. and Hall, R.D. (2010) Plant molecular stress responses face climate change. *Trends in Plant Science* 15(12), 664–674.

Ainsworth, E.A. and Ort, D.R. (2010) How do we improve crop production in a warming world? *Plant Physiology* 154(2), 526–530.

Akmakjian, G.Z. and Bailey-Serres, J. (2022) Gene regulatory circuitry of plant–environment interactions: Scaling from cells to the field. *Current Opinion in Plant Biology* 65, 102122.

Albert, N.W. and Allan, A.C. (2021) MYB genes involved in domestication and crop improvement. *Annual Plant Reviews Online* 4, 199–242.

Alonso, S., Gautam, K., Iglesias-Moya, J., Martínez, C. and Jamilena, M. (2024) Crosstalk between ethylene, Jasmonate and ABA in response to salt stress during germination and early plant growth in Cucurbita pepo. *International Journal of Molecular Sciences* 25(16), 8728.

Anwar, M., Yu, W., Yao, H., Zhou, P., Allan, A.C. and Zeng, L. (2019) NtMYB3, an R2R3-MYB from narcissus, regulates flavonoid biosynthesis. *International Journal of Molecular Sciences* 20(21), 5456.

Anwar, M., Chen, L., Xiao, Y., Wu, J., Zeng, L. *et al.* (2021) Recent advanced metabolic and genetic engineering of phenylpropanoid biosynthetic pathways. *International Journal of Molecular Sciences* 22(17), 9544.

Anwar, M., Duan, S., Ma, M., Chen, X., Wu, L. and Zeng, L. (2023) NataMYB4, a flower specific gene, regulates the flavonoid biosynthesis in Chinese narcissus. *Scientia Horticulturae* 318, 112101.

Anwar, M., Wang, J., Li, J., Altaf, M.M. and Hu, Z. (2024) MYB transcriptional factors affects upstream and downstream MEP pathway and triterpenoid biosynthesis in Chlamydomonas reinhardtii. *PRO* 12(3), 487.

Babu, M.M., Iyer, L.M., Balaji, S. and Aravind, L. (2006) The natural history of the WRKY–GCM1 zinc fingers and the relationship between transcription factors and transposons. *Nucleic Acids Research* 34(22), 6505–6520.

Ben Saad, R., Safi, H., Ben Hsouna, A., Brini, F. and Ben Romdhane, W. (2019) Functional domain analysis of LmSAP protein reveals the crucial role of the zinc-finger A20 domain in abiotic stress tolerance. *Protoplasma* 256, 1333–1344.

Bhatnagar-Mathur, P., Vadez, V. and Sharma, K.K. (2008) Transgenic approaches for abiotic stress tolerance in plants: Retrospect and prospects. *Plant Cell Reports* 27, 411–424.

Birkenbihl, R.P., Kracher, B., Roccaro, M. and Somssich, I.E. (2017) Induced genome-wide binding of three Arabidopsis WRKY transcription factors during early MAMP-triggered immunity. *The Plant Cell* 29(1), 20–38.

Bohnert, H.J., Gong, Q., Li, P. and Ma, S. (2006) Unraveling abiotic stress tolerance mechanisms–getting genomics going. *Current Opinion in Plant Biology* 9(2), 180–188.

Chang, Y., Yu, R., Feng, J., Chen, H., Eri, H. and Gao, G. (2020) NAC transcription factor involves in regulating bacterial wilt resistance in potato. *Functional Plant Biology* 47(10), 925–936.

Chaudhary, M.T., Majeed, S., Rana, I.A., Ali, Z., Jia, Y. *et al.* (2024) Impact of salinity stress on cotton and opportunities for improvement through conventional and biotechnological approaches. *BMC Plant Biology* 24(1), 20.

Chen, T., Li, W., Hu, X., Guo, J., Liu, A. and Zhang, B. (2015) A cotton MYB transcription factor, GbMYB5, is positively involved in plant adaptive response to drought stress. *Plant and Cell Physiology* 56(5), 917–929.

Chen, Q., Bao, C., Xu, F., Ma, C., Huang, L., Guo, Q. and Luo, M. (2021) Silencing GhJUB1L1 (JUB1-like 1) reduces cotton (Gossypium hirsutum) drought tolerance. *PLoS One* 16(11), e0259382.

Chen, H., Shi, Y., An, L., Yang, X., Liu, J. *et al.* (2024a) Overexpression of SlWRKY6 enhances drought tolerance by strengthening antioxidant defense and stomatal closure via ABA signaling in Solanum lycopersicum L. *Plant Physiology and Biochemistry* 213, 108855.

Chen, Y., Li, A., Yun, P., Chen, Q., Pan, D. *et al.* (2024b) Genome-wide analysis of MYB transcription factor family and AsMYB1R subfamily contribution to ROS homeostasis regulation in Avena sativa under PEG-induced drought stress. *BMC Plant Biology* 24(1), 632.

Cheng, B., Feng, X., Buitrago, S., Wan, K., Xu, Y. *et al.* (2024) The wheat TaSAMS10 improves survival rate under waterlogging stress by inducing ethylene synthesis and early flowering. *Plant Growth Regulation* 102(2), 379–393.

Corbineau, F., Xia, Q., Bailly, C. and El-Maarouf-Bouteau, H. (2014) Ethylene, a key factor in the regulation of seed dormancy. *Frontiers in Plant Science* 5, 539.

Dastmalchi, K., Chira, O., Rodriguez, M.P., Yoo, B., Serra, O., Figueras, M. and Stark, R.E. (2021) A chemical window into the impact of RNAi silencing of the StNAC103 gene in potato tuber periderms: Soluble metabolites, suberized cell walls, and antibacterial defense. *Phytochemistry* 190, 112885.

Davletova, S., Schlauch, K., Coutu, J. and Mittler, R. (2005) The zinc-finger protein Zat12 plays a central role in reactive oxygen and abiotic stress signaling in Arabidopsis. *Plant Physiology* 139(2), 847–856.

Ding, Z.J., Yan, J.Y., Xu, X.Y., Yu, D.Q., Li, G.X., Zhang, S.Q. and Zheng, S.J. (2014) Transcription factor WRKY 46 regulates osmotic stress responses and stomatal movement independently in a rabidopsis. *The Plant Journal* 79(1), 13–27.

Dubos, C., Stracke, R., Grotewold, E., Weisshaar, B., Martin, C. and Lepiniec, L. (2010) MYB transcription factors in Arabidopsis. *Trends in Plant Science* 15(10), 573–581.

Francis, B., Aravindakumar, C., Brewer, P.B. and Simon, S. (2023) Plant nutrient stress adaptation: A prospect for fertilizer limited agriculture. *Environmental and Experimental Botany* 105431.

Frugier, F., Poirier, S., Satiat-Jeunemaître, B., Kondorosi, A. and Crespi, M. (2000) A Krüppel-like zinc finger protein is involved in nitrogen-fixing root nodule organogenesis. *Genes and Development* 14(4), 475–482.

Ge, S., Han, X., Xu, X., Shao, Y., Zhu, Q. *et al.* (2020) WRKY15 suppresses tracheary element differentiation upstream of VND7 during xylem formation. *The Plant Cell* 32(7), 2307–2324.

Goel, S. and Madan, B. (2014) *Genetic Engineering of Crop Plants for Abiotic Stress Tolerance. Emerging Technologies and Management of Crop Stress Tolerance*. pp. 99–123. Elsevier: Amsterdam, Netherlands.

Gong, Q., Zhou, M., Li, X. and Guo, Y. (2024) Transcription factor MYB8 regulates iron deficiency stress response in Arabidopsis. *Plant Science* 340, 111973.

Guerin, C., Roche, J., Allard, V., Ravel, C., Mouzeyar, S. and Bouzidi, M.F. (2019) Genome-wide analysis, expansion and expression of the NAC family under drought and heat stresses in bread wheat (T. Aestivum L.). *PLoS One* 14(3), e0213390.

Han, J., Dai, J., Chen, Z., Li, W., Li, X. *et al.* (2024) Overexpression of a 'BETA'MYB factor gene, VhMYB15, increases salinity and drought tolerance in Arabidopsis thaliana. *International Journal of Molecular Sciences* 25(3), 1534.

He, X., Zhu, L., Xu, L., Guo, W. and Zhang, X. (2016) GhATAF1, a NAC transcription factor, confers abiotic and biotic stress responses by regulating phytohormonal signaling networks. *Plant Cell Reports* 35, 2167–2179.

He, H., Yang, M., Li, S., Zhang, G., Ding, Z. *et al.* (2023) Mechanisms and biotechnological applications of transcription factors. *Synthetic and Systems Biotechnology* 8(4), 565–577.

Hirayama, T. and Shinozaki, K. (2010) Research on plant abiotic stress responses in the post-genome era: Past, present and future. *The Plant Journal* 61(6), 1041–1052.

Hou, X.-M., Zhang, H.-F., Liu, S.-Y., Wang, X.-K., Zhang, Y.-M. *et al.* (2020) The NAC transcription factor CaNAC064 is a regulator of cold stress tolerance in peppers. *Plant Science* 291, 110346.

Hu, Y. and Yu, D. (2014) BRASSINOSTEROID INSENSITIVE2 interacts with ABSCISIC ACID INSENSITIVE5 to mediate the antagonism of brassinosteroids to abscisic acid during seed germination in Arabidopsis. *The Plant Cell* 26(11), 4394–4408.

Huang, J., Liu, F., Chao, D., Xin, B., Liu, K. *et al.* (2022) The WRKY transcription factor OsWRKY54 is involved in salt tolerance in rice. *International Journal of Molecular Sciences* 23(19), 11999.

Hussain, R.M., Ali, M., Feng, X. and Li, X. (2017) The essence of NAC gene family to the cultivation of drought-resistant soybean (Glycine max L. Merr.) cultivars. *BMC Plant Biology* 17, 1–11.

Iglesias-Moya, J., Cebrián, G., Garrido, D., Martínez, C. and Jamilena, M. (2023) The ethylene receptor mutation etr2b reveals crosstalk between ethylene and ABA in the control of Cucurbita pepo germination. *Physiologia Plantarum* 175(1), e13864.

Ju, L., Jing, Y., Shi, P., Liu, J., Chen, J. *et al.* (2019) JAZ proteins modulate seed germination through interaction with ABI 5 in bread wheat and Arabidopsis. *New Phytologist* 223(1), 246–260.

Junaid, M. and Gokce, A. (2024) Global agricultural losses and their causes. *Bulletin of Biological and Allied Sciences Research* 2024(1), 66–66.

Kasuga, M., Miura, S., Shinozaki, K. and Yamaguchi-Shinozaki, K. (2004) A combination of the Arabidopsis DREB1A gene and stress-inducible rd29A promoter improved drought-and low-temperature stress tolerance in tobacco by gene transfer. *Plant and Cell Physiology* 45(3), 346–350.

Khoso, M.A., Hussain, A., Ritonga, F.N., Ali, Q., Channa, M.M. *et al.* (2022) WRKY transcription factors (TFs): Molecular switches to regulate drought, temperature, and salinity stresses in plants. *Frontiers in Plant Science* 13, 1039329.

Kim, J.C., Lee, S.H., Cheong, Y.H., Yoo, C.M., Lee, S.I. *et al.* (2001) A novel cold-inducible zinc finger protein from soybean, SCOF-1, enhances cold tolerance in transgenic plants. *The Plant Journal* 25(3), 247–259.

Kobata, T. (2024) *Abiotic Factors Affect Plant Growth. Responses of Plants to Soil Flooding*. pp. 3–17. Springer: Singapore.

Kopecká, R., Kameniarová, M., Černý, M., Brzobohatý, B. and Novák, J. (2023) Abiotic stress in crop production. *International Journal of Molecular Sciences* 24(7), 6603.

Kouhen, M., Dimitrova, A., Scippa, G.S. and Trupiano, D. (2023) The course of mechanical stress: Types, perception, and plant response. *Biology* 12(2), 217.

Kumar, M., Kesawat, M.S., Ali, A., Lee, S.-C., Gill, S.S. and Kim, H.U. (2019) Integration of abscisic acid signaling with other signaling pathways in plant stress responses and development. *Plants* 8(12), 592.

Kumar, T., Wang, J.-G., Xu, C.-H., Lu, X., Mao, J. *et al.* (2024) Genetic engineering for enhancing sugarcane tolerance to biotic and abiotic stresses. *Plants* 13(13), 1739.

Lallawmkimi, M.C., Veda, D.S., Yadav, A., Kumar, M. and Rout, A. (2024) Innovative approaches in crop genetic engineering for sustainable agriculture: A review. *Journal of Advances in Biology and Biotechnology* 27(8), 615–631.

Li, J., Brader, G. and Palva, E.T. (2004) The WRKY70 transcription factor: A node of convergence for jasmonate-mediated and salicylate-mediated signals in plant defense. *The Plant Cell* 16(2), 319–331.

Li, S., Fu, Q., Chen, L., Huang, W. and Yu, D. (2011) Arabidopsis thaliana WRKY25, WRKY26, and WRKY33 coordinate induction of plant thermotolerance. *Planta* 233, 1237–1252.

Li, C., Ng, C.K.-Y. and Fan, L.-M. (2015) MYB transcription factors, active players in abiotic stress signaling. *Environmental and Experimental Botany* 114, 80–91.

Li, J., Han, G., Sun, C. and Sui, N. (2019) Research advances of MYB transcription factors in plant stress resistance and breeding. *Plant Signaling and Behavior* 14(8), 1613131.

Li, W., Pang, S., Lu, Z. and Jin, B. (2020) Function and mechanism of WRKY transcription factors in abiotic stress responses of plants. *Plants* 9(11), 1515.

Li, P., Li, X. and Jiang, M. (2021) CRISPR/Cas9-mediated mutagenesis of WRKY3 and WRKY4 function decreases salt and me-JA stress tolerance in Arabidopsis thaliana. *Molecular Biology Reports* 48(8), 5821–5832.

Li, X., Wang, Q., Guo, C., Sun, J., Li, Z. *et al.* (2022) NtNAC053, a novel NAC transcription factor, confers drought and salt tolerances in tobacco. *Frontiers in Plant Science* 13, 817106.

Li, S., Khoso, M.A., Wu, J., Yu, B., Wagan, S. and Liu, L. (2024a) Exploring the mechanisms of WRKY transcription factors and regulated pathways in response to abiotic stress. *Plant Stress* 12, 100429.

Li, X., Li, C., Shi, L., Lv, G., Li, X. *et al.* (2024b) Jasmonate signaling pathway confers salt tolerance through a NUCLEAR FACTOR-Y trimeric transcription factor complex in Arabidopsis. *Cell Reports* 43(3).

Lippuner, V., Cyert, M.S. and Gasser, C.S. (1996) Two classes of plant cDNA clones differentially complement yeast calcineurin mutants and increase salt tolerance of wild-type yeast. *Journal of Biological Chemistry* 271(22), 12859–12866.

Liu, X., Hu, P., Huang, M., Tang, Y., Li, Y., Li, L. and Hou, X. (2016) The NF-YC–RGL2 module integrates GA and ABA signalling to regulate seed germination in Arabidopsis. *Nature Communications* 7(1), 12768.

Liu, S., Zhang, C., Guo, F., Sun, Q., Yu, J. *et al.* (2022) A systematical genome-wide analysis and screening of WRKY transcription factor family engaged in abiotic stress response in sweet potato. *BMC Plant Biology* 22(1), 616.

Liu, C., Pan, Z., Wang, X., Gao, J., He, J. *et al.* (2024a) Overexpression of phosphatidylserine synthase IbPSS1 enhances salt tolerance by stimulating ethylene signaling-dependent lignin synthesis in sweet potato roots. *Plant Physiology and Biochemistry* 212, 108727.

Liu, L., Li, S., Tang, F., Li, P., Liu, J. *et al.* (2024b) MaMYBR30, a novel 1R-MYB, plays important roles in plant development and abiotic stress resistance. *Plants* 13(13), 1794.

Lv, M., Luo, W., Ge, M., Guan, Y., Tang, Y., Chen, W. and Lv, J. (2022) A group I WRKY gene, TaWRKY133, negatively regulates drought resistance in transgenic plants. *International Journal of Molecular Sciences* 23(19), 12026.

Mao, G., Meng, X., Liu, Y., Zheng, Z., Chen, Z. and Zhang, S. (2011) Phosphorylation of a WRKY transcription factor by two pathogen-responsive MAPKs drives phytoalexin biosynthesis in Arabidopsis. *The Plant Cell* 23(4), 1639–1653.

Mei, J., Mu, R., Niu, Q., Zhu, H., Chen, R. *et al.* (2024) A MYB transcription factor GhTT2 of Gossypium hirsutum regulates proanthocyanidin accumulation and improves osmotic tolerance in Arabidopsis. *Plant Cell, Tissue and Organ Culture (PCTOC)* 157(2), 39.

Mittler, R. and Blumwald, E. (2010) Genetic engineering for modern agriculture: Challenges and perspectives. *Annual Review of Plant Biology* 61(1), 443–462.

Niu, B., Zhang, Z., Zhang, J., Zhou, Y. and Chen, C. (2021) The rice LEC1-like transcription factor OsNF-YB9 interacts with SPK, an endosperm-specific sucrose synthase protein kinase, and functions in seed development. *The Plant Journal* 106(5), 1233–1246.

Niu, X., Lu, H., Fan, Y., Wang, W., Yuan, Y. *et al.* (2022) Manipulation of the transcription factor SlNAC1 for improved tolerance to abiotic stress in tomato. *Plant, Cell and Environment* 45(12), 3537–3550.

Ooka, H., Satoh, K., Doi, K., Nagata, T., Otomo, Y. *et al.* (2003) Comprehensive analysis of NAC family genes in Oryza sativa and Arabidopsis thaliana. *DNA Research* 10(6), 239–247.

Oshunsanya, S.O., Nwosu, N.J. and Li, Y. (2019) Abiotic stress in agricultural crops under climatic conditions. In: *Sustainable Agriculture, Forest and Environmental Management* pp. 71–100. Springer: Singapore.

Pabo, C.O., Peisach, E. and Grant, R.A. (2001) Design and selection of novel Cys2His2 zinc finger proteins. *Annual Review of Biochemistry* 70(1), 313–340.

Pan, J., Wang, H., You, Q., Cao, R., Sun, G. and Yu, D. (2023) Jasmonate-regulated seed germination and crosstalk with other phytohormones. *Journal of Experimental Botany* 74(4), 1162–1175.

Peleg, Z., Apse, M.P. and Blumwald, E. (2011) Engineering salinity and water-stress tolerance in crop plants: Getting closer to the field. *Advances in Botanical Research Elsevier.* 57, 405–443.

Peng, X., Zhao, Y., Li, X., Wu, M., Chai, W. *et al.* (2015) Genomewide identification, classification and analysis of NAC type gene family in maize. *Journal of Genetics* 94, 377–390.

Peng, Z.-Z., Li, H.-L., Liu, G.-S., Su, X., Zhao, X. *et al.* (2024) Tomato NAC transcription factor NOR-like1 positively regulates tomato fruit softening. *Postharvest Biology and Technology* 213, 112923.

Pu, X., Yang, L., Liu, L., Dong, X., Chen, S. *et al.* (2020) Genome-wide analysis of the MYB transcription factor superfamily in Physcomitrella patens. *International Journal of Molecular Sciences* 21(3), 975.

Puranik, S., Sahu, P.P., Srivastava, P.S. and Prasad, M. (2012) NAC proteins: Regulation and role in stress tolerance. *Trends in Plant Science* 17(6), 369–381.

Qin, H., Xiao, M., Li, Y. and Huang, R. (2024) Ethylene modulates rice root plasticity under abiotic stresses. *Plants* 13(3), 432.

Qiu, J.L., Fiil, B.K., Petersen, K., Nielsen, H.B., Botanga, C.J. *et al.* (2008) Arabidopsis MAP kinase 4 regulates gene expression through transcription factor release in the nucleus. *The EMBO Journal* 27(16), 2214–2221.

Rai, G.K., Khanday, D.M., Choudhary, S.M., Kumar, P., Kumari, S. *et al.* (2024) Unlocking nature's stress buster: Abscisic acid's crucial role in defending plants against abiotic stress. *Plant Stress* 11, 100359.

Rani, V., Rana, S., Muthamilarasan, M., Joshi, D. and Yadav, D. (2024a) Expression profiling of nuclear factor-Y (NF-Y) transcription factors during dehydration and salt stress in finger millet reveals potential candidate genes for multiple stress tolerance. *Planta* 259(6), 136.

Rani, V., Singh, V.K., Joshi, D., Singh, R. and Yadav, D. (2024b) Molecular docking insights into nuclear factor Y (NF-Y) transcription factor and pyrabactin resistance 1 (PYL) receptor proteins reveal abiotic stress regulation in finger millet. *Crop Design* 3(1), 100051.

Reguera, M., Peleg, Z. and Blumwald, E. (2012) Targeting metabolic pathways for genetic engineering abiotic stress-tolerance in crops. *Biochimica et Biophysica Acta (BBA)-Gene Regulatory Mechanisms* 1819(2), 186–194.

Rushton, P.J., Somssich, I.E., Ringler, P. and Shen, Q.J. (2010) WRKY transcription factors. *Trends in Plant Science* 15(5), 247–258.

Sehgal, A., Sita, K., Siddique, K.H., Kumar, R., Bhogireddy, S. *et al.* (2018) Drought or/and heat-stress effects on seed filling in food crops: Impacts on functional biochemistry, seed yields, and nutritional quality. *Frontiers in Plant Science* 9, 1705.

Seo, P.J., Kim, M.J., Song, J.-S., Kim, Y.-S., Kim, H.-J. and Park, C.-M. (2010) Proteolytic processing of an Arabidopsis membrane-bound NAC transcription factor is triggered by cold-induced changes in membrane fluidity. *Biochemical Journal* 427(3), 359–367.

Shang, Y., Yan, L., Liu, Z.-Q., Cao, Z., Mei, C. *et al.* (2010) The Mg-chelatase H subunit of Arabidopsis antagonizes a group of WRKY transcription repressors to relieve ABA-responsive genes of inhibition. *The Plant Cell* 22(6), 1909–1935.

Shang, X., Yu, Y., Zhu, L., Liu, H., Chai, Q. and Guo, W. (2020) A cotton NAC transcription factor GhirNAC2 plays positive roles in drought tolerance via regulating ABA biosynthesis. *Plant Science* 296, 110498.

Shu, K., Liu, X.-D., Xie, Q. and He, Z.-H. (2016) Two faces of one seed: Hormonal regulation of dormancy and germination. *Molecular Plant* 9(1), 34–45.

Singh, B.K., Delgado-Baquerizo, M., Egidi, E., Guirado, E., Leach, J.E., Liu, H. and Trivedi, P. (2023) Climate change impacts on plant pathogens, food security and paths forward. *Nature Reviews Microbiology* 21(10), 640–656.

So, K., Wang, J., Sun, S., Che, H. and Zhang, Y. (2024) Comprehensive analysis of MYB gene family and their expression under various stress conditions in Lilium pumilum. *Scientia Horticulturae* 327, 112764.

Song, J., Xie, X., Chen, C., Shu, J., Thapa, R.K. *et al.* (2021) LEAFY COTYLEDON1 expression in the endosperm enables embryo maturation in Arabidopsis. *Nature Communications* 12(1), 3963.

Song, P., Yang, R., Jiao, K., Guo, B., Zhang, L. *et al.* (2024) FvMYB108, a MYB gene from Fragaria vesca, positively regulates cold and salt tolerance of Arabidopsis. *International Journal of Molecular Sciences* 25(6), 3405.

Souer, E., van Houwelingen, A., Kloos, D., Mol, J. and Koes, R. (1996) The no apical meristem gene of petunia is required for pattern formation in embryos and flowers and is expressed at meristem and primordia boundaries. *Cell* 85(2), 159–170.

Su, J., Zhan, N., Cheng, X., Song, S., Dong, T., Ge, X. and Duan, H. (2024) Genome-wide analysis of cotton MYB transcription factors and the functional validation of GhMYB in response to drought stress. *Plant and Cell Physiology* 65(1), 79–94.

Sun, H., Hu, M., Li, J., Chen, L., Li, M. *et al.* (2018) Comprehensive analysis of NAC transcription factors uncovers their roles during fiber development and stress response in cotton. *BMC Plant Biology* 18, 1–15.

Tan, X., Zhang, H., Yang, Z., Wei, Z., Li, Y., Chen, J. and Sun, Z. (2022) NF-YA transcription factors suppress jasmonic acid-mediated antiviral defense and facilitate viral infection in rice. *PLoS Pathogens* 18(5), e1010548.

Tang, Y., Bao, X., Zhi, Y., Wu, Q., Guo, Y. *et al.* (2019) Overexpression of a MYB family gene, OsMYB6, increases drought and salinity stress tolerance in transgenic rice. *Frontiers in Plant Science* 10, 168.

Tang, F., Jiao, B., Zhang, M., He, M., Su, R., Luo, K. and Lan, T. (2024) PtoMYB031, the R2R3 MYB transcription factor involved in secondary cell wall biosynthesis in poplar. *Frontiers in Plant Science* 14, 1341245.

Tao, J.-J., Chen, H.-W., Ma, B., Zhang, W.-K., Chen, S.-Y. and Zhang, J.-S. (2015) The role of ethylene in plants under salinity stress. *Frontiers in Plant Science* 6, 1059.

Trishla, V.S. and Kirti, P.B. (2021) Structure-function relationship of Gossypium hirsutum NAC transcription factor, GhNAC4 with regard to ABA and abiotic stress responses. *Plant Science* 302, 110718.

Trotta, A., Rahikainen, M., Konert, G., Finazzi, G. and Kangasjärvi, S. (2014) Signalling crosstalk in light stress and immune reactions in plants. *Philosophical Transactions of the Royal Society B: Biological Sciences* 369(1640), 20130235.

Tuteja, N. (2007) Abscisic acid and abiotic stress signaling. *Plant Signaling and Behavior* 2(3), 135–138.

Ulas, A., Aydin, A., Ulas, F., Yetisir, H. and Miano, T.F. (2020) Cucurbita rootstocks improve salt tolerance of melon scions by inducing physiological, biochemical and nutritional responses. *Horticulturae* 6(4), 66.

Wang, L., Hua, D., He, J., Duan, Y., Chen, Z., Hong, X. and Gong, Z. (2011) Auxin response Factor2 (ARF2) and its regulated homeodomain gene HB33 mediate abscisic acid response in Arabidopsis. *PLoS Genetics* 7(7), e1002172.

Wang, J., Li, G., Li, C., Zhang, C., Cui, L. *et al.* (2021) NF-Y plays essential roles in flavonoid biosynthesis by modulating histone modifications in tomato. *New Phytologist* 229(6), 3237–3252.

Wang, H., Cheng, X., Yin, D., Chen, D., Luo, C., Liu, H. and Huang, C. (2023) Advances in the research on plant WRKY transcription factors responsive to external stresses. *Current Issues in Molecular Biology* 45(4), 2861–2880.

Wang, B., Xiong, C., Peng, Z., Luo, Z., Wang, X., Peng, S. and Yu, Z. (2024a) Genome-wide analysis of R2R3-MYB transcription factors in poplar and functional validation of PagMYB147 in defense against Melampsora magnusiana. *Planta* 260(2), 1–18.

Wang, C., Lei, J., Jin, X., Chai, S., Jiao, C., Yang, X. and Wang, L. (2024b) A sweet potato MYB transcription factor IbMYB330 enhances tolerance to drought and salt stress in transgenic tobacco. *Genes* 15(6), 693.

Wang, T., Ma, X., Chen, Y., Wang, C., Xia, Z. *et al.* (2024c) SlNAC3 suppresses cold tolerance in tomatoes by enhancing ethylene biosynthesis. *Plant, Cell and Environment* 47(8), 3132–3146.

Wang, Z., Peng, Z., Khan, S., Qayyum, A., Rehman, A. and Du, X. (2024d) Unveiling the power of MYB transcription factors: Master regulators of multi-stress responses and development in cotton. *International Journal of Biological Macromolecules* 133885.

Xia, L., Sun, S., Han, B. and Yang, X. (2023) NAC domain transcription factor gene GhNAC3 confers drought tolerance in plants. *Plant Physiology and Biochemistry* 195, 114–123.

Xie, Z., Zhang, Z.-L., Hanzlik, S., Cook, E. and Shen, Q.J. (2007) Salicylic acid inhibits gibberellin-induced alpha-amylase expression and seed germination via a pathway involving an abscisic-acid-inducible WRKY gene. *Plant Molecular Biology* 64, 293–303.

Xiong, H., Li, J., Liu, P., Duan, J., Zhao, Y. *et al.* (2014) Overexpression of OsMYB48-1, a novel MYB-related transcription factor, enhances drought and salinity tolerance in rice. *PLoS One* 9(3), e92913.

Xu, Q., He, Q., Li, S. and Tian, Z. (2014) Molecular characterization of StNAC2 in potato and its overexpression confers drought and salt tolerance. *Acta Physiologiae Plantarum* 36, 1841–1851.

Xu, Y.-P., Xu, H., Wang, B. and Su, X.-D. (2020) Crystal structures of N-terminal WRKY transcription factors and DNA complexes. *Protein and Cell* 11(3), 208–213.

Yamaguchi-Shinozaki, K. and Shinozaki, K. (2006) Transcriptional regulatory networks in cellular responses and tolerance to dehydration and cold stresses. *Annual Review of Plant Biology* 57(1), 781–803.

Yamaguchi-Shinozaki, K. and Shinozaki, K. (2007) Improving plant drought, salt and freezing tolerance by gene transfer of a single stress-inducible transcription factor. In: *Novartis Foundation Symposium 236-Rice Biotechnology: Improving Yield, Stress Tolerance and Grain Quality: Rice Biotechnology: Improving Yield, Stress Tolerance and Grain Quality: Novartis Foundation Symposium 236*. Wiley Online Library: USA.

Yamasaki, K., Kigawa, T., Inoue, M., Tateno, M., Yamasaki, T. *et al.* (2005) Solution structure of an Arabidopsis WRKY DNA binding domain. *The Plant Cell* 17(3), 944–956.

Yu, T.F., Liu, Y., Fu, J.D., Ma, J., Fang, Z.W. *et al.* (2021) The NF-Y-PYR module integrates the abscisic acid signal pathway to regulate plant stress tolerance. *Plant Biotechnology Journal* 19(12), 2589–2605.

Zhao, F., Ma, J., Li, L., Fan, S., Guo, Y. *et al.* (2016) GhNAC12, a neutral candidate gene, leads to early aging in cotton (Gossypium hirsutum L). *Gene* 576(1), 268–274.

Zhao, Q., Jing, W., Fu, X., Yang, R., Zhu, C. *et al.* (2024) TSPO-induced degradation of the ethylene receptor RhETR3 promotes salt tolerance in rose (Rosa hybrida). *Horticulture Research* 11(4), uhae040.

Zheng, L., Gao, S., Bai, Y., Zeng, H. and Shi, H. (2024) NF-YC15 transcription factor activates ethylene biosynthesis and improves cassava disease resistance. *Plant Biotechnology Journal* 22(9), 2424–2434.

Zhou, X., Lei, D., Yao, W., Li, S., Wang, H. *et al.* (2024) A novel R2R3-MYB transcription factor PbMYB1L of Pyrus bretschneideri regulates cold tolerance and anthocyanin accumulation. *Plant Cell Reports* 43(2), 34.

Zhu, N., Cheng, S., Liu, X., Du, H., Dai, M. *et al.* (2015) The R2R3-type MYB gene OsMYB91 has a function in coordinating plant growth and salt stress tolerance in rice. *Plant Science* 236, 146–156.

Zou, Y., Liu, Y., Li, W., Cao, Q., Wang, X. *et al.* (2024) Ethylene is the key phytohormone to enhance arsenic resistance in Arabidopsis thaliana. *Ecotoxicology and Environmental Safety* 281, 116644.

12 Case Studies in Crop Improvement

Hamza Sohail, Iqra Noor and Xiaodong Yang*
School of Horticulture and Landscape Architecture, Yangzhou University, Yangzhou, Jiangsu 225009, China

Abstract
The challenge of ensuring global food security amid climate change and population growth has driven advances in crop improvement strategies. Genomic tools like CRISPR/Cas9 and next-generation sequencing have enabled researchers to identify and modify key genes associated with desirable traits, resulting in resilient, high-yielding, and nutritionally-enhanced crops. Case studies in rice, cotton, and tomato reveal how specific genes enhance resistance to stresses, improve fruit quality, and accelerate breeding through innovations like speed breeding. These developments not only increase crop productivity but also support sustainable agriculture by reducing chemical inputs and enhancing resilience to environmental changes. This chapter highlights the revolutionary potential of functional genomics for modern agriculture and offers case studies of recent crop enhancements for a range of crops.

Keywords: genome editing, CRISPR/Cas, abiotic stressors, biotic stressors, climate change

12.1 Introduction

In recent years, advancing crop improvement strategies have become critical to ensuring global food security, especially in the face of a growing population and climate change (Janni *et al.*, 2024). One key technological breakthrough in this area is the use of multiomics approaches, encompassing genomics, transcriptomics, proteomics, and metabolomics.

These approaches offer a comprehensive understanding of the molecular mechanisms underlying various plant traits (Khan *et al.*, 2024). For example, high-throughput sequencing technologies have enabled the assembly of numerous plant genomes, revealing crucial genome architecture and variations essential for adaptation and breeding efforts. Functional genomics, combined with these sequencing technologies, plays a pivotal role in validating genomic regions linked to desirable traits. Tools like CRISPR/Cas9 have facilitated precise genome editing, introducing targeted mutations that enhance crop traits. Additionally, the integration of phenomics, which uses advanced phenotyping techniques, provides detailed data on traits like canopy temperature and plant height, critical for validating genetic variations (Singh *et al.*, 2024).

Bioinformatics and systems biology approaches have also transformed crop improvement by

* Email: yxd@yzu.edu.cn

 Molecular Biology of Plant Stress Responses (Muhammad Waseem and Pingwu Liu eds)
DOI: 10.1079/9781800628823.0012

enabling the integration and analysis of large-scale omics data, helping researchers construct comprehensive models of plant growth and development. These approaches have been particularly effective in identifying key regulatory networks and pathways for genetic improvement (Jamil *et al.*, 2020).

This chapter highlights the importance of integrating multiomics, advanced phenotyping, and bioinformatics in modern plant breeding. Together, these technologies have significantly enhanced our understanding of plant biology and facilitated the development of crop varieties resilient to biotic and abiotic stresses. As these technologies evolve, they promise to accelerate crop improvement efforts, contributing to global food security and sustainable agriculture.

12.2 Tools for Crop Improvement

12.2.1 Gene banks as a path to food security

Gene banks play a crucial role in the conservation and utilization of plant genetic resources, which are essential for breeding programs and ensuring food security. These institutions are responsible for the global collection, maintenance, characterization, evaluation, documentation, and distribution of plant genetic resources. By providing access to a diverse range of germplasm, gene banks support the development of crop varieties that can withstand various stresses, combat malnutrition, and improve public health (Aribi, 2024). Advanced technologies, such as high-throughput phenotypic assessments and multiomics tools, have accelerated the identification of valuable traits and genes, thereby enhancing breeding efforts for improved cultivars. In summary, gene banks are indispensable for preserving the genetic resources that drive food security and agricultural innovation (Chao *et al.*, 2024).

12.2.2 Bioinformatics

Bioinformatics is pivotal in crop improvement, enabling the analysis and management of vast genetic data. Advances in next-generation sequencing (NGS) technologies have greatly enhanced our understanding of genome architecture and the genetic basis of economically important traits (Arif *et al.*, 2024). By utilizing bioinformatics tools, researchers can analyze large datasets generated from genome sequencing, transcriptomics, proteomics, and metabolomics. These tools aid in the development of pangenomes, identification of superior haplotypes, and genome editing, thereby accelerating the breeding of climate-resilient and high-yielding crop varieties (Razzaq *et al.*, 2021a). Sophisticated computational tools and databases designed to store and retrieve biological information have enabled the shift from traditional phenotype-based analysis to genomics-driven approaches. The integration of artificial intelligence and machine learning algorithms into bioinformatics has opened new avenues for predictive analysis and efficient handling of big data, improving the accuracy and efficiency of breeding programs. As the volume of biological data continues to grow, the development of more advanced bioinformatics approaches will be crucial for addressing the challenges of sustainable agriculture and global food security (Xu *et al.*, 2022).

12.2.3 Evolution in the genotyping platforms for plant breeding

Genotyping is essential to crop improvement, enhancing the precision and efficiency of selecting desirable traits. Traditional phenotypic selection methods in plant breeding are often labor-intensive, time-consuming, and limited in detecting genetic variations at early stages (Nyirahabimana and Solmaz, 2024). Genotyping enables breeders to identify and select plants with desirable genetic traits at the molecular level, significantly accelerating the breeding process. This genomic approach helps identify single nucleotide polymorphisms (SNPs) and other molecular markers linked to key traits such as yield, pest and disease resistance, and environmental stress tolerance. Breeders can use these markers to implement marker-assisted selection (MAS) and genomic selection (GS) to develop superior crop varieties more efficiently (Aziz and Masmoudi, 2025).

Additionally, the ability of genotyping to rapidly screen large populations ensures the selection of only the best-performing plants for further breeding cycles, thereby improving overall crop productivity and sustainability (Ogutu *et al.*, 2024). Integrating genotyping platforms, such as SNP arrays and genotyping-by-sequencing, into breeding programs is crucial for meeting growing food demands and ensuring food security in the face of changing climatic conditions (Sivabharathi *et al.*, 2024).

12.2.4 Speed breeding

Speed breeding (SB) is a transformative approach in crop improvement that significantly reduces the time required to develop new plant varieties. SB accelerates the breeding cycle by manipulating growing conditions such as photoperiod, light intensity, temperature, soil moisture, and nutrition, enabling the growth of multiple generations of crops annually. Studies have demonstrated that this approach triggers early flowering and seed set, significantly reducing the time required for each breeding generation. For instance, while traditional breeding methods typically produce one to two generations per year, speed breeding can generate up to six to eight generations annually for crops like wheat and barley (He *et al.*, 2024; Imam *et al.*, 2024).

The implementation of SB involves advanced technologies like LED lighting, initially developed by NASA for growing plants in space. NASA adopted this technology for terrestrial use to provide the optimal light spectrum and duration, resulting in faster plant development (Kumar and Walia, 2024). The combination of extended photoperiods, controlled temperatures, and optimized humidity levels creates an environment where plants can grow continuously, free from the seasonal interruptions of conventional agriculture. Various crops, including wheat, barley, chickpea, pea, and canola, have successfully applied SB, significantly enhancing the efficiency of breeding programs and enabling the rapid development of climate-resilient and high-yielding crop varieties (Sajja *et al.*, 2024).

12.2.5 Multi-omics technologies

Omics technologies have revolutionized crop improvement by offering comprehensive insights into the genetic, molecular, and biochemical foundations of crop traits. The integration of genomics, transcriptomics, proteomics, metabolomics, and phenomics provides a multi-dimensional understanding of the mechanisms governing phenotypic traits, stress responses, and overall crop development (Kaur *et al.*, 2021; Razzaq *et al.*, 2021b; Yang *et al.*, 2021). Genomics tools, bolstered by advancements in NGS, enable the identification and analysis of genetic variations across different crop species, leading to the discovery of key traits related to yield, quality, and stress resistance. These genomic insights are crucial for MAS and GS, accelerating the breeding process and improving crop performance under diverse environmental conditions (Edwards *et al.*, 2013; Mir *et al.*, 2019).

Transcriptomics offers detailed gene expression profiles across various developmental stages and under different stress conditions, revealing regulatory networks and pathways vital for crop adaptation and resilience. RNA sequencing has been instrumental in identifying differentially expressed genes and understanding gene regulatory mechanisms in response to biotic and abiotic stresses. This knowledge aids in developing crops with enhanced stress tolerance and productivity by targeting specific gene expression patterns and regulatory pathways (Bashir *et al.*, 2019; Vishwanath *et al.*, 2024). Similarly, proteomics and metabolomics provide insights into the functional proteins and metabolic pathways involved in stress responses and trait development, offering a comprehensive view of the biochemical processes underlying crop improvement (Eldakak *et al.*, 2013; Razzaq *et al.*, 2019).

Phenomics, the high-throughput phenotyping of crop traits, complements these omics approaches by linking genetic and molecular data with observable phenotypic traits. Advanced phenotyping techniques, including imaging and sensor-based methods, allow for precise measurement of traits such as yield, quality, and stress resistance (Cobb *et al.*, 2013; Rahman *et al.*, 2015). The integration of phenomics with other omics data facilitates the identification of key phenotypic traits and supports the development of climate-smart crops (Fig. 12.1). By leveraging the combined power of multiomics, researchers can enhance crop breeding programs, ensuring the development of resilient and high-yielding crop varieties that can meet

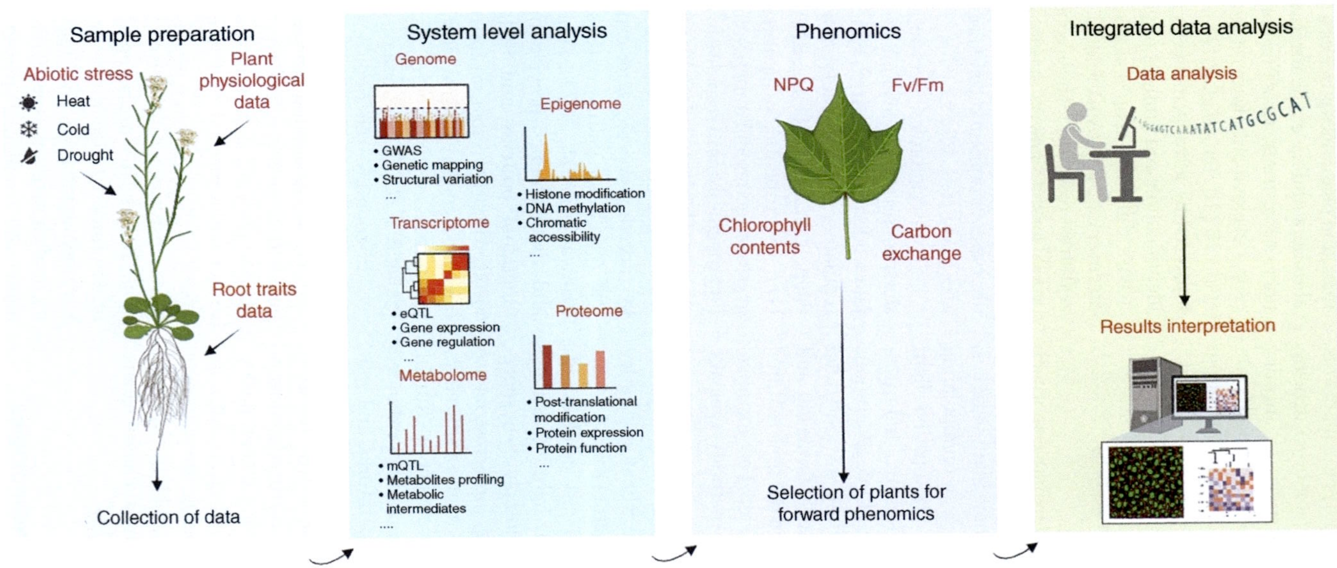

Fig. 12.1. A step-by-step procedure explaining the role of multiomics. (Figure author's own.)

the growing global food demand and adapt to changing environmental conditions (Razzaq *et al.*, 2021a).

12.2.6 NGS technologies

Sequence-based breeding has significantly transformed crop improvement strategies by utilizing modern genomic tools and technologies. The NGS technology has provided extensive genomic data, including genome assemblies and high-density genetic maps. These resources have facilitated the identification of genetic markers associated with various agronomic traits in crops. By enabling the rapid sequencing of large populations and detecting genomic variations, sequence-based breeding enhances the precision and efficiency of selecting desirable traits. This approach supports continuous population improvement and accelerates genetic gains, which are essential for developing superior crop varieties (Pandey *et al.*, 2016; Ruperao *et al.*, 2023).

Sequencing-based trait mapping plays a crucial role in this process, combining classical genetics with advanced sequencing technologies to map traits directly to specific genomic regions. Techniques such as bulked segregant analysis and its advanced forms, like QTL-Seq and MutMap, have proven effective in pinpointing genes responsible for important traits. These approaches streamline the identification of causal mutations and facilitate the breeding of crops with enhanced yield, quality, and stress resistance. By integrating high-throughput sequencing and genotyping platforms, sequence-based breeding enables the efficient selection of breeding lines with superior alleles or haplotypes, ultimately leading to the development of improved crop varieties (Soumya and Sankar, 2024).

12.2.7 Forward breeding technologies

Forward breeding plays a crucial role in crop improvement by utilizing genomic tools and technologies to enhance selection accuracy and efficiency. This approach leverages MAS and marker-assisted backcrossing to accelerate the breeding process, particularly for traits that are challenging to manage through traditional methods. Forward breeding is especially effective in developing crops that require specific environmental conditions or host developmental stages, or when pyramiding multiple monogenic traits or several quantitative trait loci for single traits (Cobb *et al.*, 2019; Naqvi *et al.*, 2022).

The application of forward breeding in crops like wheat, potato, and groundnut has demonstrated significant improvements in yield and resistance to biotic and abiotic stresses (Das *et al.*, 2015). For instance, the CIMMYT Global Wheat Program (GWP) (Fig. 12.2) utilizes advanced genetics and genomics tools to develop higher-yielding, more productive wheat varieties that are resistant to diseases like rust and fusarium head blight (Reynolds and Braun, 2022). Similarly, in potato breeding, MAS facilitates the selection of disease-resistant varieties, reducing the need for extensive phenotypic evaluations and saving significant time and costs. The integration of genomic resources, such as whole-genome

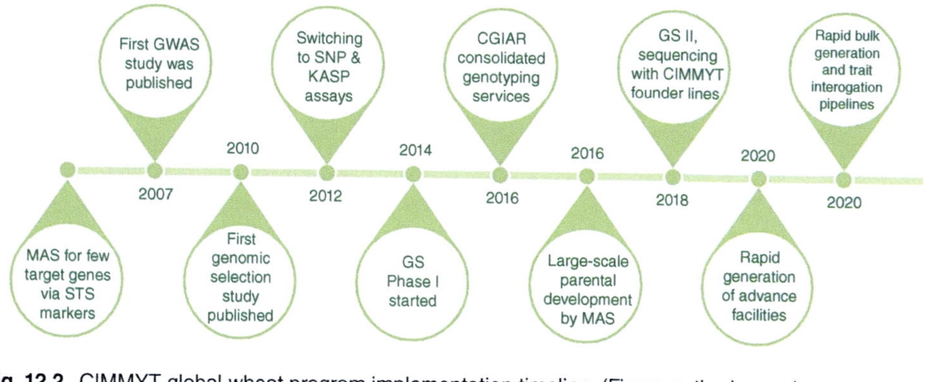

Fig. 12.2. CIMMYT global wheat program implementation timeline. (Figure author's own.)

sequencing and SNP genotyping platforms, into forward breeding strategies has enabled the development of crops with enhanced nutritional quality and climate resilience (Razzaq *et al.*, 2021a).

12.2.8 Genomic selection

Genomic selection (GS) involves developing prediction models using a training population that has been both genotyped and phenotyped (Akdemir and Isidro-Sánchez, 2019). These models are then used to estimate the genomic estimated breeding values (GEBVs) of individuals in a breeding population. By focusing on genetic potential rather than observed traits alone, GS enables more efficient selection cycles and accelerates the development of improved crop varieties. This approach is particularly advantageous for traits governed by multiple genes, such as yield, disease resistance, and abiotic stress tolerance, where traditional breeding methods may be less effective. The integration of GS with other advanced breeding techniques, such as SB and high-throughput phenotyping, further enhances its effectiveness, enabling faster and more precise crop improvement (Habier *et al.*, 2007; Contaldi *et al.*, 2021).

The success of GS in crop breeding is supported by several key factors, including the accuracy of prediction models, the density of genetic markers, and the genetic diversity within the training population (Voss-Fels *et al.*, 2019). High-density genotyping platforms and advancements in computational tools have significantly improved the accuracy of GEBVs, making GS a powerful tool in modern plant breeding programs. Additionally, the ability to rapidly cycle through generations and select the best candidates based on genetic predictions ensures continuous genetic gain and the development of high-performing crop varieties that can address the challenges of food security and climate change (Varshney *et al.*, 2016; Kim and Kwak, 2020).

12.2.9 Genetic engineering

Genetic engineering has revolutionized crop improvement by enabling precise modifications that enhance various plant traits. One significant application of genetic engineering is the development of genetically modified (GM) crops that exhibit beneficial attributes such as resistance to biotic and abiotic stresses, herbicide tolerance, pest resistance, and improved nutritional profiles. For example, herbicide-resistant GM crops, such as glyphosate-tolerant soybean and maize, have been developed to withstand specific herbicides, allowing farmers to control weeds more effectively without harming the crops. This trait is achieved by modifying the plant's genome to either degrade the herbicide or alter the target enzyme, rendering it insensitive to the herbicide's effects (Bonny, 2016; Wang *et al.*, 2017).

Additionally, genetic engineering facilitates biofortification, which enhances the nutritional value of crops. By inserting genes responsible for synthesizing essential vitamins and minerals, crops like golden rice, enriched with vitamin A, have been created (Zhu *et al.*, 2013; Sheoran *et al.*, 2022). This approach addresses malnutrition by ensuring that staple crops contain higher levels of vital nutrients. Furthermore, advancements in genome editing technologies, including CRISPR/Cas, have made it possible to introduce or modify traits with high precision, significantly improving crop yield, quality, and resistance to biotic and abiotic stresses. Overall, genetic engineering holds enormous promise for achieving food security and enhancing the nutritional quality of food crops (Wang *et al.*, 2022).

12.3 Recent Advancements in Abiotic Stress Resistance

Abiotic stress, including factors such as drought, salinity, cold, heat, and heavy metals, significantly limits plant growth and productivity, thereby threatening global food security (Bera *et al.*, 2022). The rapid population growth over the past half century has necessitated a deeper investigation into the environmental constraints contributing to reduced crop production. Historical estimates, like those by Boyer in the early 1980s, indicated that environmental factors could reduce crop yields by up to 70%. Various abiotic stresses affect approximately 96.5% of global rural land, causing irreversible damage to cultivated lands and adversely impacting crop production and quality (Imran *et al.*, 2021). In an increasingly changing environment, these abiotic stresses pose considerable challenges to agricultural sustainability. While traditional breeding and transgenic technologies have been employed to mitigate these risks, modern molecular tools have significantly advanced our ability to develop stress-resistant crops. CRISPR/Cas-mediated genome editing has shown immense potential in creating crop varieties with enhanced resistance to abiotic stressors (Zhang *et al.*, 2018; Şimşek *et al.*, 2024). Furthermore, evolving genome editing techniques such as prime editing and base editing, along with the production of mutant libraries, have accelerated the development of modern cultivars adapted to stressful conditions. These approaches not only improve crop yield and quality but also address biosafety concerns by producing transgene-free genetically edited plants, which can more easily meet regulatory norms.

12.3.1 Heat stress

Global warming, driven by increased concentrations of greenhouse gases in the atmosphere, is causing a rise in global temperatures. This phenomenon is leading to more frequent and intense heatwaves that necessitate heat-tolerant future crops. The scientists are working to generate heat-stress-tolerant crops, and in recent years, progress has been made in model crops like *Arabidopsis*. Recent studies on the impact of heat stress on *Arabidopsis*, rice, wheat, and tomato reveal several crucial regulatory mechanisms that enhance thermotolerance. THERMO-WITH ABA-RESPONSE1 (TWA1) acts as a temperature-sensing transcriptional co-regulator that interacts with JASMONATE-ASSOCIATED MYC-LIKE (JAM) transcription factors (TFs) and TOPLESS (TPL) proteins at elevated temperatures, facilitating the upregulation of heat-shock TF A2 (HSFA2) and heat shock proteins (HSPs) to enhance the plant's heat stress response (Bohn *et al.*, 2024). Additionally, X-box Binding Protein 1-Like1 (NFXL1) directly regulates DEHYDRATION-RESPONSIVE ELEMENT-BINDING PROTEIN 2A (DREB2A), which is crucial for reproductive thermotolerance by activating heat stress-responsive genes, including HSFA3 and HSP17.6, thereby underscoring its essential role during the reproductive stage (Zhu *et al.*, 2023). Another study highlights that m^{6}A RNA methylation marks transposon RNA, suppressing transposon mobility by inhibiting virus-like particle assembly and extrachromosomal DNA production, with the heat-activated retroelement *Onsen* being selectively demethylated by AtALKBH9B in stress granules (SGs) to allow mobilization (Fan *et al.*, 2023). Furthermore, AtCYP18-2, a homolog of PPIL1 in *Arabidopsis*, is shown to be crucial for heat tolerance by regulating pre-messenger RNA (mRNA) splicing; under heat stress, AtCYP18-2 interacts with spliceosome components and small nuclear RNAs U5 and U6, affecting the splicing of heat stress-responsive genes and ensuring the production of functional proteins like HSP101 and HSFA2 necessary for the heat-stress response (Lee *et al.*, 2023).

In rice, ONAC023 is a key regulator that enhances drought and heat tolerance by inducing the expression of stress-responsive genes involved in water transport, reactive oxygen species (ROS) homeostasis, and alternative splicing, facilitated by the remorin protein OSREM1.5 (Chang *et al.*, 2024). Additionally, N-terminal acetylation mediated by the OsNAA15 and OsNAA10 proteins, forming the N-terminal acetyltransferase (NatA) complex, is essential for thermoresponsive growth. The OsNAA15.1 splicing variant promotes growth under high temperatures by reducing ROS through the acetylation and degradation of glycolate oxidases GLO1 and GLO5 (Li *et al.*, 2024c). In wheat, the natural variation of the STKc_GSK3 kinase gene TaSG-D1E286K enhances heat tolerance by stabilizing the phosphorylation of TaPIF4, while sequence variations in the TaPIF4 promoter introduced from European germplasm negatively impact heat tolerance (Cao *et al.*, 2024). The alternative splicing of the TaHSFA6e gene results in two functional transcripts, with TaHSFA6e-III significantly improving thermotolerance by promoting SG disassembly and efficient translation re-initiation during heat-stress recovery (Wen *et al.*, 2023). In tomato, the plastid isoforms of triose phosphate isomerase, SlTPI1 and SlTPI2, exhibit differential responses to heat stress, with SlTPI2 showing higher thermotolerance regulated by heat shock factors (HSFs) and SlTPI1 being suppressed by WRKY TFs (Chen *et al.*,

Table 12.1. Recent advancements in heat stress tolerance. (Table author's own.)

Plant	Gene	Function	Reference
Arabidopsis	TWA1	Increases heat tolerance by regulating HSFA2 and HSPs.	Bohn *et al.* (2024)
	NFXL1	Enhances thermotolerance by regulating DREB2A and activating HSFA3 and HSP17.6.	Zhu *et al.* (2023)
	AtALKBH9B	AtALKBH9B demethylates the heat-activated retroelement Onsen in SGs, enabling its mobilization.	Fan *et al.* (2023)
	AtCYP18-2	Regulates pre-mRNA splicing and production of HSP101 and HSFA2.	Lee *et al.* (2023)
Rice	ONAC023	Enhances heat tolerance by inducing the expression of stress-responsive genes involved in water transport, ROS homeostasis, and alternative splicing.	Chang *et al.* (2024)
	OsNAA15 and OsNAA10	Promotes heat tolerance by forming the NatA complex; reduces ROS by acetylation and degradation of GLO1 and GLO5.	Li *et al.* (2024f)
Wheat	TaSG-D1E286K	Enhances heat tolerance by improving the phosphorylation and stability of TaPIF4.	Cao *et al.* (2024)
	TaHSFA6e	Regulates TaHSP70 gene expression with TaHSFA6e-III isoform, enhancing thermotolerance by promoting SG disassembly and efficient translation.	Wen *et al.* (2023)
Tomato	SlTPI2 and SlTPI1	SlTPI2 enhances thermotolerance and photosynthetic efficiency; SlTPI1 is suppressed by WRKY TFs, contributing to differential heat stress responses.	Chen *et al.* (2023)

2023). These interconnected findings highlight the complex regulatory networks in *Arabidopsis*, rice, wheat, and tomato that contribute to enhanced thermotolerance and provide potential genetic targets for developing heat-resilient crop varieties (Table 12.1).

12.3.2 Drought stress

Drought is a major abiotic stressor that has a negative impact on the growth and development of plants. Plants adapt to drought scenarios by modulating several metabolic and physiological pathways, which ultimately alleviate the negative effects of drought. Due to the intensification of global warming, many studies have been done in recent years to develop drought-tolerant genotypes capable of withstanding drought stress without compromising production (Sohail *et al.*, 2024).

Recent studies on drought stress in *Arabidopsis* reveal that the overexpression of the long non-coding RNA DANA1 enhances drought tolerance by interacting with DANA1-INTERACTING PROTEIN 1 (DIP1) to promote histone deacetylation of drought-responsive genes CYP707A1 and CYP707A2, thereby improving drought resistance (Cai *et al.*, 2024). Additionally, phloem-localized metacaspase, AtMC3, overexpression enhances drought resistance by promoting vascular tissue differentiation and maintaining effective long-distance transport, with knock-out mutants showing impaired responses to drought and reduced sensitivity to abscisic acid (ABA) (Pitsili *et al.*, 2023). Furthermore, the Regulator of Flowering and Stress (RFS) gene family enhances drought tolerance by delaying flowering in *Brachypodium* and *Arabidopsis*. Overexpression of BdRFS increases biomass and regulates lipid metabolism, specifically digalactosyldiacylglycerol

(DGDG) and phosphatidylcholine (PC) levels, impacting flowering and stress responses. (Ying *et al.*, 2022).

In rice, the SNAC1-OsERF103-OsSDG705 module is crucial for drought response. OsERF103 interacts with SNAC1 to regulate drought-related genes like OsbZIP23 and recruits OsSDG705 to enhance H3K4me3 deposition, boosting gene expression under drought conditions (Yang *et al.*, 2024). Moreover, in another study, the role of cysteine-rich receptor-like kinases (CRKs) in the rice stress response was investigated. The study highlights the expansion of CRK genes on chromosome 7 and uses CRISPR/Cas9 to create mutants for stress condition analysis. Findings show that different CRK genes enhance rice's tolerance to drought, cold, heat, and pathogens, where OsCRK29 was specifically found to regulate drought tolerance (Ye *et al.*, 2024). In wheat, overexpression of TaGW2 promotes degradation of the TaARR12 TF, which represses drought resistance genes. Ultimately, TaGW2 improves water use efficiency and reduces water loss, which boosts drought tolerance (Li *et al.*, 2023d). Moreover, the TF TabHLH27 enhances drought resistance in wheat by promoting root growth and improving water use efficiency. It regulates genes that help the plant respond to drought stress, leading to better survival and yield under water-limited conditions (Wang *et al.*, 2024a).

In cotton, GhCIPK6D1 increases drought sensitivity by promoting stomatal opening and increasing water loss; overexpression of GhCIPK6D1 leads to higher drought sensitivity, while its knock-out enhances drought tolerance by reducing stomatal aperture and water loss (Sun *et al.*, 2024b). Similarly, SlMYB92 and SlASFT enhance drought resistance in tomatoes by promoting suberin deposition in the root exodermis, which reduces water loss and improves water retention under drought conditions (Cantó-Pastor *et al.*, 2024). In *Populus*, the PtrAREB1-2-COMPASS pentamer enhances drought resistance by recruiting the COMPASS complex to drought-responsive genes, increasing their expression through H3K4me3 modification, thereby improving drought tolerance and biomass production (Zhang *et al.*, 2023a). These findings highlight diverse genetic pathways and mechanisms in *Arabidopsis*, rice, wheat, cotton, tomato, and *Populus* that contribute to enhanced drought tolerance (Table 12.2).

12.3.3 Cold stress

Cold stress is a harmful abiotic factor that damages plants at morphological, physiological, and biochemical levels. Understanding how plants adapt to and manage cold stress is crucial for identifying key genes or metabolites that contribute to plant tolerance under these adverse conditions (Liu *et al.*, 2023a). Recent studies have extensively screened genes associated with cold stress to enhance crop resilience. Inducer of CBF Expression 1 (ICE1) enhances *Arabidopsis'* cold tolerance by regulating cold-responsive genes like CBF3, which are crucial for cold tolerance. ICE1 also interacts with NPR1 to activate PR1 through the salicylic acid (SA) signaling pathway, boosting plant immunity against pathogens. This dual role helps *Arabidopsis* withstand cold stress and pathogen attacks (Li *et al.*, 2024b). Moreover, under cold stress, MYB43 is degraded by PRL1, allowing ICE1 to activate CBF genes, enhancing cold tolerance in *Arabidopsis* (Zheng *et al.*, 2023).

In rice, the NAC-type TF OsNAC5 enhances cold tolerance by activating ABA Insensitive 5 (OsABI5). This activation stabilizes OsABI5, boosting the expression of cold-responsive (COR) genes. Consequently, rice can respond quickly and effectively to cold stress, improving germination and seedling growth in low temperatures (Li *et al.*, 2023c). Moreover, OsKASI-2 enhances chilling tolerance in rice by increasing unsaturated fatty acids, preserving membrane fluidity, and maintaining ROS homeostasis, thereby protecting cell integrity and supporting chloroplast function under cold stress (Zhang *et al.*, 2024c). Similarly, *OsTIL1* enhances chilling tolerance in rice by protecting cell membranes from oxidative damage by ROS and maintaining 18:3-glycerolipid biosynthesis, preserving membrane damage under cold stress (Table 12.3) (Ji *et al.*, 2023).

In tomato, SlNAC3 negatively impacts cold tolerance through activation of ethylene (ET) biosynthetic genes such as SlACO1, SlACS2, and SlACS4, leading to increased ET production, which in turn reduces cold tolerance in tomato plants (Wang *et al.*, 2024e). Furthermore, the non-expresser of pathogenesis-related (PR) genes 1 (SlNPR1) confers chilling tolerance to tomatoes through the upregulation of key genes involved in ferulic acid (FA) synthesis, such as

Table 12.2. Recent advancements in drought stress tolerance. (Table author's own.)

Plant	Gene	Function	Reference
Arabidopsis	DANA1	Enhances drought tolerance by promoting histone deacetylation of CYP707A1 and CYP707A2.	Cai *et al.* (2024)
	AtMC3	Increases drought tolerance through vascular differentiation and long-distance transport.	Pitsili *et al.* (2023)
	BdRFS	Increases drought resistance by delaying flowering and increasing biomass and lipids.	Ying *et al.* (2022)
Rice	OsERF103	OsERF103 increases drought resistance by interacting with SNAC1 to regulate drought tolerance genes.	Yang *et al.* (2024)
	OsCRK29	Increases drought resistance and seed-setting rates.	Ye *et al.* (2024)
Wheat	TaGW2	Increases drought resistance by promoting the degradation of TaARR12 and reducing water loss.	Li *et al.* (2024e)
	TabHLH27	Increases drought resistance through improved root growth and water use efficiency.	Wang *et al.* (2024a)
Cotton	GhCIPK6D1	Increases drought sensitivity by promoting stomatal opening and water loss.	Sun *et al.* (2024b)
Tomato	SlMYB92 and SlASFT	Increases drought resistance by promoting suberin deposition.	Cantó-Pastor *et al.* (2024)
Populus	PtrAREB1-2–COMPASS	Increases drought resistance by improving biomass production	Zhang *et al.* (2023a)

Table 12.3. Recent advancements in cold stress tolerance. (Table author's own.)

Plant	Gene	Function	Reference
Arabidopsis	ICE1	Increases the cold tolerance by regulating cold tolerance gene CBF3.	Li *et al.* (2024d)
	MYB43	Increases cold tolerance through PRL1 mediated MYB43 degradation.	Zheng *et al.* (2023)
Rice	OsNAC5	Increases cold tolerance by activating the OsABI5 and COR genes.	Li *et al.* (2024c)
	OsKASI-2	Increases cold tolerance by maintaining ROS and chloroplast membrane integrity.	Zhang *et al.* (2024c)
	OsTIL1	Increases cold tolerance by reducing ROS and membrane damage.	Ji *et al.* (2023)
Tomato	SlNAC3	Decreases cold tolerance through activation of ethylene (ET) biosynthesis genes.	Wang *et al.* (2024e)
	SlNPR1	Increases cold tolerance by regulating FA biosynthesis.	Shu *et al.* (2022)
Citrus	PtrERF110	Increases cold tolerance by regulating sterols and sugars.	Khan *et al.* (2024)
	PtrZAT12	Increases cold tolerance through the regulation of COR genes.	Zhang *et al.* (2023b)
Kiwi fruit	AcePosF21	Increases cold tolerance by regulating AsA and AceGGP3.	Liu *et al.* (2023a)

PAL, C4H, C3H, and COMT, under low-temperature conditions. The increased FA levels help scavenge ROS, thereby protecting the plant cells from oxidative stress and improving overall cold resistance (Shu *et al.*, 2022). In citrus, overexpression of PtrERF110 enhances cold tolerance by regulating sugar and sterol biosynthesis, maintaining cell membrane stability during cold stress (Khan *et al.*, 2024). Similarly, overexpression of the C2H2-type zinc-finger protein (ZFP) PtrZAT12 enhances chilling tolerance in citrus by acting downstream of CBF1, binding to COR gene promoters, and activating genes that mitigate cold damage (Zhang *et al.*, 2023b). Another study on kiwifruit reveals that AcePosF21 enhances chilling tolerance by regulating ascorbic acid (Asc) biosynthesis and activating the expression *of GDP-L-galactose phosphorylase 3 (AceGGP3)* which ultimately increases cold tolerance by reducing oxidative damage (Liu *et al.*, 2023a). These findings underscore the potential for genetic manipulation to develop cold-resistant crops.

12.3.4 Salinity stress

Salinity stress causes significant damage to plants by disrupting their physiological and biochemical processes. It leads to ion toxicity, osmotic stress, and oxidative damage, severely affecting plant growth and development. As a result, salinity stress hampers seed germination, root development, and overall plant productivity. Research has focused on understanding the molecular mechanisms behind these responses to develop strategies for enhancing plant tolerance to salinity stress (Geng *et al.*, 2022; Sohail *et al.*, 2022). In *Arabidopsis*, the TF DIVARICATA1 (DIV1) enhances salinity tolerance by upregulating germination-associated genes like DELAY OF GERMINATION 1-LIKE 3 (DOGL3) and GA-STIMULATED ARABIDOPSIS 4 (GASA4), aiding the plant's response to salinity (Zhang *et al.*, 2024a). Moreover, PP2C.D6 and PP2C.D7 contribute to salinity tolerance by interacting with the salt overly sensitive 1 (SOS1) protein. These interactions occur through the C-terminal region of SOS1, inhibiting its activity. This inhibition helps regulate sodium ion (Na^+) transport and maintain ion homeostasis under salt stress conditions, enhancing the plant's ability to tolerate high salinity (Fu *et al.*, 2022). Similarly, phosphatidic acid (PA) enhances salinity tolerance in *Arabidopsis* by regulating Na^+ and potassium (K^+) homeostasis. PA binds to SOS2, activating the Na^+/H^+ antiporter SOS1 for Na^+ efflux and promoting SOS2-mediated phosphorylation of calcium-binding protein 8 (SCaBP8), which alleviates its inhibition of the K^+ transporter AKT1, thus balancing Na^+ and K^+ levels under salt stress (Li *et al.*, 2023b). Furthermore, NIGT1.4 plays a crucial role in salinity tolerance in *Arabidopsis* by promoting primary root elongation under salt stress. It achieves this by interacting with ABA signaling pathways and directly inducing the expression of ERF1, which modulates auxin activity and the expression of functional genes essential for maintaining root growth under saline conditions (Hu *et al.*, 2023).

In rice, OsHAK9 enhances salinity tolerance by preventing gibberellin (GA) degradation through OsGA2ox7 mediation, maintaining higher GA levels, and balancing K^+/Na^+ ions during seed germination under salt stress (Zeng *et al.*, 2024). Moreover, OsBBX17 regulates salinity stress tolerance in rice by interacting with OsMPK1, which phosphorylates OsBBX17, reducing its DNA-binding activity. This derepresses the transcription of OsHAK2 and OsHAK7, enhancing Na^+/K^+ homeostasis and promoting salinity tolerance (Shen *et al.*, 2023b). In tomato, SlWRKY81 enhances salinity stress tolerance by regulating spermidine synthesis and maintaining Na^+/K^+ homeostasis through the interaction with SlJAZ1 in the jasmonic acid (JA) pathway (Shang *et al.*, 2024). Similarly, SlSOS2 improves salinity stress tolerance in tomatoes by enhancing Na^+/K^+ homeostasis in roots, which reduces Na^+ toxicity under salt stress conditions (Hong *et al.*, 2022). In wheat, TaSPL6-D improves salinity tolerance by activating the Na^+-selective transporter TaHKT1;5-D, enhancing Na^+ exclusion from shoots, and maintaining yield under salt stress. This regulation helps reduce Na^+ toxicity in leaves, ensuring better growth performance and yield stability in saline conditions (Wang *et al.*, 2024d). Furthermore, GhUBC10-2 negatively regulates salinity tolerance in cotton by promoting the degradation of GhGSTU17, which is involved in ROS scavenging, thereby impairing the plant's ability to cope with salt stress (Sun *et al.*, 2024c).

The functional studies of these genes provide crucial insights into the molecular mechanisms underlying salinity stress tolerance in various crops. By understanding the specific roles of such genes, researchers can develop targeted genetic modifications to enhance salinity tolerance. These advancements pave the way for breeding or engineering crops with improved ion homeostasis, stress-responsive signaling pathways, and enhanced antioxidant capabilities. Ultimately, this knowledge will contribute to the development of resilient crop varieties, ensuring sustainable agricultural productivity in saline environments, and addressing food security challenges in the face of climate change (Table 12.4).

12.3.5 Heavy metal stress

Heavy metals negatively impact plants by inducing oxidative stress through the generation of ROS. This oxidative stress results in cellular damage, including lipid peroxidation, protein denaturation, and DNA damage, which disrupts vital physiological processes such as photosynthesis and respiration. Heavy metals also interfere with nutrient uptake and enzyme activities, leading to nutrient deficiencies and metabolic imbalances. Consequently, plants exhibit stunted growth, chlorosis, reduced biomass, and lower crop yields, ultimately compromising plant health and productivity (Noor *et al.*, 2022).

Generating heavy metal-tolerant crops is crucial for sustainable agriculture. These crops ensure productivity in contaminated soils, provide safer food by reducing metal accumulation, and aid in environmental remediation by stabilizing and extracting heavy metals from polluted areas. This approach supports food security, soil health, and ecosystem stability (Mitra *et al.*, 2022). In *Arabidopsis*, Trichome Birefringence (TBR) allelic variation enhances zinc (Zn) toxicity tolerance by modulating pectin methylesterification in root cell walls, which facilitates Zn sequestration and improves root growth under high Zn conditions (Zhong *et al.*, 2024). Moreover, the C_2H_2 ZFP ZAT17 negatively regulates cadmium (Cd) tolerance by upregulating Cd transport genes like *IRT1 and NRT2.1* and affecting the alternative splicing of Cd-responsive genes through interactions with the MOS4-associated complex (MAC). This regulatory mechanism suggests ZAT17 plays a crucial role in modulating Cd stress responses by influencing gene expression and splicing (Feng *et al.*, 2024). Similarly, overexpression of *CaCOMT1* enhances Cd tolerance in *Arabidopsis* by promoting cell wall lignin synthesis, which increases Cd fixation in root cell

Table 12.4. Recent advancements in salinity stress tolerance. (Table author's own.)

Plant	Gene	Function	Reference
Arabidopsis	DIV1	Increases salinity tolerance by upregulating DOGL3 and GASA4.	Zhang *et al.* (2024a)
	PP2C.D6 and PP2C.D7	Increases salinity tolerance by interacting with SOS1 and regulating Na^+ contents.	Fu *et al.* (2022)
	NIGT1.4	Increases salinity tolerance by enhancing root elongation.	Hu *et al.* (2023)
Rice	OsHAK9	Increases salinity tolerance through GA and K^+/Na^+ homeostasis.	Zeng *et al.* (2024)
	OsBBX17	Increases salinity tolerance through Na^+/K^+ homeostasis.	Shen *et al.* (2023b)
Tomato	SlWRKY81	Increases salinity tolerance by regulating spermidine and Na^+/K^+.	Shang *et al.* (2024)
	SlSOS2	Increases salinity tolerance by reducing Na^+ contents in roots.	Hong *et al.* (2022)
Wheat	TaSPL6-D	Increases salinity tolerance by enhancing Na^+ exclusion from shoots.	Wang *et al.* (2024d)
Cotton	GhUBC10-2	Decreases salinity tolerance by altering the ROS scavenging ability of plants.	Sun *et al.* (2024c)

walls and reduces Cd translocation to aboveground parts, thereby mitigating Cd toxicity (Shen *et al.*, 2023a).

In rice, ZFP36 enhances aluminum (Al) tolerance through an ABA-dependent pathway. Under Al stress, ZFP36 expression increases, enhancing root elongation and antioxidant defense by regulating genes like *OsAPX1* and *OsALS1*. It interacts with *OsPDK1*, which modulates ZFP36 activity, contributing to Al detoxification (Su *et al.*, 2024). Furthermore, *OsYSL15* enhances chromium (Cr) tolerance in rice by facilitating Cr(III) uptake. Its expression increases under Cr stress, and knockdown reduces Cr absorption, indicating its crucial role in Cr tolerance (Li *et al.*, 2024a). In another study, *OsAKR1* enhanced Cd tolerance, and its overexpression improved Cd resistance by modulating ROS scavenging and reducing cytotoxic aldehyde levels, while loss-of-function mutants showed increased Cd sensitivity and higher ROS levels (Guo *et al.*, 2024). The genes *OsRBOHA* and *OsRBOHI* play crucial roles in Cd tolerance in rice by mediating ROS signaling. These genes are upregulated in response to Cd stress, leading to an increase in ROS production, which triggers systemic signaling and root stress avoidance mechanisms. Knock-out mutants of these genes show significantly reduced ROS signaling and a decreased ability to avoid local Cd stress, demonstrating their essential roles in maintaining root architecture under Cd stress. Similarly, *OsCOPT7*, plays a crucial role in controlling copper (Cu) translocation from roots to shoots. Knock-out mutants of *OsCOPT7* exhibit increased Cu accumulation in roots but decreased Cu concentrations in shoots and grains, enhancing Cu tolerance and reducing Cu toxicity effects (Tang *et al.*, 2024).

Cotton and tobacco are two economically important crops. In cotton, *GhRCD1* overexpression increases Cd tolerance through the GhbHLH12-GhMYB44-GhHMA1 pathway, while *GhRCD1* knock-out results in increased Cd sensitivity and ROS accumulation (Wei *et al.*, 2024). While in tobacco, under cd stress, overexpression of *NtARF11* leads to inhibition of *NtNRT1.1* expression with lower Cd concentrations in both shoots and roots and increases the plant's biomass and chlorophyll content. Additionally, *NtARF11* boosts antioxidant enzyme activities, reducing ROS and improving overall stress resilience (Jia *et al.*, 2024). Functional studies of heavy metal tolerance genes unlock the potential to revolutionize crop resilience against metal toxicity. By pinpointing and engineering key metal tolerance genes, scientists can cultivate crops that excel in metal sequestration, minimize harmful metal uptake, and bolster antioxidant defenses. This approach not only enhances food security and soil health but also promotes ecosystem stability, forging a path toward a more resilient agricultural future (Table 12.5).

12.4 Recent Advancements in Biotic Stress Resistance

In the era of climate change, biotic stressors such as insects and diseases pose significant threats to crop production. Rising temperatures, altered precipitation patterns, and increased atmospheric carbon dioxide (CO_2) levels create favorable conditions for the proliferation and spread of pests and pathogens. These changes can exacerbate the vulnerability of crops, leading to a higher incidence and severity of infestations and infections. The resulting crop damage not only reduces yields but also impacts food quality and security. Addressing these challenges requires integrated pest management strategies, resistant crop varieties, and innovative agricultural practices to safeguard crop production in a changing climate (Devi *et al.*, 2022; Subedi *et al.*, 2023).

12.4.1 Disease resistance

Diseases significantly impact crop plants by reducing yield and quality. Pathogens such as bacteria, viruses, and fungi can cause a variety of symptoms, including stunted growth, chlorosis, necrosis, and wilting, which impede photosynthesis and nutrient uptake. In rice, for example, diseases like rice blast and bacterial blight lead to severe yield losses by attacking the plant's vascular system and disrupting water and nutrient transport. The rice stripe mosaic virus (RSMV) not only reduces yield but also causes delayed heading, creating a conducive environment for vector insects to overwinter, thus promoting further disease spread. These adverse effects necessitate robust disease management strategies

Table 12.5. Recent advancements in salinity stress tolerance. (Table author's own.)

Plant	Gene	Function	Reference
Arabidopsis	TBR	Increases Zn toxicity tolerance by sequestration in roots.	Zhong *et al.* (2024)
	ZAT17	Decreases Cd tolerance by upregulating Cd transporters.	Feng *et al.* (2024)
	CaCOMT1	Increases Cd tolerance through the promotion of lignin synthesis.	Shen *et al.* (2023a)
Rice	ZFP36	Increases Al tolerance by upregulating Al transporter genes.	Su *et al.* (2024)
	OsYSL15	Increases Cr tolerance.	Li *et al.* (2024b)
	OsAKR1	Increases Cd tolerance by ROS-scavenging and reducing cytotoxic aldehydes.	Guo *et al.* (2024)
	OsRBOHA and OsRBOHI	Increases Cd tolerance by mediating ROS signaling and root stress avoidance.	Wang *et al.* (2023)
	OsCOPT7	Increases Cu tolerance by controlling Cu translocation from roots to shoots and reducing Cu contents.	Tang *et al.* (2024)
Cotton	GhRCD1	Increases Cd tolerance by ROS homeostasis and interacting with GhbHLH12.	Wei *et al.* (2024)
Tobacco	NtARF11	Increases Cd tolerance by inhibiting NtNRT1.1 expression and reducing Cd uptake.	Jia *et al.* (2024)

to ensure food security and sustainable agricultural productivity. Gene functional studies can identify key resistance genes and pathways, enabling the development of disease-resistant crop varieties through genetic engineering and breeding, thus enhancing crop resilience and productivity (Chen *et al.*, 2024; Liu *et al.*, 2024; Zhao *et al.*, 2024).

In rice, HAF1, an E3 ubiquitin ligase, enhances RSMV resistance by regulating the degradation of Hd1 and ELF3. The RSMV P6 protein disrupts HAF1 function, stabilizing Hd1 and ELF3, and delaying rice heading, thus increasing susceptibility to RSMV (Chen *et al.*, 2024). Moreover, OsSWEET genes in rice are integral to managing bacterial blight resistance. Knockdown of OsSWEET genes, such as OsSWEET11a, OsSWEET14, and OsSWEET15, enhances bacterial blight resistance by preventing this efflux, thus limiting bacterial growth and lesion formation. Therefore, manipulating the expression of OsSWEET genes is a viable strategy for breeding rice varieties with improved resistance to bacterial blight (Wang *et al.*, 2024f). Similarly, the *OsS5H* gene in rice encodes a SA 5-hydroxylase enzyme, which is involved in converting SA into 2,5-dihydroxybenzoic acid (2,5-DHBA). Knock-out mutants of *OsS5H* show enhanced resistance to both bacterial blight and rice blast due to increased SA levels and upregulation of defense-related genes like *OsWRKY45* (Liu *et al.*, 2023b). Furthermore, the *OsMYBxoc1* gene in rice regulates resistance to bacterial leaf streak caused by *Xanthomonas oryzae pv. oryzicola* (*Xoc*). Overexpression of *OsMYBxoc1* represses the transcription of the iron transport gene *OsNRAMP5*, which reduces iron availability and thus inhibits pathogen proliferation (Zhang *et al.*, 2024b). Additionally, *OsHXK5* enhances the susceptibility to bacterial leaf streak (BLS) in rice by compromising the plant's defense responses. Overexpression of *OsHXK5* increases susceptibility to *Xoc* infection, while disruption of its promoter's effector binding element through gene editing confers resistance to the disease (Wang *et al.*, 2024c).

In wheat, *TaMKP1* negatively regulates disease resistance by dephosphorylating key MAPK phosphatases (*TaMPK3/4/6*), thus suppressing immune signaling. CRISPR/Cas9 knock-out of *TaMKP1* enhances resistance to fungal pathogens, including rust and powdery mildew, while also improving wheat's agronomic traits like plant height and grain yield (Liu *et al.*, 2024). In another study examining wheat, *TaCIPK14* knockdown enhanced resistance to stripe rust by increasing ROS and elevating the expression of defense-related genes like *TaPR1*, *TaPR2*, and *TaPR5*. Additionally, knock-out plants exhibit broad-spectrum resistance to multiple races of the stripe rust pathogen without compromising key agronomic traits (He *et al.*, 2022).

Cotton is a major fiber crop globally, and recent studies have identified multiple resistance genes in this crop. *MYB33* in cotton plays a dual role in balancing plant growth and defense against *Verticillium dahliae* infection. When *MYB33* is overexpressed, it enhances plant height by promoting cell elongation, but this comes at the cost of reduced resistance to *Verticillium dahliae* as it suppresses the expression of defense-related genes like *GhDFR1*. On the other hand, knocking out *MYB33* results in shorter plants with enhanced resistance to *Verticillium dahliae*, as it allows the activation of defense mechanisms (Guang *et al.*, 2023). Similarly, *GbPP2C80* interacts with *GbWAKL14* to modulate ROS production through the MAPK pathway, weakening the plant's immune response. Knocking out either gene enhances resistance to *Verticillium* and *Fusarium* wilts by increasing ROS levels and defense signaling (Zhao *et al.*, 2024).

In tomato, the ET response factor *SlERF.C1* is a crucial regulator in tomato defense against *Botrytis cinerea* (*B. cinerea*). It enhances resistance by activating the transcription of defense-related *PR* genes, such as *PR1* and *PR1a*, by directly binding to their promoters. Overexpression of *SlERF.C1* significantly boosts resistance to *B. cinerea*, while its knock-out results in increased susceptibility, highlighting its essential role in the plant's immune response (Deng *et al.*, 2024). Moreover, *SlbZIP68* enhances tomatoes' resistance to late blight by activating defense-related genes in the SA/JA pathways and improving the ROS-scavenging system. Knock-out of *SlbZIP68* reduces this resistance, leading to increased susceptibility to late blight disease (Yang *et al.*, 2023). In tobacco, *UbEF1B* and *CCR4/NOT3* are key host factors that interact with geminivirus replication proteins. Knock-out of these genes using CRISPR/Cas9 confers high-efficiency, broad-spectrum resistance to multiple geminiviruses without affecting plant growth. This resistance is achieved by disrupting the virus's ability to replicate, making these genes promising targets for developing virus-resistant crops (Li *et al.*, 2023a).

In grapes, *enhanced disease resistance 1* (*EDR1*) negatively regulates resistance to powdery mildew. CRISPR-induced chimeric mutations in the *VviEDR1* gene enhance resistance by promoting defense mechanisms such as increased callose deposition, H_2O_2 accumulation, and activation of SA and ET signaling pathways. Notably, these CRISPR mutants, particularly chimeric lines, show improved resistance without compromising growth. While homozygous and bi-allelic mutants exhibited severe growth defects and spontaneous lesions leading to plant death (Yu *et al.*, 2024). Overall, the integration of functional studies related to resistance genes offers a promising avenue for developing robust, disease-resistant varieties (Table 12.6).

12.4.2 Insect pest resistance

Insect pests pose a major threat to both rice and cotton crops, leading to significant economic losses due to reduced yields. The damage caused by these pests includes physical harm to plant tissues, disruption of normal plant growth, and reduced seed production. Molecular studies provide insights into potential genetic strategies to enhance crop resistance against these pests, offering a promising avenue for sustainable agricultural practices (Li *et al.*, 2023e; Teng *et al.*, 2024). In cotton, overexpression of specific genes (GhTPS14 and GhCYP82L1) researchers enhanced (3E)-4,8-dimethyl-1,3,7-nonatriene (DMNT) emissions. DMNT attracts natural enemies like parasitoid wasps, which target the mirid bugs, thereby boosting the plant's indirect defense. Additionally, DMNT creates an ecological trap that lures the bugs but inhibits their growth and increases their mortality due to its toxicity at higher concentrations. Thus, manipulating DMNT emissions in cotton presents a promising strategy for effectively controlling mirid bugs (Teng *et al.*, 2024). Moreover, *GhMLP423* plays a crucial role in cotton's resistance against insects by enhancing systemic-acquired resistance (SAR) through the activation of SA and PR genes. It induces wound-responsive signaling, including Ca^{2+} flux and ROS accumulation, which together strengthen the plant's defense mechanisms against both chewing and sap-sucking insects (Sun *et al.*, 2023). Similarly, *GhHAM* plays a crucial role in cotton's resistance against insects by regulating the formation

Table 12.6. Recent advancements in disease tolerance. (Table author's own.)

Plant	Gene	Function	Reference
Rice	HAF1	Increase RSMV resistance by delaying rice heading.	Chen *et al.* (2024)
	OsSWEET	Knock-out increases bacterial blight resistance by preventing and limiting bacterial growth.	Wang *et al.* (2024f)
	OsS5H	Knock-out increases resistance to bacterial blight and rice blast by regulating SA levels.	Liu *et al.* (2023b)
	OsMYBxoc1	Increase *Xoc* resistance by regulating *OsNRAMP5*-mediated iron contents.	Zhang *et al.* (2024b)
	OsHXK5	Knock-out increases resistance to bacterial leaf streak.	Wang *et al.* (2024c)
Wheat	*TaMKP1*	Knock-out increases resistance to fungal pathogens and improves yield.	Liu *et al.* (2024)
	TaCIPK14	Knock-out increases ROS production and defense gene expression.	He *et al.* (2022)
Cotton	*MYB33*	Knock-out restricts growth while boosting resistance against *Verticillium dahliae*.	Guang *et al.* (2023)
	GbPP2C80	Knock-out increases resistance to *Verticillium* and *Fusarium* wilts	Zhao *et al.* (2024)
Tomato	*SlERF.C1*	Increase resistance against *B. cinerea* through *PR* gene activation.	Deng *et al.* (2024)
	SlbZIP68	Increase resistance to late blight by increasing ROS scavenging.	Yang *et al.* (2023)
Tobacco	*UbEF1B and CCR4/NOT3*	Knock-out increases resistance to multiple geminiviruses.	Li *et al.* (2023a)
Grapes	*VviEDR1*	Knock-out increases resistance to powdery mildew through H_2O_2 and cellulose deposition.	Yu *et al.* (2024)

of pigment glands, which are responsible for producing gossypol and other secondary metabolites that deter pests. Knock-out of *GhHAM* leads to reduced gossypol levels and increased susceptibility to pests like cotton bollworms and aphids. *GhHAM* functions upstream of the GoPGF-GhJUB1 module, directly influencing gland formation and terpene production, thus contributing to broad-spectrum pest resistance in cotton (Long *et al.*, 2024).

In rice, *OsPCF5* enhances resistance against brown planthopper (BPH) by interacting with MYB proteins (OsMYB22, OsMYB30, and OsMYB30C). Overexpression of *OsPCF5* leads to increased BPH resistance, while its regulation is controlled by OsmiR319, which negatively affects BPH resistance. *OsPCF5* works through the phenylpropanoid pathway to strengthen plant defenses (Sun *et al.*, 2024a). Similarly, knock-out of *OsWRKY71* in rice impairs *Bph15*-mediated resistance, leading to increased susceptibility to BPH, as it regulates key defense-related genes and pathways involved in the plant's response to BPH infestation (Li *et al.*, 2023e). Furthermore, overexpression of *Osmotin1* decreases phytophagous mite *Schizotetranychus oryzae* numbers and mitigates seed loss, while its knock-out leads to higher mite infestation and greater seed reduction (Keil *et al.*, 2023). Functional studies of resistant genes have provided a strong foundation for developing pest-resistant crops. By understanding the specific roles these genes play in enhancing plant defenses, such as producing protective compounds, regulating defense pathways, and activating resistance mechanisms. Researchers can now design crops that are more resilient to insect attacks. These insights pave the way for genetic engineering approaches that reduce reliance on chemical pesticides, contributing to sustainable agriculture (Table 12.7).

Table 12.7. Recent advancements in insect pest tolerance. (Table author's own.)

Plant	Gene	Function	Reference
Cotton	GhTPS14 and GhCYP82L1	Increases resistance against mirid bugs by increasing DMNT emissions.	Teng *et al.* (2024)
	GhMLP423	Increases resistance against chewing and sap-sucking insects by activating the SA and PR genes.	Sun *et al.* (2023)
	GhHAM	Knock-out increased sensitivity against bollworms and aphids by decreasing gossypol levels.	Long *et al.* (2024)
Rice	*OsPCF5*	Increases resistance to BPH by interacting with MYB proteins and activating the phenylpropanoid pathway.	Sun *et al.* (2024a)
	Osmotin1	Increases resistance against mites and decrease seed loss.	Keil *et al.* (2023)
	OsWRKY71	Knock-out decreased resistance against BPH.	Li *et al.* (2023b)

12.5 Recent Advancements in Crop Quality Traits

Genetic engineering and functional studies of crop quality trait-related genes have the potential to revolutionize the cultivation of fruits and vegetables. By understanding and manipulating specific genes that control traits like size, taste, nutritional content, and resistance to environmental stressors, scientists can develop varieties that not only yield better-quality produce but also withstand pests and diseases more effectively. These advances not only improve the overall quality and marketability of produce but also contribute to sustainable agricultural practices by reducing the need for chemical inputs and enhancing resilience against environmental stresses. As a result, these advancements not only benefit consumers but also promote sustainable agricultural practices, ensuring that crops are more resilient and environmentally friendly (Jin *et al.*, 2023; Dong *et al.*, 2024).

In tomatoes, Expansin *SlExp1* and endoglucanase *SlCel2* synergistically promote tomato fruit softening by regulating cell wall disassembly. *SlExp1* acts as a cell wall loosening agent, while *SlCel2* hydrolyzes components of the cell wall matrix. Together, their co-expression during fruit ripening enhances the breakdown of pectin and xyloglucan, leading to a softer fruit texture (Su *et al.*, 2023). In another study, SlBEL11 prevented premature fruit drop in tomatoes by regulating flavonoid biosynthesis and fine-tuning auxin efflux. It is highly expressed in the fruit abscission zone (FAZ), and its downregulation leads to impaired flavonoid biosynthesis, particularly quercetin, which is crucial for maintaining proper auxin transport. This disruption in auxin balance causes premature fruit drop, highlighting the role of SlBEL11 in stabilizing auxin levels to prevent early fruit abscission (Dong *et al.*, 2024). Moreover, SlYTH2, an m^{6}A reader, negatively regulates tomato fruit aroma by repressing the translation of aroma-related genes. Knock-out of SlYTH2 enhances the production of aroma volatiles by increasing the translation efficiency of key genes involved in volatile synthesis, leading to a richer fruit aroma without affecting other aspects of fruit quality. In rice, OsSG34 plays a crucial role in regulating rice grain shape and appearance by modulating lipid metabolism. It affects triglyceride content, leading to changes in cell proliferation and expansion, resulting in slender grains with improved transparency and reduced chalkiness. (Jin *et al.*, 2023). Similarly, miR5504 regulates rice grain shape and appearance by influencing cell expansion, leading to changes in grain width and thickness (Wang *et al.*, 2024b).

12.5 Future Perspectives

In the future, the continued integration of multi-omics technologies, advanced phenotyping, and bioinformatics will be crucial in driving the next wave of crop improvement. As climate change intensifies, the need for crops that can withstand extreme environmental conditions, such as drought, salinity, and heat, will become even more critical.

Future research will likely focus on refining genome editing techniques like CRISPR/Cas9 to develop crops with precise modifications that enhance resilience, yield, and nutritional content. Additionally, the advancement of predictive modeling through artificial intelligence and machine learning will allow for more accurate identification of key genetic traits and the rapid development of crop varieties tailored to specific environments.

References

Aziz, A.M., and Masmoudi, K. (2025). Molecular Breakthroughs in Modern Plant Breeding Techniques. *Horticultural Plant Journal* 11(1), 15–41.

Akdemir, D. and Isidro-Sánchez, J. (2019) Design of training populations for selective phenotyping in genomic prediction. *Sci Rep* 9, 1446.

Aribi, M. M. (2024). Plant gene banks: conservation of genetic resources. In: *Sustainable utilization and conservation of plant genetic diversity* (pp. 753-775). Singapore: Springer Nature Singapore.

Arif, M., Men, S., Nawaz, A. F., Li, X., Xu, L. *et al.* (2024). Bcl-2-associated athanogene (BAG) co-chaperones: key players in multiple abiotic and biotic stress tolerance in plants. *Journal of Plant Growth Regulation*, 1–14.

Bashir, W., Anwar, S., Zhao, Q., Hussain, I. and Xie, F. (2019). Interactive effect of drought and cadmium stress on soybean root morphology and gene expression. *Ecotoxicology and Environmental Safety*, 175, 90-101. https://doi.org/10.1016/j.ecoenv.2019.03.042

Bera, K., Dutta, P., Sadhukhan, S. (2022). Plant Responses Under Abiotic Stress and Mitigation Options Towards Agricultural Sustainability. In: Roy, S., Mathur, P., Chakraborty, A.P., Saha, S.P. (eds) Plant Stress: Challenges and Management in the New Decade. *Advances in Science, Technology & Innovation*. Springer, Cham. https://doi.org/10.1007/978-3-030-95365-2_1

Bohn, L., Huang, J., Weidig, S., Yang, Z., Heidersberger, C. *et al.* (2024). The temperature sensor TWA1 is required for thermotolerance in Arabidopsis. *Nature*, 629(8014), 1126–1132. https://doi.org/10.1038/s41586-024-07424-x

Bonny S. (2016). Genetically Modified Herbicide-Tolerant Crops, Weeds, and Herbicides: Overview and Impact. *Environmental Management*, 57(1), 31–48. https://doi.org/10.1007/s00267-015-0589-7

Cai, J., Zhang, Y., He, R., Jiang, L., Qu, Z., Gu, J., Yang, J., Legascue, M. F., Wang, Z., Ariel, F., Adelson, D. L., Zhu, Y., & Wang, D. (2024). LncRNA DANA1 promotes drought tolerance and histone deacetylation of drought responsive genes in Arabidopsis. EMBO Reports, 25(2), 796–812. https://doi.org/10.1038/s44319-023-00030-4

Cantó-Pastor, A., Kajala, K., Shaar-Moshe, L., Manzano, C., Timilsena, P. *et al.* (2024). A suberized exodermis is required for tomato drought tolerance. Nature Plants, 10(1), 118–130. https://doi.org/10.1038/s41477-023-01567-x

Cao, J., Qin, Z., Cui, G., Chen, Z., Cheng, X. *et al.* (2024). Natural variation of STKc_GSK3 kinase TaSG-D1 contributes to heat stress tolerance in Indian dwarf wheat. *Nature Communications*, 15(1). https://doi.org/10.1038/s41467-024-46419-0

Chang, Y., Fang, Y., Liu, J., Ye, T., Li, X. *et al.* (2024). Stress-induced nuclear translocation of ONAC023 improves drought and heat tolerance through multiple processes in rice. *Nature Communications*, 15(1). https://doi.org/10.1038/s41467-024-50229-9

Chao, J., Li, T., Yin, H., & Wang, Z. (2024). Adoption of multiple sustainable agricultural practices among farmers in northwest, China. *Cogent Food & Agriculture*, 10(1), 2300189.

Chen, C., Zhang, M., Ma, X., Meng, Q., & Zhuang, K. (2023). Differential heat-response characteristics of two plastid isoforms of triose phosphate isomerase in tomato. *Plant Biotechnology Journal*, 22(3), 650–661. https://doi.org/10.1111/pbi.14212

Chen, S., Zhong, X., Wang, Z., Chen, B., Huang, X. *et al.* (2024). Rice stripe mosaic virus hijacks rice heading-related gene to promote the overwintering of its insect vector. *Journal of Integrative Plant Biology*, 66(9), 2000–2016. https://doi.org/10.1111/jipb.13722

Cobb, J. N., Declerck, G., Greenberg, A., Clark, R., & McCouch, S. (2013). Next-generation phenotyping: requirements and strategies for enhancing our understanding of genotype-phenotype relationships and its relevance to crop improvement. *TAG. Theoretical and applied genetics. Theoretische und angewandte Genetik*, 126(4), 867–887. https://doi.org/10.1007/s00122-013-2066-0

Cobb, J. N., Juma, R. U., Biswas, P. S., Arbelaez, J. D., Rutkoski, J., Atlin, G., Hagen, T., Quinn, M., & Ng, E. H. (2019). Enhancing the rate of genetic gain in public-sector plant breeding programs: lessons from the breeder's equation. *TAG. Theoretical and applied genetics. Theoretische und angewandte Genetik*, 132(3), 627–645. https://doi.org/10.1007/s00122-019-03317-0

Contaldi, F., Cappetta, E., Esposito, S. (2021). Practical Workflow from High-Throughput Genotyping to Genomic Estimated Breeding Values (GEBVs). In: Tripodi, P. (eds) Crop Breeding. Methods in Molecular Biology, vol 2264. Humana, New York, NY. https://doi.org/10.1007/978-1-0716-1201-9_9

Das, P., Nutan, K. K., Lata, S., & Pareek, A. (2015). Understanding salinity responses and adopting 'omics-based' approaches to generate salinity tolerant cultivars of rice. *Frontiers in Plant Science*, 6, 155215. https://doi.org/10.3389/fpls.2015.00712

Deng, H., Pei, Y., Xu, X., Du, X., Xue, Q., Gao, Z., Shu, P., Wu, Y., Liu, Z., Jian, Y., Wu, M., Wang, Y., Li, Z., Pirrello, J., Bouzayen, M., Deng, W., Hong, Y., & Liu, M. (2024). Ethylene-MPK8-ERF.C1-PR module confers resistance against Botrytis cinerea in tomato fruit without compromising ripening. *New Phytologist*, 242(2), 592–609. https://doi.org/10.1111/nph.19632

Devi, B., Kumar, M. N., Chutia, M., & Bhattacharyya, N. (2022). Abiotic and biotic stress challenges of Cassava in changing climate and strategies to overcome: A review. *Scientia Horticulturae*, 305, 111432. https://doi.org/10.1016/j.scienta.2022.111432

Dong, X., Liu, X., Cheng, L., Li, R., Ge, S., Wang, S., Cai, Y., Liu, Y., Meng, S., Jiang, C., Shi, C., Li, T., Fu, D., Qi, M., & Xu, T. (2024). SlBEL11 regulates flavonoid biosynthesis, thus fine-tuning auxin efflux to prevent premature fruit drop in tomato. *Journal of Integrative Plant Biology*, 66(4), 749–770. https://doi.org/10.1111/jipb.13627

Edwards, D., Batley, J., & Snowdon, R. J. (2013). Accessing complex crop genomes with next-generation sequencing. *TAG. Theoretical and applied genetics. Theoretische und angewandte Genetik*, 126(1), 1–11. https://doi.org/10.1007/s00122-012-1964-x

Eldakak, M., Milad, S. I., Nawar, A. I., & Rohila, J. S. (2013). Proteomics: A biotechnology tool for crop improvement. *Frontiers in Plant Science*, 4, 41649. https://doi.org/10.3389/fpls.2013.00035

Feng, Q., Zhao, L., Jiang, S., Qiu, Y., Zhai, T., Yu, S., Yang, W., & Zhang, S. (2024). The C2H2 family protein ZAT17 engages in the cadmium stress response by interacting with PRL1 in Arabidopsis. Journal of Hazardous Materials, 465, 133528. https://doi.org/10.1016/j.jhazmat.2024.133528

Fu, H., Yu, X., Jiang, Y., Wang, Y., Yang, Y., Chen, S., Chen, Q., & Guo, Y. (2022). SALT OVERLY SENSITIVE 1 is inhibited by clade D Protein phosphatase 2C D6 and D7 in Arabidopsis thaliana. The Plant Cell, 35(1), 279–297. https://doi.org/10.1093/plcell/koac283

Geng, S., Sohail, H., Cao, H., Sun, J., Chen, Z., Zhou, L., Wang, W., Ye, R., Yang, L., & Bie, Z. (2022). An efficient root transformation system for CRISPR/Cas9-based analyses of shoot–root communication in cucurbit crops. Horticulture Research, 9. https://doi.org/10.1093/hr/uhab082

Guang, H., Xiaoyang, G., Zhian, W., Ye, W., Peng, W., Linfang, S., Bingting, W., Anhong, Z., Fuguang, L., & Jiahe, W. (2023). The cotton MYB33 gene is a hub gene regulating the trade-off between plant growth and defense in Verticillium dahliae infection. Journal of Advanced Research, 61, 1–17. https://doi.org/10.1016/j.jare.2023.08.017

Guo, R., Zhang, Q., Chen, C. Z., Sun, J. Y., Tu, C. Y., He, M. X., Shen, R. F., Huang, J., & Zhu, X. F. (2024). A novel aldo-keto reductase gene, OsAKR1, from rice confers higher tolerance to cadmium stress in rice by an in vivo reactive aldehyde detoxification. Journal of Hazardous Materials, 470, 134212. https://doi.org/10.1016/j.jhazmat.2024.134212

Habier, D., Fernando, R. L., & Dekkers, J. C. (2007). The impact of genetic relationship information on genome-assisted breeding values. *Genetics*, 177(4), 2389–2397. https://doi.org/10.1534/genetics.107.081190

He, F., Wang, C., Sun, H., Tian, S., Zhao, G., Liu, C., Wan, C., Guo, J., Huang, X., Zhan, G., Yu, X., Kang, Z., & Guo, J. (2022). Simultaneous editing of three homoeologues of TaCIPK14 confers broad-spectrum resistance to stripe rust in wheat. Plant Biotechnology Journal, 21(2), 354–368. https://doi.org/10.1111/pbi.13956

He, R., Ju, J., Liu, K., Song, J., Zhang, S., Zhang, M. *et al.* (2024). Technology of plant factory for vegetable crop speed breeding. *Frontiers in Plant Science*, 15, 1414860.

Hong, Y., Guan, X., Wang, X., Kong, D., Yu, S., Wang, Z., Yu, Y., Chao, Z., Liu, X., Huang, S., Zhu, J., Zhu, G., & Wang, Z. (2022). Natural variation in SlSOS2 promoter hinders salt resistance during tomato domestication. Horticulture Research, 10(1). https://doi.org/10.1093/hr/uhac244

Hu, Y., Zeng, L., Lv, X., Guo, J., Li, X., Zhang, X., Wang, D., Wang, J., Bi, J., Julkowska, M. M., & Li, B. (2023). NIGT1.4 maintains primary root elongation in response to salt stress through induction of ERF1 in Arabidopsis. The Plant Journal, 116(1), 173–186. https://doi.org/10.1111/tpj.16369

Imam, Z., Sultana, R., Parveen, R., Swapnil, Singh, D., Sinha, S., & Sahoo, J. P. (2024). Understanding the concept of speed breeding in crop improvement: opportunities and challenges towards global food security. *Tropical Plant Biology*, 17(1), 1-23.

Imran, Q. M., Falak, N., Hussain, A., Mun, B., & Yun, B. (2021). Abiotic Stress in Plants; Stress Perception to Molecular Response and Role of Biotechnological Tools in Stress Resistance. *Agronomy*, 11(8), 1579. https://doi.org/10.3390/agronomy11081579

Ji, L., Zhang, Z., Liu, S., Zhao, L., Li, Q., Xiao, B., Suzuki, N., Burks, D. J., Azad, R. K., & Xie, G. (2023). The OsTIL1 lipocalin protects cell membranes from reactive oxygen species damage and maintains the 18:3-containing glycerolipid biosynthesis under cold stress in rice. The Plant Journal, 117(1), 72–91. https://doi.org/10.1111/tpj.16470

Jia, H., Zhu, Z., Zhan, J., Luo, Y., Yin, Z., Wang, Z., Yan, X., Shao, H., & Song, Z. (2024). NtARF11 positively regulates cadmium tolerance in tobacco by inhibiting expression of the nitrate transporter NtNRT1.1. Journal of Hazardous Materials, 473, 134719. https://doi.org/10.1016/j.jhazmat.2024.134719

Jin, X., Chen, J., Khan, A., Chen, Z., Gao, R., Lu, Y., & Zheng, X. (2023). Triacylglycerol lipase, OsSG34, plays an important role in grain shape and appearance quality in rice. The Plant Journal, 117(3), 840–855. https://doi.org/10.1111/tpj.16532

Kaur, G., Vikal, Y., Kaur, L., Kalia, A., Mittal, A., Kaur, D., & Yadav, I. (2021). Elucidating the morpho-physiological adaptations and molecular responses under long-term waterlogging stress in maize through gene expression analysis. *Plant Science*, 304, 110823. https://doi.org/10.1016/j.plantsci.2021.110823

Keil, R., De Oliveira Neves, L., Da Silva, L. C. O., Lamb, T. I., Berghahn, E., Pita, F. M., Johann, L., Wang, Y., Feng, Z., Wang, G., Zuo, S., & Sperotto, R. A. (2023). Osmotin1 is involved in rice resistance to Schizotetranychus oryzae (Acari: Tetranychidae) infestation. Pest Management Science, 80(4), 2154–2161. https://doi.org/10.1002/ps.7955

Khan, M., Dahro, B., Wang, Y., Wang, M., Xiao, W., Qu, J., Zeng, Y., Fang, T., Xiao, P., Xu, X., Li, C., & Liu, J. (2024). The transcription factor ERF110 promotes cold tolerance by directly regulating sugar and sterol biosynthesis in citrus. The Plant Journal, 119(5), 2385–2401. https://doi.org/10.1111/tpj.16925

Kim, H.S., Kwak, SS. Crop biotechnology for sustainable agriculture in the face of climate crisis. *Plant Biotechnol Rep* 14, 139–141 (2020). https://doi.org/10.1007/s11816-020-00619-4

Kumar, K. P. and Walia, P. (2024). "Speed Breeding in Cereal Crops: Accelerating Genetic Improvement for Rapid Agricultural Advancement." *Journal of Experimental Agriculture International* 46 (6):465–477.

Lee, A., Park, H. J., Jo, S. H., Jung, H., Kim, H., Lee, H., Kim, Y., Jung, C., & Cho, H. S. (2023). The spliceophilin CYP18-2 is mainly involved in the splicing of retained introns under heat stress in Arabidopsis. Journal of Integrative Plant Biology, 65(5), 1113–1133. https://doi.org/10.1111/jipb.13450

Li, H., Gong, P., Xu, X., Zhou, X., & Li, F. (2023a). Knockout of the virus replication-related genes UbEF1B and CCR4/NOT3 by CRISPR/Cas9 confers high-efficiency and broad-spectrum resistance to geminiviruses in Nicotiana benthamiana. Plant Biotechnology Journal, 22(4), 793–795. https://doi.org/10.1111/pbi.14265

Li, J., Shen, L., Han, X., He, G., Fan, W., Li, Y., Yang, S., Zhang, Z., Yang, Y., Jin, W., Wang, Y., Zhang, W., & Guo, Y. (2023b). Phosphatidic acid–regulated SOS2 controls sodium and potassium homeostasis in Arabidopsis under salt stress. The EMBO Journal, 42(8). https://doi.org/10.15252/embj.2022112401

Li, R., Song, Y., Wang, X., Zheng, C., Liu, B., Zhang, H., Ke, J., Wu, X., Wu, L., Yang, R., & Jiang, M. (2023c). OsNAC5 orchestrates OsABI5 to fine-tune cold tolerance in rice. Journal of Integrative Plant Biology, 66(4), 660–682. https://doi.org/10.1111/jipb.13585

Li, S., Zhang, Y., Liu, Y., Zhang, P., Wang, X., Chen, B., Ding, L., Nie, Y., Li, F., Ma, Z., Kang, Z., & Mao, H. (2023d). The E3 ligase TaGW2 mediates transcription factor TaARR12 degradation to promote drought resistance in wheat. The Plant Cell, 36(3), 605–625. https://doi.org/10.1093/plcell/koad307

Li, X., Zhang, J., Shangguan, X., Yin, J., Zhu, L., Hu, J., Du, B., & Lv, W. (2023e). Knockout of OsWRKY71 impairs Bph15-mediated resistance against brown planthopper in rice. Frontiers in Plant Science, 14. https://doi.org/10.3389/fpls.2023.1260526

Li, J., Chen, W., Xu, K., Xie, W., Qi, H., Tang, Y., Wang, S., Deng, T., Morel, J., & Qiu, R. (2024a). Fe(III) transporter OsYSL15 may play a key role in the uptake of Cr(III) in rice (Oryza sativa L.). Journal of Hazardous Materials, 469, 133531. https://doi.org/10.1016/j.jhazmat.2024.133531

Li, S., He, L., Yang, Y., Zhang, Y., Han, X., Hu, Y., & Jiang, Y. (2024b). INDUCER OF CBF EXPRESSION 1 promotes cold-enhanced immunity by directly activating salicylic acid signaling. The Plant Cell, 36(7), 2587–2606. https://doi.org/10.1093/plcell/koae096

Li, X., Tang, H., Xu, T., Wang, P., Ma, F., Wei, H., Fang, Z., Wu, X., Wang, Y., Xue, Y., & Zhang, B. (2024c). N-terminal acetylation orchestrates glycolate-mediated ROS homeostasis to promote rice thermoresponsive growth. New Phytologist, 243(5), 1742–1757. https://doi.org/10.1111/nph.19928

Liu, X., Bulley, S. M., Varkonyi-Gasic, E., Zhong, C., & Li, D. (2023a). Kiwifruit bZIP transcription factorAcePosF21elicits ascorbic acid biosynthesis during cold stress. PLANT PHYSIOLOGY, 192(2), 982–999. https://doi.org/10.1093/plphys/kiad121

Liu, X., Yu, Y., Yao, W., Yin, Z., Wang, Y., Huang, Z., Zhou, J., Liu, J., Lu, X., Wang, F., Zhang, G., Chen, G., Xiao, Y., Deng, H., & Tang, W. (2023b). CRISPR/Cas9-mediated simultaneous mutation of three salicylic acid 5-hydroxylase (OsS5H) genes confers broad-spectrum disease resistance in rice. Plant Biotechnology Journal, 21(9), 1873–1886. https://doi.org/10.1111/pbi.14099

Liu, S., Zhang, F., Su, J., Fang, A., Tian, B., Yu, Y., Bi, C., Ma, D., Xiao, S., & Yang, Y. (2024). CRISPR-targeted mutagenesis of mitogen-activated protein kinase phosphatase 1 improves both immunity and yield in wheat. Plant Biotechnology Journal, 22(7), 1929–1941. https://doi.org/10.1111/pbi.14312

Long, L., Xu, F., Yuan, M., Shang, S., Song, H., Zhao, J., Hu, G., Zhang, Z., Zhao, X., Ma, J., Hussain, A., Wang, P., Cai, Y., Jin, S., & Gao, W. (2024). GhHAM regulates GoPGF-dependent gland development and contributes to broad-spectrum pest resistance in cotton. The Plant Journal, 119(2), 879–894. https://doi.org/10.1111/tpj.16803

Mir, R. A., Bhat, K. A., Shah, A. A., and Zargar, S. M. (2019). Role of Silicon in Abiotic Stress Tolerance of Plants. *Improving Abiotic Stress Toler. Plants* 2020, 271–290. doi: 10.1201/9780429027505-15

Mitra, M., Agarwal, P., & Roy, S. (2022). Plant response to heavy metal stress: an insight into the molecular mechanism of transcriptional regulation. Elsevier eBooks (pp. 337–367). https://doi.org/10.1016/b978-0-323-90613-5.00004-2

Naqvi, R. Z., Siddiqui, H. A., Mahmood, M. A., Najeebullah, S., Ehsan, A., Azhar, M., Farooq, M., Amin, I., Asad, S., Mukhtar, Z., Mansoor, S., & Asif, M. (2022). Smart breeding approaches in post-genomics era for developing climate-resilient food crops. *Frontiers in Plant Science*, 13, 972164. https://doi.org/10.3389/fpls.2022.972164

Noor, I., Sohail, H., Sun, J., Nawaz, M. A., Li, G., Hasanuzzaman, M., & Liu, J. (2022). Heavy metal and metalloid toxicity in horticultural plants: Tolerance mechanism and remediation strategies. Chemosphere, 303, 135196. https://doi.org/10.1016/j.chemosphere.2022.135196

Nyirahabimana, F., & Solmaz, İ. (2024). Haploid induction through ovary culture in cucumber. *In Vitro Cellular & Developmental Biology-Plant*, 60(1), 122-130.

Ogutu, E. A., Madahana, S. L., Bhavani, S., & Macharia, G. (2024). Genotypex environment interaction: trade-offs between the agronomic performance and stability of durum (Triticum turgidum) wheat to stem-rust resistance in Kenya. *Frontiers in Plant Science*, 15, 1427483.

Pandey, A., Sharma, M., & Pandey, G. K. (2016). Emerging Roles of Strigolactones in Plant Responses to Stress and Development. *Frontiers in Plant Science*, 7, 165491. https://doi.org/10.3389/fpls.2016.00434

Pitsili, E., Rodriguez-Trevino, R., Ruiz-Solani, N., Demir, F., Kastanaki, E., Dambire, C., De Pedro-Jové, R., Vercammen, D., Salguero-Linares, J., Hall, H., Mantz, M., Schuler, M., Tuominen, H., Van Breusegem, F., Valls, M., Munné-Bosch, S., Holdsworth, M. J., Huesgen, P. F., Rodriguez-Villalon, A., & Coll, N. S. (2023). A phloem-localized Arabidopsis metacaspase (AtMC3) improves drought tolerance. New Phytologist, 239(4), 1281–1299. https://doi.org/10.1111/nph.19022

Rahman, A., Mostofa, M. G., Alam, M. M., Nahar, K., Hasanuzzaman, M., & Fujita, M. (2015). Calcium Mitigates Arsenic Toxicity in Rice Seedlings by Reducing Arsenic Uptake and Modulating the Antioxidant Defense and Glyoxalase Systems and Stress Markers. *BioMed research international*, 2015, 340812. https://doi.org/10.1155/2015/340812

Razzaq, A., Sadia, B., Raza, A., Khalid Hameed, M., and Saleem, F. (2019). Metabolomics: a way forward for crop improvement. *Metabolites* 9:303. doi: 10.3390/metabo9120303

Razzaq, A., Kaur, P., Akhter, N., Wani, S. H., & Saleem, F. (2021a). Next-Generation Breeding Strategies for Climate-Ready Crops. *Frontiers in Plant Science*, 12, 620420. https://doi.org/10.3389/fpls.2021.620420

Razzaq, A., Wani, S. H., Saleem, F., Yu, M., Zhou, M., & Shabala, S. (2021b). Rewilding crops for climate resilience: economic analysis and de novo domestication strategies. *Journal of Experimental Botany*, 72(18), 6123-6139.

Razzaq, S., Zhou, B., Adil, M., Ullah, Z., Guo, H., Zia-Ur-Rehman, M., Xiaopeng, C., Qian, L. W., & Lu, H. (2024). Cadmium Stress Alleviation: Interplay of Micronutrients and Enzymatic/Non-enzymatic Species in Maize by Organic and Inorganic Amendments. *Water, Air, and Soil Pollution*, 235(5), Article 305.

Reynolds, M.P., Braun, HJ. (2022). Wheat Improvement. In: Reynolds, M.P., Braun, HJ. (eds) Wheat Improvement. Springer, Cham. https://doi.org/10.1007/978-3-030-90673-3_1

Ruperao, P., Gandham, P., Odeny, D. A., Mayes, S., Selvanayagam, S., Thirunavukkarasu, N., Das, R. R., Srikanda, M., Gandhi, H., Habyarimana, E., Manyasa, E., Nebie, B., Deshpande, S. P., & Rathore, A.

(2023). Exploring the sorghum race level diversity utilizing 272 sorghum accessions genomic resources. *Frontiers in Plant Science*, 14, 1143512. https://doi.org/10.3389/fpls.2023.1143512

Sajja, S.B., Mathew, A., Pasupuleti, J., T., R. (2024). Speed Breeding to Accelerate Crop Improvement. In: Priyadarshan, P.M., Jain, S.M., Penna, S., Al-Khayri, J.M. (eds) Digital Agriculture. Springer, Cham. https://doi.org/10.1007/978-3-031-43548-5_14

Shang, C., Liu, X., Chen, G., Li, G., Hu, S., Zheng, H., Ge, L., Long, Y., Wang, Q., & Hu, X. (2024). Sl-WRKY81 regulates Spd synthesis and Na+/K+ homeostasis through interaction with SlJAZ1 mediated JA pathway to improve tomato saline-alkali resistance. The Plant Journal, 118(6), 1774–1792. https://doi.org/10.1111/tpj.16709

Shen, C., Huang, B., Hu, L., Yuan, H., Huang, Y., Wang, Y., Sun, Y., Li, Y., Zhang, J., & Xin, J. (2023a). Comparative transcriptome analysis and Arabidopsis thaliana overexpression reveal key genes associated with cadmium transport and distribution in root of two Capsicum annuum cultivars. Journal of Hazardous Materials, 465, 133365. https://doi.org/10.1016/j.jhazmat.2023.133365

Shen, T., Xu, F., Chen, D., Yan, R., Wang, Q., Li, K., Zhang, G., Ni, L., & Jiang, M. (2023b). A B-box transcription factor OsBBX17 regulates saline-alkaline tolerance through the MAPK cascade pathway in rice. New Phytologist, 241(5), 2158–2175. https://doi.org/10.1111/nph.19480

Sheoran S, Kaur Y, Kumar S, Shukla S, Rakshit S and Kumar R (2022) Recent Advances for Drought Stress Tolerance in Maize (*Zea mays* L.): Present Status and Future Prospects. *Front. Plant Sci.* 13:872566. doi: 10.3389/fpls.2022.872566

Shu, P., Li, Y., Xiang, L., Sheng, J., & Shen, L. (2022). SlNPR1 modulates chilling stress resistance in tomato plant by alleviating oxidative damage and affecting the synthesis of ferulic acid. Scientia Horticulturae, 307, 111486. https://doi.org/10.1016/j.scienta.2022.111486

Şimşek Ö, Dalda Şekerci A, Isak MA, Bulut F, İzgü T, Tütüncü M, Dönmez D. Optimizing micropropagation and rooting protocols for diverse lavender genotypes: a synergistic approach integrating machine learning techniques. Horticulturae. 2024;10(1):52.

Singh, H., Kumar, N., & Kumar, A. (2024). Enhancing resource use efficiency in crops through plant functional traits. In *Plant Functional Traits for Improving Productivity* (pp. 97-117). Singapore: Springer Nature Singapore.

Sivabharathi, R. C., Rajagopalan, V. R., Suresh, R., Sudha, M., Karthikeyan, G., Jayakanthan, M., & Raveendran, M. (2024). Haplotype-based breeding: A new insight in crop improvement. *Plant Science*, 346, 112129.

Sohail, H., Noor, I., Nawaz, M. A., Ma, M., Shireen, F., Huang, Y., Yang, L., & Bie, Z. (2022). Genome-wide identification of plasma-membrane intrinsic proteins in pumpkin and functional characterization of CmoPIP1-4 under salinity stress. Environmental and Experimental Botany, 202, 104995. https://doi.org/10.1016/j.envexpbot.2022.104995

Sohail, H., Noor, I., Hasanuzzaman, M., Geng, S., Wei, L., Nawaz, M. A., Huang, Y., Yang, L., & Bie, Z. (2024). CmoPIP1-4 confers drought tolerance in pumpkin by altering hydrogen sulfide signaling. Plant Physiology and Biochemistry, 208, 108443. https://doi.org/10.1016/j.plaphy.2024.108443

Soumya, S. L., & Sankar, S. M. (2024). Perspectives of MUTMAP and its variants towards Next-Generation Plant Breeding. In *Apple Academic Press eBooks* (pp. 219–242). https://doi.org/10.1201/9781003361862-9

Su, G., Lin, Y., Wang, C., Lu, J., Liu, Z., He, Z., Shu, X., Chen, W., Wu, R., Li, B., Zhu, C., Rose, J. K. C., Grierson, D., Giovannoni, J. J., Shi, Y., & Chen, K. (2023). Expansin SlExp1 and endoglucanase Sl-Cel2 synergistically promote fruit softening and cell wall disassembly in tomato. The Plant Cell, 36(3), 709–726. https://doi.org/10.1093/plcell/koad291

Su, N., Gong, Y., Hou, X., Liu, X., Shabala, S., Demidchik, V., Yu, M., Jiang, M., & Huang, L. (2024). Zinc finger protein ZFP36 and pyruvate dehydrogenase kinase PDK1 function in ABA-mediated aluminum tolerance in rice. The Crop Journal, 12(5), 1483–1495. https://doi.org/10.1016/j.cj.2024.06.004

Subedi, B., Poudel, A., & Aryal, S. (2023). The impact of climate change on insect pest biology and ecology: Implications for pest management strategies, crop production, and food security. Journal of Agriculture and Food Research, 14, 100733. https://doi.org/10.1016/j.jafr.2023.100733

Sun, L., Alariqi, M., Wang, Y., Wang, Q., Xu, Z., Zafar, M. N. *et al.* (2023). Construction of host plant Insect-Resistance mutant library by High-Throughput CRISPR/CAS9 system and identification of a Broad-Spectrum insect Resistance gene. Advanced Science, 11(4). https://doi.org/10.1002/advs.202306157

Sun, B., Shen, Y., Zhu, L., Yang, X., Liu, X., Li, D., Zhu, M., Miao, X., & Shi, Z. (2024a). OsmiR319-OsPCF5 modulate resistance to brown planthopper in rice through association with MYB proteins. BMC Biology, 22(1). https://doi.org/10.1186/s12915-024-01868-3

Sun, W., Xia, L., Deng, J., Sun, S., Yue, D., You, J., Wang, M., Jin, S., Zhu, L., Lindsey, K., Zhang, X., & Yang, X. (2024b). Evolution and subfunctionalization of CIPK6 homologous genes in regulating cotton drought resistance. Nature Communications, 15(1). https://doi.org/10.1038/s41467-024-50097-3

Sun, Y., Tian, Z., Zuo, D., Wang, Q., & Song, G. (2024c). GhUBC10-2 mediates GhGSTU17 degradation to regulate salt tolerance in cotton (Gossypium hirsutum). Plant Cell & Environment, 47(5), 1606–1624. https://doi.org/10.1111/pce.14839

Tang, Z., Li, Y., Zhang, Z., Huang, X., & Zhao, F. (2024). OsCOPT7 is a copper exporter at the tonoplast and endoplasmic reticulum and controls Cu translocation to the shoots and grain of rice. Plant Cell & Environment, 47(6), 2163–2177. https://doi.org/10.1111/pce.14867

Teng, D., Liu, D., Khashaveh, A., Lv, B., Sun, P., Geng, T., Cui, H., Wang, Y., & Zhang, Y. (2024). Engineering DMNT emission in cotton enhances direct and indirect defense against mirid bugs. Journal of Advanced Research. https://doi.org/10.1016/j.jare.2024.05.022

Varshney, R. K., Singh, V. K., Hickey, J. M., Xun, X. U., Marshall, D. F., Wang, J., ... Ribaut, J.-M. (2016). Analytical and decision support tools for genomics-assisted breeding. *Trends in Plant Science*, 21, 354–363.

Vishwanath S, Rajangam J, Rajadurai K R, Gnanasekaran M, Anitha T, & Ravi R. (2024). Enhancing floral diversity: A review of mutation breeding techniques in flower crops. *Plant Science Today*, 11(4).

Voss-Fels, K.P., Stahl, A., Wittkop, B. *et al.* Breeding improves wheat productivity under contrasting agrochemical input levels. *Nat. Plants* 5, 706–714 (2019).

Wang, H., Prentice, I. C., Keenan, T. F., Davis, T. W., Wright, I. J., Cornwell, W. K., Evans, B. J., & Peng, C. (2017). Towards a universal model for carbon dioxide uptake by plants. *Nature Plants*, 3(9), 734-741. https://doi.org/10.1038/s41477-017-0006-8

Wang Y, Kang Y, Zhong M, Zhang L, Chai X, Jiang X, Yang X. Effects of Iron Deficiency Stress on Plant Growth and Quality in Flowering Chinese Cabbage and Its Adaptive Response. *Agronomy*. 2022; 12(4):875. https://doi.org/10.3390/agronomy12040875

Wang, H., Zhao, X., Xuan, W., Wang, P., & Zhao, F. (2023). Rice roots avoid asymmetric heavy metal and salinity stress via an RBOH-ROS-auxin signaling cascade. Molecular Plant, 16(10), 1678–1694. https://doi.org/10.1016/j.molp.2023.09.007

Wang, D., Zhang, X., Cao, Y., Batool, A., Xu, Y., Qiao, Y., Li, Y., Wang, H., Lin, X., Bie, X., Zhang, X., Jing, R., Dong, B., Tong, Y., Teng, W., Liu, X., & Xiao, J. (2024a). TabHLH27 orchestrates root growth and drought tolerance to enhance water use efficiency in wheat. Journal of Integrative Plant Biology, 66(7), 1295–1312. https://doi.org/10.1111/jipb.13670

Wang, H., Wang, X., Li, Y., Cui, Y., Yan, X., Gao, J., Ouyang, J., & Li, S. (2024b). Pleiotropic effects of MIR5504 underlying plant height, grain yield and quality in rice. Plant and Cell Physiology, 65(5), 781–789. https://doi.org/10.1093/pcp/pcae015

Wang, J., Liao, Z., Jin, X., Liao, L., Zhang, Y., Zhang, R., Zhao, X., Qin, H., Chen, J., He, Y., Zhuang, C., Tang, J., & Huang, S. (2024c). Xanthomonas oryzae pv. oryzicola effector Tal10a directly activates rice OsHXK5 expression to facilitate pathogenesis. The Plant Journal, 119(5), 2423–2436. https://doi.org/10.1111/tpj.16929

Wang, M., Cheng, J., Wu, J., Chen, J., Liu, D., Wang, C., Ma, S., Guo, W., Li, G., Di, D., Zhang, Y., Han, D., Kronzucker, H. J., Xia, G., & Shi, W. (2024d). Variation in TaSPL6-D confers salinity tolerance in bread wheat by activating TaHKT1;5-D while preserving yield-related traits. Nature Genetics, 56(6), 1257–1269. https://doi.org/10.1038/s41588-024-01762-2

Wang, T., Ma, X., Chen, Y., Wang, C., Xia, Z., Liu, Z., Gao, L., & Zhang, W. (2024e). SlNAC3 suppresses cold tolerance in tomatoes by enhancing ethylene biosynthesis. Plant Cell & Environment, 47(8), 3132–3146. https://doi.org/10.1111/pce.14933

Wang, X., Ju, Y., Wu, T., Kong, L., Yuan, M., Liu, H., Chen, X., & Chu, Z. (2024f). The clade III subfamily of OsSWEETs directly suppresses rice immunity by interacting with OsHMGB1 and OsHsp20L. Plant Biotechnology Journal, 22(8), 2186–2200. https://doi.org/10.1111/pbi.14338

Wei, X., Geng, M., Yuan, J., Zhan, J., Liu, L., Chen, Y., Wang, Y., Qin, W., Duan, H., Zhao, H., Li, F., & Ge, X. (2024). GhRCD1 promotes cotton tolerance to cadmium by regulating the GhbHLH12–GhMYB44–GhHMA1 transcriptional cascade. Plant Biotechnology Journal, 22(7), 1777–1796. https://doi.org/10.1111/pbi.14301

Wen, J., Qin, Z., Sun, L., Zhang, Y., Wang, D., Peng, H., Yao, Y., Hu, Z., Ni, Z., Sun, Q., & Xin, M. (2023). Alternative splicing of TaHSFA6e modulates heat shock protein–mediated translational regulation in response to heat stress in wheat. New Phytologist, 239(6), 2235–2247. https://doi.org/10.1111/nph.19100

Xu, H., Liu, Q., Wang, S., Yang, G., & Xue, S. (2022). A global meta-analysis of the impacts of exotic plant species invasion on plant diversity and soil properties. *Science of the total environment*, 810, 152286.

Yang, S., Zhang, J., Geng, Y., Tang, Z., Wang, J., Guo, F., Meng, J., Wang, Q., Wan, S., & Li, X. (2021). Transcriptome analysis reveals the mechanism of improving erect-plant-type peanut yield by single-seeding precision sowing. *PeerJ*, 9, e10616. https://doi.org/10.7717/peerj.10616

Yang, W., Liu, C., Fu, Q., Jia, X., Deng, L., Feng, C., Wang, Y., Yang, Z., Yang, H., & Xu, X. (2023). Knockout of SlbZIP68 reduces late blight resistance in tomato. Plant Science, 336, 111861. https://doi.org/10.1016/j.plantsci.2023.111861

Yang, L., Xu, L., Guo, J., Li, A., Qi, H., Wang, J., & Song, S. (2024). SNAC1-OsERF103-OsSDG705 module mediates drought response in rice. New Phytologist, 241(6), 2480–2494. https://doi.org/10.1111/nph.19552

Ye, T., Wang, H., An, C., Tu, H., Zhang, L., Hu, D., Xiong, H., & Xiong, L. (2024). An expanded cysteine-rich receptor-like kinase gene cluster functionally differentiates in drought, cold, heat, and pathogen stress responses in rice. Plant Biotechnology Journal, 22(10), 2672–2674. https://doi.org/10.1111/pbi.14381

Ying, S., Scheible, W., & Lundquist, P. K. (2022). A stress-inducible protein regulates drought tolerance and flowering time in Brachypodium and Arabidopsis. PLANT PHYSIOLOGY, 191(1), 643–659. https://doi.org/10.1093/plphys/kiac486

Yu, X., Guo, Y., Yang, Q., Yu, H., Lu, M., Zhao, L., Jin, Z., Xu, X., Feng, J., & Wen, Y. (2024). Chimeric mutations in grapevine ENHANCED DISEASE RESISTANCE1 improve resistance to powdery mildew without growth penalty. PLANT PHYSIOLOGY, 195(3), 1995–2015. https://doi.org/10.1093/plphys/kiae169

Zeng, P., Xie, T., Shen, J., Liang, T., Yin, L., Liu, K., He, Y., Chen, M., Tang, H., Chen, S., Shabala, S., Zhang, H., & Cheng, J. (2024). Potassium transporter OsHAK9 regulates seed germination under salt stress by preventing gibberellin degradation through mediating OsGA2ox7 in rice. Journal of Integrative Plant Biology, 66(4), 731–748. https://doi.org/10.1111/jipb.13642

Zhang, J., Xie, M., Tuskan, G. A., Muchero, W., & Chen, J. (2018). Recent Advances in the Transcriptional Regulation of Secondary Cell Wall Biosynthesis in the Woody Plants. *Frontiers in Plant Science*, 9, 408857. https://doi.org/10.3389/fpls.2018.01535

Zhang, B., Wang, Z., Dai, X., Gao, J., Zhao, J., Ma, R., Chen, Y., Sun, Y., Ma, H., Li, S., Zhou, C., Wang, J. P., & Li, W. (2023a). A COMPASS histone H3K4 trimethyltransferase pentamer transactivates drought tolerance and growth/biomass production in Populus trichocarpa. New Phytologist, 241(5), 1950–1972. https://doi.org/10.1111/nph.19481

Zhang, Y., Xiao, W., Wang, M., Khan, M., & Liu, J. (2023b). A C2H2-type zinc finger protein ZAT12 of Poncirus trifoliata acts downstream of CBF1 to regulate cold tolerance. The Plant Journal, 117(5), 1317–1329. https://doi.org/10.1111/tpj.16562

Zhang, D., He, T., Wang, X., Zhou, C., Chen, Y., Wang, X., Wang, S., He, S., Guo, Y., Liu, Z., & Chen, M. (2024a). Transcription factor DIVARICATA1 positively modulates seed germination in response to salinity stress. PLANT PHYSIOLOGY, 195(4), 2997–3009. https://doi.org/10.1093/plphys/kiae231

Zhang, H., Sun, B., Wu, W., Li, Y., Yin, Z., Lu, C., Zhao, H., Kong, L., & Ding, X. (2024b). The MYB transcription factor OsMYBxoc1 regulates resistance to Xoc by directly repressing transcription of the iron transport gene OsNRAMP5 in rice. Plant Communications, 5(6), 100859. https://doi.org/10.1016/j.xplc.2024.100859

Zhang, L., Wang, S., Bai, B., Chen, Y., Xiang, Z., Chen, C., Kuang, X., Yang, Y., Fu, J., Chen, L., & Mao, D. (2024c). OsKASI-2 is required for the regulation of unsaturation levels of membrane lipids and chilling tolerance in rice. Plant Biotechnology Journal, 22(8), 2157–2172. https://doi.org/10.1111/pbi.14336

Zhao, N., Guo, A., Wang, W., Li, B., Wang, M., Zhou, Z., Jiang, K., Aierxi, A., Wang, B., Adjibolosoo, D., Xia, Z., Li, H., Cui, Y., Kong, J., & Hua, J. (2024). GbPP2C80 Interacts with GbWAKL14 to Negatively Co-Regulate Resistance to Fusarium and Verticillium wilt via MPK3 and ROS Signaling in Sea Island Cotton. Advanced Science, 11(30). https://doi.org/10.1002/advs.202309785

Zheng, P., Cao, L., Zhang, C., Fang, X., Wang, L., Miao, M., Tang, X., Liu, Y., & Cao, S. (2023). The transcription factor MYB43 antagonizes with ICE1 to regulate freezing tolerance in Arabidopsis. New Phytologist, 238(6), 2440–2459. https://doi.org/10.1111/nph.18882

Zhong, K., Zhang, P., Wei, X., Platre, M. P., He, W., Zhang, L., Małolepszy, A., Cao, M., Hu, S., Tang, S., Li, B., Hu, P., & Busch, W. (2024). Natural variation of TBR confers plant zinc toxicity tolerance through root cell wall pectin methylesterification. Nature Communications, 15(1). https://doi.org/10.1038/s41467-024-50106-5

Zhu, J., Zhang, L., Li, W., Han, S., Yang, W., & Qi, L. (2013). Reference Gene Selection for Quantitative Real-time PCR Normalization in Caragana intermedia under Different Abiotic Stress Conditions. *PLOS ONE*, 8(1), e53196. https://doi.org/10.1371/journal.pone.0053196

Zhu, Q., Zhang, L., & Liu, J. (2023). NFXL1 functions as a transcriptional activator required for thermotolerance at reproductive stage in Arabidopsis. Journal of Integrative Plant Biology, 66(1), 54–65. https://doi.org/10.1111/jipb.13604

13 Future Perspectives and Emerging Technologies of Molecular Responses to Plant Stress

Muhammad Awais Ibrahim[1], Javed Abbas[1], Asma Ajmal[1], Muhammad Muzamil Sattar[1], Javed Iqbal[2], Banzeer Ahsan Abbasi[3*], Ghulam Murtaza[4], Rashid Iqbal[5,6] and Sana Basharat[7]

[1]Department of Plant Sciences, Faculty of Biological Sciences, Quaid-i-Azam University, Islamabad, Pakistan; [2]Department of Botany, Bacha Khan University, Charsadda 24420, Khyber Pakhtunkhwa, Pakistan; [3]Department of Botany, Rawalpindi Women University, Rawalpindi 46300, Pakistan; [4]School of Agriculture, Yunnan University, Kunming, Yunnan 650504, China; [5]Department of Agronomy, Faculty of Agriculture and Environment, The Islamia University of Bahawalpur 63100, Pakistan; [6]Department of Life Sciences, Western Caspian University, Baku, Azerbaijan; [7]College of Tropical Agriculture and Forestry, Hainan University, Danzhou, Hainan, China

Abstract

Plants continuously face a variety of environmental challenges, roughly classified into biotic and abiotic stresses, which significantly threaten their growth and productivity. Biotic stress, including pathogens and herbivorous insects, compromise physical integrity, while inorganic stresses, such as drought, salinity, extreme temperatures, and heavy metals, disrupt cellular homeostasis and induce oxidative stress. Understanding the complex molecular mechanisms underlying stress responses is particularly crucial in the context of climate change. This chapter examines complex signaling routes, transcriptional networks, hormonal regulation, and epigenetic modifications that enable plants to adapt to many stress conditions. Crosstalk between stress routes often results in the incidence of cross-tolerances, affecting the plant's defense strategies. The role of secondary metabolites, such as flavonoids and terpenoids, is emphasized due to their antioxidants and antimicrobial properties that contribute to increased resistance. Additionally, emerging epigenetic mechanisms—such as DNA methylation and small RNAs—regulate gene expression and facilitate stress memory, enabling plants to "remember" previous stress events and enhance their subsequent stress responses. Technological progress in genomics, transcriptomics, proteomics, metabolic, and systems biology provides powerful tools to understand plant stress reactions at many biological levels. In addition, the chapter highlights state-of-the-art breeding techniques, including CRISPER/Cas genome editing, marker-assisted selection (MAS), and speed breeding, which are important for engineering stress-revolutionary crops. Integration of machine learning, bio-infection science, and advanced imaging techniques accelerates the identity of stress biomarkers and facilitates accurate agriculture. The conclusion explores the socioeconomic effects together with ethical aspects of molecular technologies as they relate to sustainable farming systems' accessibility. The complete framework in this chapter provides methods through which molecular information can be used, along with biotechnology approaches, for enhancing crop adaptation through food security enhancements during uncertain environmental times.

* Email: benazirahsanabbasi786@gmail.com

DOI: 10.1079/9781800628823.0013

Keywords: biotic and abiotic stresses, climate change, epigenetic mechanisms, gene expression, CRISPR-Cas9 genome editing, food security

13.1 Introduction

Plants acquire ongoing awareness about environmental threats that create negative impacts on their growth as well as their growth and productivity patterns. The adverse effects of these biotic and abiotic stresses need complex molecular reactions to manage them. Pathogens including bacteria, fungi and viruses, along with herbivorous insects and nematodes, constitute biotic stresses that harm plant physical structures (Gassmman *et al.*, 2016). The abiotic environmental factors that create stress include drought, salinity, harsh temperatures, and toxic heavy metal exposure. These factors trigger oxidative damage and prevent cells from maintaining normal equilibrium (Jin *et al.*, 2024). Given the increasing frequency of these stresses due to climate change, understanding the stress reactions of the plant at the molecular level is important for improving agricultural flexibility (Kaur *et al.*, 2024). The molecular system that controls plant reactions for stress includes complex signaling tract, including hormonal regulation, transcriptional control, and epigenetic modifications. The crosstalk in plants between biotic and inorganic stress signaling is important in shaping the defense strategies, as some stress routes overlap, causing cross-tolerance phenomena (Nejat and Mantri, 2017). For example, drought stress is shown to affect the immunity of the plant, sometimes increasing the resistance of pathogens, while in other times the sensitivity increases depending on the specific stress conditions (Pandey and Senthil-Kumar, 2019). The capacity of plants to integrate signs from many stresses determines their overall adaptive ability and survival.

Understanding these molecular reactions has significant implications for permanent agriculture. As the world's population is expected to exceed nine billion by 2050, an increase in food production is a global challenge, decreasing the environmental impact (Dresselhaus and Hückelhoven, 2018). Breeding and biotechnology intervention can increase crop tolerance for stress-nature genes and many tension targeting routes, which can ensure food security. The newly-developed gene editing technologies, including Advance and CRISPR-Cas9 along with genetic engineering plant-microbe interaction, offer research opportunities to enhance flexibility (Dangi *et al.*, 2018). Through the combination of transcriptomic and proteomic and metabolomic techniques we gain a better comprehension of plants under stress, thus creating paths for improving target crops. Scientific interest in how secondary metabolites function as plant protectors during stress reactions continues to grow because these compounds fundamentally shield the plant system from damage. Plants produce flavonoid, alkaloid, and terpenoid metabolites in response to biotic and innovative stress leading to increased resistance (Umar *et al.*, 2021).

Secondary metabolites serve as antioxidants enough to deliver antimicrobial defense mechanisms while simultaneously functioning as signaling molecules during plant adaptive processes (Tajima *et al.*, 2020). The mechanisms of stress regulation in secondary metabolism from this plant enable opportunities for new engineering strategies to enhance flexibility. DNA serves as an important regulatory factor that controls histone modifications, along with small RNA and plant stress reactions. Plants achieve heritable changes in gene expression through DNA amendments that do not modify DNA sequence fundamental structure, hence enabling plants to retain memory of past stressors while enhancing their tolerance (Apostolova, 2023). Scientific research has uncovered the plant capability of recognizing future stressors in their environment (Saijo and Loo, 2019). The insights gained from these discoveries help epigenome-ornament technologies produce stress-tolerant crops.

13.1.1 Objective of the chapter

The main focus of this chapter is to connect plant stress reactions to modern molecular approaches that enhance crop resistance throughout technological plant evolution. The chief goal is to examine modern research findings about signaling routes, secondary metabolites, and epigenetic modifications for building foundations of innovative agricultural strategies. By emphasizing interdisciplinary approaches such as genetic engineering, systems biology, and computational modeling, we want to present a roadmap that can be used to translate molecular knowledge into practical applications. Finally, this chapter aims to contribute to global efforts in durable food production and inform as to how scientific progress can be used to reduce the effects of biological and abiotic stress on crops (Table 13.1).

13.2 Heat Stress Perception and Signaling Pathways in Plants

Heat stress is an important environmental challenge that disrupts plant growth, growth, and productivity. Plants perceive heat stress through various molecular sensors, which start a signaling cascade to activate protective reactions. The perception of heat stress occurs mainly on the plasma membrane, where temperature-sensitive proteins and lipid components detect changes in thermal conditions (Dündar and Topak, 2025). Calcium channels, such as cyclic nucleotide-gated channels, play a pivotal role in heat sensing by facilitating the influx of Ca^{2+} ions, which function as secondary messengers in cellular signaling pathways (Borgin and Guihar, 2021). Additionally, reactive oxygen species (ROS) act as signaling molecules, triggering redox-dependent routes that control the heat shock factors (HSFs) and other protective mechanisms (Fortunato *et al.*, 2023). These complex signaling networks ensure that plants can respond to rapid heat stress and thus reduce cellular damage. Once considering heat stress, the plants activate HSFs, which reflect the transcription of heat shock proteins (HSPs) to prevent protein scratching and aggregation (Table 13.2). HSPs act as molecular chaperones, helping folding and stabilization under excessive temperature (Haider *et al.*, 2021). These HSP70 and HSP90 families play an important role in maintaining cellular processes during heat stress (Borgin and Guihur, 2021). In addition, epigenetic amendments, such as histone acetylation and DNA methylation, contribute to heat stress memory, allowing plants to maintain the physical impression of pre-stress risk and maintain a rapid response to subsequent heat waves (Zheng *et al.*, 2024).

The implementation of transcriptional modifications, together with epigenetic regulatory elements, gives plants greater ability to resist high temperatures. The functioning of hormonal signals serves as a fundamental process for heat stress optimization. Studies have proven

Table 13.1. Type of plant stress and their effects on plants. (Table author's own.)

Stress type	Examples	Effects on plants
Biotic stress	Pathogenic microbes (bacteria, fungi, viruses, nematodes), herbivorous insects	Disrupts physiological functions, induces defense responses
Abiotic stress	Drought, salinity, extreme temperatures, heavy metals	Causes oxidative damage, disrupts cellular homeostasis

Table 13.2. Molecular mechanisms showing responses of plant during stress. (Table author's own.)

Mechanism	Role in stress response	Examples
Signaling pathways	Transduces stress signals through messengers	Calcium signaling, ROS signaling
Hormonal regulation	Modulates plant growth and defense	ABA, ET, SA
Epigenetic modifications	Regulates gene expression without changing DNA sequence	DNA methylation, histone modifications
Secondary metabolites	Provides antioxidant and antimicrobial properties	Flavonoids, terpenoids

that ethylene (ET) serves as the primary plant hormone, which regulates thermotolerance through its ability to control gene expression while organizing various stress response functions (Poór *et al.*, 2021). The heat-resistance function of brassinosteroids (BRs) occurs when these molecules connect to HSFs to enhance stress-existing gene expression (Albertos *et al.*, 2022). Plants that cross these defense routes between abscisic acid (ABA), salicylic acid (SA), and nitric oxide (NO) in stock collections develop stronger protection systems (Rai *et al.*, 2019). The hormonal behavior creates a stable plant reaction toward heat stress to maintain growth efficiency and survival capacities.

Plant molecular biologists have discovered during recent studies how microRNA functions in regulating heat stress responses. Post-transcriptional regulators, such as microRNAs (miRNAs), target specific messenger RNAs (mRNAs) for degradation or translational inhibition, thereby modulating heat stress signaling pathways (Ding *et al.*, 2020). The process of heat stress optimization includes chloroplast retrograde signaling as a significant mechanism for nuclear signaling that originates from damaged chloroplasts to trigger safety gene activation (Sun and Guo, 2016). The research demonstrates plant heat stress reactions contain multiple factors while showing genetic and biotechnological approaches develop heat tolerance in agricultural crops (Table 13.2).

13.3 Role of Stress-responsive Genes, Transcription Factors (TFs), and Secondary Metabolites

Plants have developed an elaborate stress coping mechanism using stress-disciplinary genes in combination with transcription factors (TFs) and secondary metabolites. Stress-disciplinary genes are important in starting signaling cascades that mediate physical and biochemical adaptation for biotic and inorganic stresses (Yoon *et al.*, 2020; Pan *et al.*, 2022). Among the various molecular components involved in stress tolerance, late embryogenesis abundant (LEA) proteins, dehydrins, and HSPs have been identified as key contributors to stress plasticity. These proteins play protective roles by stabilizing cellular structures, preventing protein aggregation, and enhancing cellular resilience under dehydration and heat stress conditions (Ramesh *et al.*, 2020). Additionally, antioxidant enzymes such as superoxide dismutase (SOD), catalase (CAT), and ascorbate peroxidase (APX), encoded by specific genes, are essential for scavenging ROS, thereby mitigating oxidative stress and preventing cellular damage (Ying *et al.*, 2025) (see Fig. 13.1). The knowledge of the regulation of these genes is useful in increasing plant flexibility by genetic engineering and selective reproduction (Sarkar *et al.*, 2020).

TFs serve as master regulators of gene expression in stress reactions by changing binding and transcriptional activity for certain DNA sequences. Many TF families such as NAC, WRKY, AP2/ERF, MyB, and bZIP have been reported as playing an important role in plant stress tolerance (Hoang et *al.*, 2017). For example, NAC TFS controlled the expression of the stress-inducible gene participating in drought and salt stress resistance through adjusting the ABA-reliant pathway (Wu *et al.*, 2020). In this context, WRKY TFs contribute to redox regulation by enhancing disease resistance and inducing the expression of defense-related genes (Jin *et al.*, 2017). In addition, AP2/ERF TFs are also important in ET signaling and change the reactions of drought, and cold and heat stress (Guo *et al.*, 2021; Hrmova and Hussain, 2021). The complex crosstalk among these TFs facilitates broad adaptive responses to various abiotic and biotic stresses in plants (Nazar *et al.*, 2017).

Secondary metabolites represent another crucial component of plant stress responses, functioning as protective agents against both biotic and abiotic stressors. Compounds such as flavonoids, alkaloids, terpenoids, and phenolics exhibit antioxidant, antimicrobial, and signaling roles, contributing significantly to the plant's adaptive defense mechanisms (Zong *et al.*, 2016; Fernando, 2020). The use of these metabolites through metabolic engineering has high potential for the production of stress-flexible crops (Goel *et al.*, 2019). By leveraging these insights, plant scientists are developing novel strategies to enhance crop productivity and resilience in the face of global climate change (Wang *et al.*, 2017).

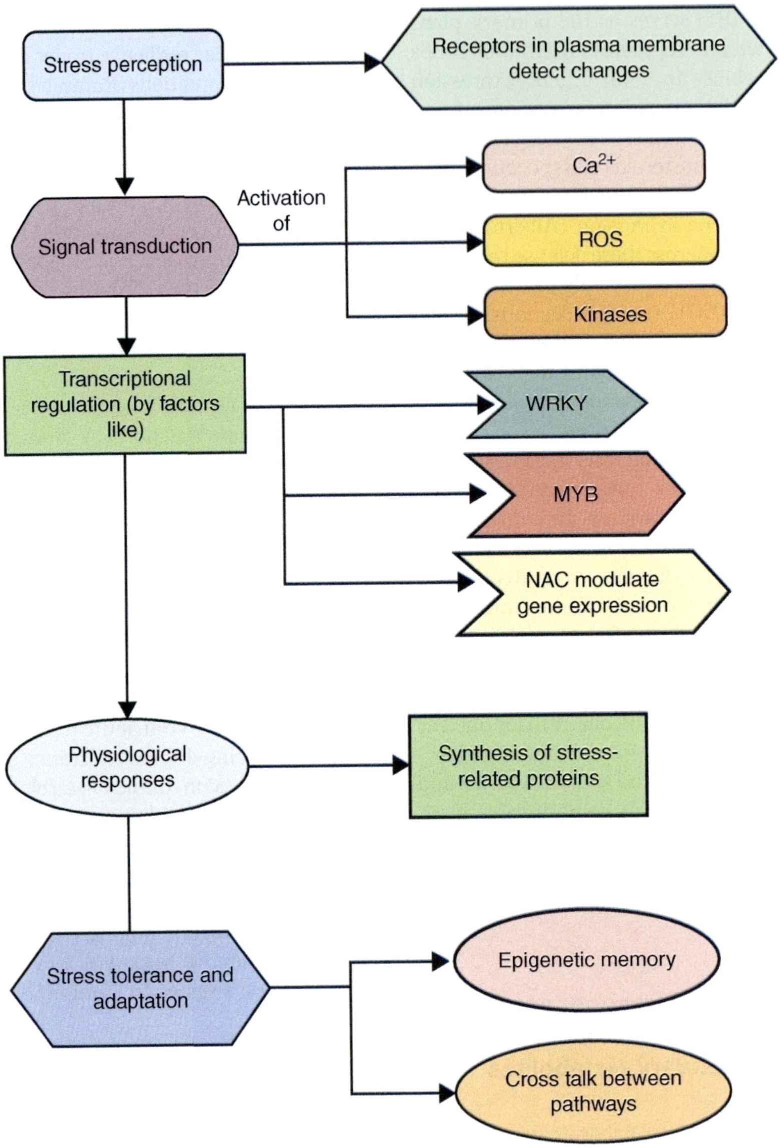

Fig. 13.1. Molecular response to plant stress. (Figure author's own.)

13.4 Stress Memory and Epigenetic Modifications

Plants demonstrate extraordinary plasticity under environmental stress, in which stress is an important contributor to stress memory in epigenetic amendments. DNA methylation, histone modification, and small RNA regulation are epigenetic mechanisms that allow plants to "remember" previous stress and react better to recurring stress (Kumar *et al.*, 2018). Such modifications can be transient or stable, and some epigenetic scars remain in generations, which affect stress tolerance in progeny (Mozgová *et al.*, 2019). Researchers have found in *Arabidopsis* that the exposure was reiterated by exposure epigenetic priming for rapidly adapted reactions due to drought stress (Ramakrishnan *et al.*,

2022). This capacity for stress memory retention is an important implication for the improvement of crops, as it provides a non-native route towards increasing stress flexibility (Saeed *et al.*, 2022). One of the critical molecular components of stress memory is DNA, an epigenetic mark that modifies gene expression, but not the sequence of DNA (Kumar and Mohapatra, 2024). Methylation modifications in stress-melodious genes can be conveyed to tolerance in later generations by replacing chromatin access (Turgut-Kara *et al.*, 2020). Histone also controls stress memory by replacing chromatin structure and transcriptional activity (Shanker *et al.*, 2020). For example, a well-recorded epigenetic stress memory form, includes histone modifications, which silences the flowering locus C (FLC) genes. This guarantees the flowers under the seasonal conditions (Ramakrishnan *et al.*, 2022).

In addition to chromatin modifications, small RNAs (sRNAs) play an important role in epigenetic regulation by guiding DNA methylation and post-transcriptional gene silencing (Lämke and Bäurle, 2017). The role of miRNAs and small intervention RNAs (siRNAs) in stress adaptation is well documented and studies show that these molecules regulate major genes involved in drought and salt tolerance (Zheng *et al.*, 2022). In addition, species of various plant have seen transgenerational epigenetic inheritance, where stress-inspired epigenetic scars are transmitted to progeny, which reflect increased resistance to abiotic stress (Tricker, 2015). However, the stability of these changes remains an area of active research, in which the debate arises whether the stress memory is permanent or can be reset in future generations. Application of epigenetics in agriculture provides promising avenues to develop stress-flexible crops without genetic amendments. Taking advantage of the knowledge of stress memory, researchers can use epigenome editing techniques to increase plant adaptability for environmental changes (Miryeganeh, 2021). CRISPR-based epigenetic editing allows for accurate amendments in recent advances in DNA methylation patterns, which provides a tool to manipulate stress-tension-existed genes without changing genetic sequences (Bulgakov, 2024).

Understanding mutual activity between chromatin remodeling, small RNA and DNA methylation will be necessary to use epigenetic memory to improve crop flexibility in the face of climate change (Nishio *et al.*, 2023). Future research will need to address how to maintain desirable epigenetic amendments across several generations, reducing unexpected effects on plant growth and development (Table 13.3).

13.5 Transcriptomics, Proteomics, and Metabolomics Integration in Plant Stress Response

The integration of transcriptomics, proteomics, and metabolomics offers a comprehensive understanding of plant responses to environmental stress across multiple biological levels. Transcriptomics enables the identification of key regulatory genes activated under stress conditions (Wang *et al.*, 2020). Proteomics, through large-scale analysis of protein expression and function, reveals stress-induced changes in protein abundance and post-translational modifications (PTMs) that govern cellular responses (Wang, 2019). Metabolomics, on the other hand, examines stress-induced biochemical alterations and identifies key metabolites involved in plant adaptation (Ghatak *et al.*, 2018). The convergence of these omics platforms has led to significant advancements in stress tolerance research, including the identification of stress biomarkers and critical metabolic pathways that can be targeted to improve crop performance (Soda *et al.*, 2015).

Table 13.3. Role of omics in plant stress research. (Table author's own.)

Omics approach	Application in stress response	Examples
Transcriptomics	Identifies stress-related genes	RNA sequencing, microarrays
Proteomics	Analyzes stress-induced proteins	Mass spectrometry, protein assays
Metabolomics	Studies metabolic changes under stress	GC-MS, LC-MS
Epigenomics	Investigates stress memory mechanisms	DNA methylation analysis

Recent progresses in omics integration have shown how different layers of biological regulation are coordinated with each other to determine stress reactions. For example, transcriptomic analysis has identified genes related to drought tolerance, while protective research has identified the post-disconnection modifications required for protein functions during water scarcity conditions (Joshi *et al.*, 2022). Metabolomic profiling has also established osmoprotectant accumulation, including prolines and sugars, helping to maintain cellular homeostasis (Piasecka *et al.*, 2019). By combining these datasets, scientific stress can keep a broad map of the adaptation mechanism, which highlights both short-term and long-term plant reactions (Voelckel *et al.*, 2017).

One of the primary challenges in omics research lies in managing the vast quantities of data generated during integrative analyses, as well as in establishing accurate correlations among transcript, protein, and metabolite concentrations. Recent advances have been achieved through the development of computational models in bioinformatics and machine learning, which facilitate the prediction of stress responses under multi-oxidative conditions (Sanchez-Muñoz *et al.*, 2024). For instance, by examining the co-expression of genes and corresponding changes in protein and metabolite levels, researchers have identified key regulatory hubs involved in stress response pathways (Jamil *et al.*, 2020).

The development of a public omics database has created a more convenient way to compare data between species and bring researchers to translate the findings from model plants to agricultural relevant crops (Rurek and Smolibowski, 2024). Interestingly, the future of omics-driven stress research will hinge on the integration of multiple omics layers with environmental and phenotypic data in order to engineer plant resilience to stress. A combination of transcriptomics, proteomics, and metabolomics, with phenomics (analysis of plant trait) and lipidomic (lipid profiling) provides an integrative approach for studying plant adaptation (Phelix and Feltus, 2015). By taking advantage of these techniques, researchers can identify novel stress-unique symptoms and engineer crops that can withstand excessive climatic conditions. As omics technologies develop, their integration would be important in developing future models for plant stress reactions, eventually increasing agricultural productivity and stability (Yan *et al.*, 2024).

13.6 Systems Biology and Bioinformatics in Deciphering Complex Stress Responses

Conversely, the complexity of plant stress reactions requires a system biology approach that covers high-three-tropical omics data and computational modeling to detect complex regulatory networks. Synthesis integrates transcriptomics, proteomics and metabolomics to create forecast models for stress reactions that map the gene-to-protein-to-metabolite interaction under stress (Sahoo *et al.*, 2020). The systematic visual plant of stress responses provides the opportunity to understand the adaptation of the plant more widely by identifying the major regulators that control the adaptation mechanism (Parreira *et al.*, 2016). Thus, using bioinformatics tools to integrate the omics dataset into interactive networks provides a more detailed understanding of the mechanisms regulating the plant stress tolerance (Shafi and Zahoor, 2019).

Bioinformatics science plays an important role in managing the vast amounts of data generated during multi-science experiments. Increasing use of computational tools has made it possible to manufacture genome-wide association studies (GWAS) and quantitative locomotor mapping to find stress-related genes (Laha *et al.*, 2020). In addition, the new ML algorithm is being developed to estimate plants reactions to environmental challenges using multi-omics data sets (Ambrosino *et al.*, 2020). The combination of coupled transcriptomics and protective data with physical reactions allows the manufacture of digital twins of plants to simulate growth and stress adaptation under various environmental conditions.

Systems biology, coupled with the emerging omics tools, provide new means of constructing high-performance agriculture. Epigenomics and transcriptomics strategies have established how plants produce stress memory and hence allowed targeted genetic modification to enhance crops' tolerance under conditions of extreme stress (Shameer *et al.*, 2019). Recent advances in network biology are uncovering new information on interactions between various stress signaling networks in plants, resulting in discoveries of promising target genes for enhancing plants' stress resilience under climate change (Mosa *et al.*, 2017). In order to fully ingest and study plant stress responses more fully, systems biology must simultaneously engage with artificial intelligence (AI) and synthetic biology. With several AI-enabled bioinformatics pipelines already available, it is possible to create predictions of the early adaptation patterns to stress (Upadhyay *et al.*, 2017).

Other approaches, such as synthetic biology (e.g. designing synthetic regulatory circuits that mimic natural responses to stress), emerge as promising approaches for engineering crops that are more resilient. As computational tools constantly evolve, systems biology, coupled with AI and biotechnology, will serve as critical building blocks for development of a global climate-adaptive agricultural strategy to mitigate the effect of global climate change.

13.7 CRISPR/Cas Systems for Stress Tolerance Breeding

The CRISPR/Cas genome editing system has brought revolution in plant breeding by enabling accurate and targeted amendments to stress-related genes. Given the increasing impact of climate change, plants are often exposed to extreme environmental conditions, including drought, salinity, and temperature extremes. Traditional reproductive methods have faced challenges in the production of stress-tolerant crops because the genes involved in stress reactions are quite complex.

The CRISPR/Cas9 system can make accurate changes in planting genes, which has helped to improve how plants handle stress by replacing signs of stress and processing routes (Ahmad *et al.*, 2021a). Studies have shown that using CRISPR to remove specific elements that limit stress reactions, like the ABA signaling genes, can significantly increase the capacity of the crop to adapt to dry and salty conditions (Saikia *et al.*, 2024). The CRISPR/Cas technique also allows scientists to edit several genes at once, which is useful to replace many genes managing stress in plants (Fig. 13.2) and provides better stress resilience against different types of stress in the plant (Mushtaq *et al.*, 2018). For example, in tomatoes, stress-related genes have been improved for opposing diseases like drying and bacteria wilt (Kumar *et al.*, 2023). In rice and wheat, CRISPR/Cas has been used to increase the plants' tolerance to salt as it changes how they control salt and water within their cells (Nazir *et al.*, 2022).

Beyond the simple gene knock-out, the CRISPR/Cas system has been adapted for base editing, prime editing, and epigenome modifications, which provides more accuracy in stress tolerance reproduction (Biswas *et al.*, 2021). Adenosine deaminases-acting RNA (ADARs) editing allows for solo-nucleotide modifications without causing double-strand breaks, reducing the off-target effect. Similarly, CRISPR intervention (CRISPRi) and CRISPR activation (CRISPR) techniques enable the fine-tuned regulation of stress-swollen genes by modifying transcriptional activity (Pandita, 2021).

This progress provides great potential to develop climate-flexible crops that can withstand many environmental challenges without compromising development and yield. Stress tolerance is very far away for global food security in the application of CRISPR/Cas technology in reproduction. As regulatory structures develop, the commercialization of genome-edited crops is expected to expand, which provides a permanent solution for agriculture (Ahmad *et al.*, 2021b).

Future research should focus on integrating CRISPR to adapt to reduce unexpected mutation and increase its efficiency in reproductive programs (Wang *et al.*, 2021). Using CRISPR/Cas for targeted crop improvement, agricultural climate can become more flexible, ensuring food production stability in the coming decades.

13.8 Speed Breeding, MAS, MARS, GWAS, and Genomic Selection Techniques in Stress Tolerance Breeding

The recent progress in plant reproductive technology has greatly increased the accuracy and efficiency of breeding stress-tolerant crops. Speed

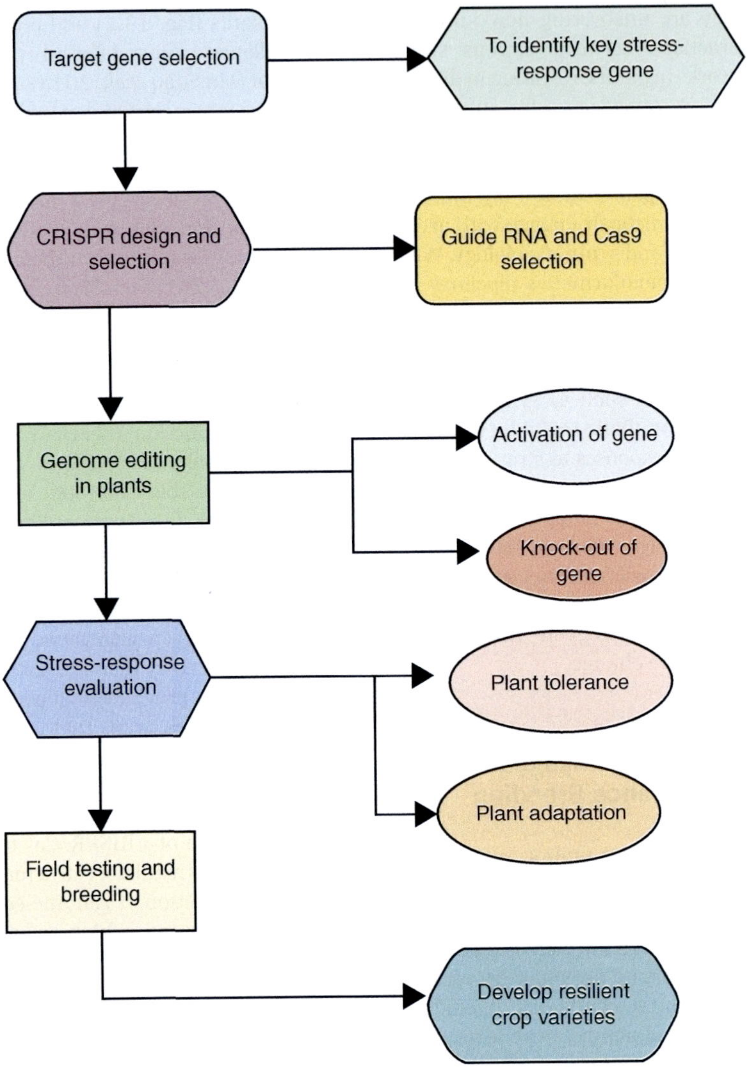

Fig. 13.2. CRISPR/Cas application in stress tolerance. (Figure author's own.)

breeding (SB), marker-assisted selection (MAS), marker-assisted recurrent selection (MARS), genome-wide association studies (GWAS), and genomic selection (GS) are the most optimistic ways to carry forward crop reforms.

SB is an interesting method that manipulates environmental conditions to accelerate generation time, reduces the breeding cycle by a stunning 50%. This means that we can soon harvest genetic benefits (Bhat *et al.*, 2023). SB is successfully coupled with other genomics devices to increase the tolerance of crops towards abiotic stress, especially in grain staples such as wheat and rice (Rana *et al.*, 2019). Additionally, MAS has reduced the reproductive process by enabling plants according to genetic markers associated with desired symptoms, which greatly reduces the time required for phenotypic evaluation (Şahin *et al.*, 2024). In addition, MARS and GS have revolutionized reproductive accuracy by allowing breeders to handle complex symptoms controlled by several genes. MARS is particularly valuable for accumulating many useful alleles in the reproductive population, as it increases drought and salt tolerance in plants such as maize and wheat. Similarly, GS appoints a high-gradation marker in the genome to estimate breeding values, allowing better area tests without the need for better selection of better genotypes (Sinha *et al.*, 2023). This method has been effective in improving flexibility of stress in especially climate-sensitive crops such as grains, where genetic variation plays an important role in adaptation in stress-response paths (Wani *et al.*, 2020).

The GWAS book is significant in the identification of stress-tolerance loci, offering a rich understanding of the genetic architecture of stress responses. Through the integration of phenotypic differences and genetic markers, GWAS has enabled the identification of genes that regulate salt tolerance in rice (Wang *et al.*, 2022) and the identification of drought-resistance in wheat (Banka *et al.*, 2024). The combination of GWAS with ML and AI-driven phenotyping has further advanced the fine-grained characteristic map, which allows dissected complex stress responses to be accurately mapped (Chang-Brahim *et al.*, 2024). Such progression is paving the way for the next generation breeding strategies, where genomic data is taken advantage of for future modeling to increase stress flexibility in crops.

The integration of these novel molecular devices is changing plant reproduction, ensuring that crops can withstand the increasing challenges caused by climate change. By combining SB, MAS, MARS, GWAS, and GS, researchers are accelerating the growth of stress-flexible varieties by reducing resource-intensive field tests (Langridge and Chalmers, 2004). With the increasing availability of genetic and phenotypic data, integrating these datasets with CRISPR-based genome editing holds significant potential to enhance reproductive efficiency. This integration may facilitate the engineering of crop varieties with improved resilience to fluctuating environmental conditions (Haqué *et al.*, 2021).

Future research should aim to expand the accessibility and optimize the efficacy of these technologies to enable their widespread adoption in agriculture.

13.9 Advanced Imaging and Non-destructive Monitoring Technologies

The combination of highly sophisticated imaging and non-destructive monitoring technologies has greatly increased our ability to identify plants' stress reactions in real time. Remote sensing methods, such as hyperspectral and multispectral imaging, have been used to identify large-scale early stress, allowing crop health mass monitoring (Skendžić *et al.*, 2023). Thermography, which measures heat radiation emitted from plant surfaces, is an effective method for detecting water and heat stress by identifying variations in canopy temperature (Elvanidi *et al.*, 2018). Additionally, spectroscopy enables the assessment of plant physiological status, including chlorophyll content and photosynthetic efficiency, through the analysis of visible and proximal reflectance. As such, it serves as a valuable tool for monitoring both biotic and abiotic stress factors.

The modern approach to determine the stress of the plant includes positron emission tomography (PET) and magnetic resonance imaging (MRI). Such methods provide special understanding of internal physiological processes, including nutritious transport in plant tissues, water expansion, and metabolic currents. PET scans make it possible to monitor the routes in real time by which carbon and nitrogen is assimilated, about which knowledge is important to appreciate how plants distribute resources in stressful conditions. Similarly, magnetic resonance imaging (MRI) has been employed to visualize root architecture and soil moisture distribution, providing valuable insights into drought-resistance mechanisms and facilitating the improvement of associated phenotypic traits (Lassalle, 2021).

Increasing the accuracy of monitoring stress comes from the integration of these imaging technologies with AI and ML. Automated classification of stress symptoms, discrimination among various stress types, and prediction of crop yield through image analysis have demonstrated significant potential in agricultural diagnostics (Merrick *et al.*, 2022). When integrated with artificial intelligence (AI) algorithms, hyperspectral imaging can detect subtle spectral variations associated with physiological changes such as phase transitions or nutrient deficiencies, enabling pre-symptomatic intervention (Zhang *et al.*, 2019).

These technologies support real-time, non-destructive evaluation of plant stress, offering substantial potential for precision agriculture. Moreover, the integration of remote sensing with molecular marker analysis enhances the identification of stress responses in known genotypes, thereby accelerating breeding programs aimed at developing climate-resilient crop varieties (Satapathy, 2020). Future research should focus on reducing costs and access obstacles for these advanced imaging techniques, which ensures their broad adoption for permanent agricultural practices.

13.10 Application of Machine Learning in Transcriptomic and Phenotypic Data Integration

The integration of ML with phenotypic and transcriptomic facts has modified the dynamics of plant stress research, making viable advanced stress reactions to create mechanisms of predictions and flexibility. The machine learning algorithm was trained on large-scale transcriptomic data to identify stress-related gene expression patterns and to distinguish different levels of stress based on gene expression profiles (Kang *et al.*, 2018). These fashion nervous networks use the inclusive deep study strategies of the network and assist vector machines (SVMs), which fall into variable changes in gene expression related to special stresses and receive excessive accuracy of stress conditions (Kang *et al.*, 2019). For example, stress GenePred, a ML set of rules based on a completely double nerve network, states that the traditional statistical processes have been improved in the detection of stress biomarkers in *Arabidopsis* (Sanchez-Muñoz *et al.*, 2024). One of the major uses of ML in transcriptomics is its application in prediction of functional genes. Unlike the targeted approaches, ML has the ability to process deep genomic data and highlight novel gene functions such as stress adaptation (Mahood *et al.*, 2020).

ML models can also reduce gene relations with regulation network data and predict the function of some genes in stress-response routes (Henry and Stinchcombe, 2025). In addition, the meta-analysis of transcriptomic datasets using machine learning has facilitated the identification of a "core stress gene set," comprising universally conserved genes involved in multi-stress response pathways (Sanchez-Muñoz *et al.*, 2024).

The combination of ML with advanced plant phenomena allows us to study how plants handle stress more effectively. Convolutional neural network (CNN), a type of ML algorithm, is used for automatic plant phenotyping. This method draws key features from images, before they are noticeable by eye (Centorame *et al.*, 2024). By using both transcriptomic and phenotypic information, researchers can estimate how genetic differences affect plant reactions in different environments. Under stress, plants sometimes increase breeding to become more optimal (Hadish *et al.*, 2023). This approach is especially beneficial to map genotype-phenotype relationships, which helps identify crop varieties with better stress tolerance. Plant stress research lies in its integration with ML's future multi-omics data, including epigenomics, proteomics, and metabolomics. Emerging AI models are being developed to create a future framework that simultaneously simulates plant reactions for several stresses, allowing the design of more flexible crop varieties (Van Dijk *et al.*, 2020). As the ML algorithm progresses, its use will grow in precise agriculture, which will provide real-time surveillance and future analysis for crop stress management. Future studies should involve enhancing model explainability and moral use of AI-based decision making in plant breeding.

13.11 Predictive Modeling for Multi-stress Resilience Breeding

Predictive modeling has emerged as a powerful tool in plant breeding, enabling researchers to identify genetic traits that enhance multi-stress tolerance. ML-based models, such as support vector machine (SVMs) and random forest algorithms, are now being used to predict stress-existent genes in many abiotic stress conditions (Meher *et al.*, 2022). For example, ASRpro computational models utilize transcriptomic data to enhance the efficiency of gene identification, specifically targeting genes associated with six distinct abiotic stress conditions relevant to reproductive development. With the inclusion of genetic, environmental and phenotypic information, these models aid the breeders to produce climate-friendly crops with more efficiency (Ly *et al.*, 2017).

The genomic-practical prediction model has also promoted multi-stress flexibility breeding by taking advantage of the genome-wide marker information to predict plant performance under various environmental conditions. The factorial regression genomic best linear unbiased predicter (FR-GBLUP) approach has successfully predicted dry and heat stress tolerance in wheat (Ly *et al.*, 2017). Similarly, the Bayesian deep learning models have been employed to forecast plant reactions for many stresses through the inclusion of multi-specific genomic information with environmental components (Montesinos-López *et al.*, 2018). This achievement allows breeders to work in silico, which significantly reduces the time and expenses of field tests.

Multi-level modeling methods are also being used to forecast plant physical and structural adaptation to environmental stresses. Gangwar *et al.* (2023) studied integrated genetic, physical, and environmental data on how wheat and oats bend due to air stress, which are indicative of reproductive strategies for housing resistance. In the same manner, hyperspectral imaging and ML have been used to forecast the dominance signature by examining the ideal signature for the water scenario (Lin *et al.*, 2022). These integrated methods enable scientists to evaluate the adaptability of the plant at numerous levels ranging from gene expression to the area's performance. As modeling techniques continue to advance, their integration with genomic selection and AI-driven phenotyping is expected to further enhance the precision of multi-stress resilience in reproductive traits. The application of multi-trait deep learning frameworks enables the simultaneous prediction of key agronomic traits, thereby improving breeding programs aimed at developing climate-resilient crop varieties (Bhatta *et al.*, 2020).

Future research should focus on refining these models by incorporating various environmental variables, such as soil microbiome interactions and climate projections, to adapt to stress adaptation strategies in various agricultural systems.

13.12 Socioeconomic Impacts of Emerging Molecular Technologies

The use of molecular technologies in agriculture has a profound impact on world food production, economic growth and rural livelihood. New reproductive methods, including genetic modification (GM), CRISPR gene editing, and molecular MAS, have given rise to high crop yields, low pesticides, and better food security (Zilberman, 2018). Research suggests that transgenic crops have increased farm revenue by reducing input costs and increasing productivity, which are major benefits for the country's smallholder farmers (Brookes, 2022). In addition, the adoption of new molecular technologies has maximized land use efficiency, by reducing agricultural expansion and slowing down greenhouse gas emissions (Saikanth *et al.*, 2024). Although these advantages exist, the socioeconomic effects of molecular technologies are not universally distributed. Political disputes and regulatory sanctions have obstructed the introduction of genetically modified organisms (GMOs) in most countries, which reduce their economic benefits (Gollin, 2020). In addition, issues related to intellectual property rights and dominance of seed markets by corporations have given rise to moral concerns about farmer dependence on biotechnology firms (Binimelis and Wickson, 2018). In other examples, technological innovation has resulted in displacement of rural jobs through extended automation in the production of crops (Ivanova, 2021). These issues are a call to take action for the policies of equilibrium that favor innovation when making equal access to technological progress.

The global regulatory landscape for molecular technologies varies widely, affecting international trade and market access to biotechnology-derived crops. While some countries have adopted genetic engineering and genome editing as a tool for sustainable agriculture, others have implemented strict regulatory obstacles due to consumer concerns and environmental ideas (Fischer and Buyel, 2020). The development of a harmonious regulatory structure can facilitate international cooperation and improve technology transfer in areas where climate most requires flexibility. Additionally, moral thoughts around gene editing and synthetic biology are the subject of debate, especially about the potential long-term ecological effects (Sadiku *et al.*, 2020).

Future research must concentrate on enhancing the capability and access to molecular technology for small farmers so that development in biotechnology is used to achieve inclusive and sustainable agricultural growth. Enhancing public-private collaboration and encouraging open communication between policy makers, scientists, and the public will be crucial in determining the ethical and socio-economic direction of agricultural biotechnology (Shikha *et al.*, 2024). In addition, combining these methods with other digital innovations like blockchain and AI, can enhance traceability, ethical trade practices, and compliance with regulations, by further evolving their socio-economic advantages.

13.13 Conclusion

In an era marked by environmental challenges and increasing global population, it is more important than ever that plant resistance and flexibility to cope with biotic and inorganic stresses are increased. This chapter has widely discussed molecular, genetic, and physiological mechanisms underlying the stress reactions of plants, stating how the signaling route, transcriptional regulation, and secondary metabolites collectively contribute to the adaptation capacity.

The integration of omics technologies like transcriptomics, proteomics, metabolomics, and system biology approaches have revolutionized our ability to decode these complex networks and provided valuable insight into major regulators of stress tolerance. Apart from that, sophisticated biotechnology tools like CRISPR/Cas genome editing, MAS, and GS have presented new opportunities in breeding climate-resilient crops with enhanced productivity. The addition of ML and cutting-edge imaging tools amplifies precise agriculture by allowing real-time monitoring of stress as well as future forecasting. Nevertheless, socio-economic and moral aspects of their adoption should be addressed in detail to facilitate equitable access, ensure sustainable development and gain the public's approval.

Lastly, efforts should be made to bridge the gap between molecular research and its practical application in order to achieve food security ahead of the threat of global climate change. Future action should aim at the harmonization of inter-disciplinary research, enhancing the public-private collaboration, and making the scientific advancements bear fruit in form of real value to farmers worldwide. Through capitalizing on the entire potential of molecular technologies as well as revolutionary reproductive techniques, we can have a sustainable agricultural system for generations to come.

References

Ahmad, A., Ashraf, S., Munawar, N., Jamil, A., Ghaffar, A. and Shahbaz, M. (2021a) CRISPR/CAS-Mediated abiotic stress tolerance in crops. In: *Springer eBooks*, (pp. 177–211). https://doi.org/DOI: 10.1007/978-981-15-7142-8_6.

Ahmad, S., Sheng, Z., Jalal, R. S., Tabassum, J., Ahmed, F. K., Hu, S., Shao, G., Wei, X., Abd-Elsalam, K. A., Hu, P. and Tang, S. (2021b) CRISPR–Cas technology towards improvement of abiotic stress tolerance in plants. In: *Elsevier eBooks*, (pp. 755–772). https://doi.org/DOI: 10.1016/b978-0-12-821910-2.00021-7.

Albertos, P., Dündar, G., Schenk, P., Carrera, S., Cavelius, P., Sieberer, T. and Poppenberger, B. (2022) Transcription factor BES1 interacts with HSFA1 to promote heat stress resistance of plants. *The EMBO Journal* 41(3). DOI: 10.15252/embj.2021108664.

Ambrosino, L., Colantuono, C., Diretto, G., Fiore, A. and Chiusano, M.L. (2020) Bioinformatics Resources for plant abiotic stress responses: State of the art and opportunities in the fast evolving -omics era. *Plants* 9(5), 591. DOI: 10.3390/plants9050591.

Apostolova, E.L. (2023) Molecular mechanisms of plant defense against abiotic stress. *International Journal of Molecular Sciences* 24(12), 10339. DOI: 10.3390/ijms241210339.

Banka, V., Pheirim, R. and Waghmode, P. (2024) Genomic innovations in wheat (Triticum aestivum L.): A comprehensive review of recent developments and future directions. *Journal of Advances in Biology and Biotechnology* 27(9), 795–806. DOI: 10.9734/jabb/2024/v27i91353.

Bhat, J.A., Feng, X., Mir, Z.A., Raina, A. and Siddique, K.H.M. (2023) Recent advances in artificial intelligence, mechanistic models, and speed breeding offer exciting opportunities for precise and accelerated genomics-assisted breeding. *Physiologia Plantarum* 175(4), e13969. DOI: 10.1111/ppl.13969.

Bhatta, M., Gutierrez, L., Cammarota, L., Cardozo, F., Germán, S. *et al.* (2020) Multi-trait genomic prediction model increased the predictive ability for agronomic and malting quality traits in barley (Hordeum vulgare L.). *G3 Genes Genomes Genetics* 10(3), 1113–1124. DOI: 10.1534/g3.119.400968.

Binimelis, R. and Wickson, F. (2018) The troubled relationship between GMOs and beekeeping: an exploration of socioeconomic impacts in Spain and Uruguay. *Agroecology and Sustainable Food Systems* 43(5), 546–578. DOI: 10.1080/21683565.2018.1514678.

Biswas, D., Saha, S.C. and Dey, A. (2021) CRISPR-Cas genome-editing tool in plant abiotic stress-tolerance. *Plant Gene* 26, 100286. DOI: 10.1016/j.plgene.2021.100286.

Bourgine, B. and Guihur, A. (2021) Heat shock signaling in land plants: from plasma membrane sensing to the transcription of small heat shock proteins. *Frontiers in Plant Science* 12, 710801. DOI: 10.3389/fpls.2021.710801.

Brookes, G. (2022) Farm income and production impacts from the use of genetically modified (GM) crop technology 1996–2020. *GM Crops and Food* 13(1), 171–195. DOI: 10.1080/21645698.2022.2105626.

Bulgakov, V.P. (2024) Chromatin modifications and memory in regulation of stress-related polyphenols: finding new ways to control flavonoid biosynthesis. *Critical Reviews in Biotechnology* 44(8), 1478–1494. DOI: 10.1080/07388551.2024.2336529.

Centorame, L., Gasperini, T., Ilari, A., Del Gatto, A. and Pedretti, E.F. (2024) An overview of machine learning applications on plant phenotyping, with a focus on sunflower. *Agronomy* 14(4), 719. DOI: 10.3390/agronomy14040719.

Chang-Brahim, I., Koppensteiner, L.J., Beltrame, L., Bodner, G., Saranti, A. *et al.* (2024) Reviewing the essential roles of remote phenotyping, GWAS and explainable AI in practical marker-assisted selection for drought-tolerant winter wheat breeding. *Frontiers in Plant Science* 15, 1319938. DOI: 10.3389/fpls.2024.1319938.

Dangi, A.K., Sharma, B., Khangwal, I. and Shukla, P. (2018) Combinatorial interactions of biotic and abiotic stresses in plants and their molecular mechanisms: Systems biology approach. *Molecular Biotechnology* 60(8), 636–650. DOI: 10.1007/s12033-018-0100-9.

Ding, Y., Huang, L., Jiang, Q. and Zhu, C. (2020) MicroRNAs as important regulators of heat stress responses in plants. *Journal of Agricultural and Food Chemistry* 68(41), 11320–11326. DOI: 10.1021/acs.jafc.0c03597.

Dresselhaus, T. and Hückelhoven, R. (2018) Biotic and abiotic stress responses in crop plants. *Agronomy* 8(11), 267. DOI: 10.3390/agronomy8110267.

Dündar, M.A. and Topak, R. (2025) Evaluation of yield, quality and water productivity of bread wheat cultivated under supplemental irrigation strategies in semi-arid conditions. *Plant and Soil*, 1–20.

Elvanidi, A., Friedel, M,, Max, J,, Katsoulas, N, and Zinkernagel, J, (2018) Remote sensing of canopy light interception and plant water deficit stress under greenhouse conditions. *DGG-Proceedings* 8(3), 1–5. DOI: 10.5288/dgg-pr-ae-2018.

Fernando, V.D. (2020) Major transcription factor families involved in salinity stress tolerance in plants. In: *Elsevier eBooks*, pp. 99–109. DOI: 10.1016/b978-0-12-819334-1.00007-1.

Fischer, R. and Buyel, J.F. (2020) Molecular farming – The slope of enlightenment. *Biotechnology Advances* 40, 107519. DOI: 10.1016/j.biotechadv.2020.107519.

Fortunato, S., Lasorella, C., Dipierro, N., Vita, F. and De Pinto, M.C. (2023). Redox signaling in plant heat stress response. *Antioxidants* 12(3), 605. DOI: 10.3390/antiox12030605.

Gangwar, T., Susko, A.Q., Baranova, S., Guala, M., Smith, K.P. and Heuschele, D.J. (2023) Multi-scale modelling predicts plant stem bending behaviour in response to wind to inform lodging resistance. *Royal Society Open Science* 10(1), 221410. DOI: 10.1098/rsos.221410.

Ghatak, A., Chaturvedi, P. and Weckwerth, W. (2018) Metabolomics in plant stress physiology. *Advances in Biochemical Engineering, Biotechnology*, 164, 187–236. DOI: 10.1007/10_2017_55.

Goel, P., Bhuria, M., Sinha, R., Sharma, T.R. and Singh, A.K. (2019) Promising transcription factors for salt and drought tolerance in plants. In: *Energy, Environment, and Sustainability*, pp. 7–50. Springer: Singapore. DOI: 10.1007/978-981-15-0690-1_2.

Gollin, D. (2020) Conserving genetic resources for agriculture: economic implications of emerging science. *Food Security* 12(5), 919–927. DOI: 10.1007/s12571-020-01035-w.

Guo, J., Sun, B., He, H., Zhang, Y., Tian, H. and Wang, B. (2021) Current understanding of BHLH transcription factors in plant abiotic stress tolerance. *International Journal of Molecular Sciences* 22(9), 4921. DOI: 10.3390/ijms22094921.

Hadish, J.A., Honaas, L.A. and Ficklin, S.P. (2023) Predicting phenotypic traits using a massive RNA-seq dataset. *bioRxiv pp. 2023–12. (Cold Spring Harbor Laboratory)*. DOI: 10.1101/2023.12.05.570195.

Haider, S., Iqbal, J., Naseer, S., Yaseen, T., Shaukat, M. *et al.* (2021) Molecular mechanisms of plant tolerance to heat stress: current landscape and future perspectives. *Plant Cell Reports* 40(12), 2247–2271. DOI: 10.1007/s00299-021-02696-3.

Haque, M.A., Rafii, M.Y., Yusoff, M.M., Ali, N.S., Yusuff, O. *et al.* (2021) Advanced breeding strategies and future perspectives of salinity tolerance in rice. *Agronomy* 11(8), 1631. DOI: 10.3390/agronomy11081631.

Henry, G.A. and Stinchcombe, J.R. (2025) Predicting Fitness-Related traits using gene expression and machine learning. *Genome Biology and Evolution* 17(2), p.evae275. DOI: 10.1093/gbe/evae275.

Hoang, X.L.T., Nhi, D.N.H., Thu, N.B.A., Thao, N.P. and Tran, L.P. (2017) Transcription Factors and Their Roles in Signal Transduction in Plants under Abiotic Stresses. *Current Genomics* 18(6), 483–497. DOI: 10.2174/1389202918666170227150057.

Hrmova, M. and Hussain, S.S. (2021) Plant transcription factors involved in drought and associated stresses. *International Journal of Molecular Sciences* 22(11), 5662. DOI: 10.3390/ijms22115662.

Ivanova, P.T. (2021) New technologies in agricultural business and their impact on employment. *Agribusiness and Rural Areas – Economy, Innovation and Growth 2021 Conference Proceedings*, 1, 134–143. University Publishing House, Science and Economics, University of Economics, Varna.

Jamil, I.N., Remali, J., Azizan, K.A., Muhammad, N.A.N., Arita, M., Goh, H. and Aizat, W.M. (2020) Systematic Multi-Omics Integration (MOI) approach in plant systems biology. *Frontiers in Plant Science* 11, 944. DOI: 10.3389/fpls.2020.00944.

Jin, C., Li, K., Xu, X., Zhang, H., Chen, H. *et al.* (2017) A Novel NAC transcription factor, PbeNAC1, of pyrus betulifolia confers cold and drought tolerance via interacting with PbeDREBs and activating the expression of stress-responsive genes. *Frontiers in Plant Science* 8, 01049. DOI: 10.3389/fpls.2017.01049.

Jin, S., Wei, M., Wei, Y. and Jiang, Z. (2024) Insights into Plant Sensory Mechanisms under Abiotic Stresses. *Plants* 13(14), 1907. DOI: 10.3390/plants13141907.

Joshi, N., Bisht, K. and Lohani, P. (2022) Alterations in transcriptome, proteome and metabolome during drought stress inplants. *Asian Journal of Microbiology Biotechnology and Environmental Sciences*, 24(2), 350–356. DOI: 10.53550/ajmbes.2022.v24i02.023.

Kang, D., Ahn, H., Lee, S., Lee, C., Hur, J., Jung, W. and Kim, S. (2018) Identifying stress-related genes and predicting stress types in Arabidopsis using logical correlation layer and CMCL loss through time-series data. In *2018 IEEE International Conference on Bioinformatics and Biomedicine (BIBM)*, 399–404. IEEE, 2018. DOI: 10.1109/bibm.2018.8621581.

Kang, D., Ahn, H., Lee, S., Lee, C. J., Hur, J., Jung, W., *et al.* (2019) StressGenePred: a twin prediction model architecture for classifying the stress types of samples and discovering stress-related genes in arabidopsis. *BMC Genomics* 20:949. doi: 10.1186/s12864-019-6283-z.

Kang, X., Zhao, L. and Liu, X. (2024) Calcium Signaling and the Response to Heat Shock in Crop Plants. *International Journal of Molecular Sciences*, 25(1), 24. https://doi.org/10.3390/ijms25010324

Kaur, A., Singh, G., Menon, S., Kumari, K., Ngairangbam, H. and Choudhary, S. (2024) Elucidating the molecular mechanisms of plant stress responses: Deciphering the impacts of abiotic and biotic stressors and developing innovative mitigation strategies for sustainable agriculture. *International Journal of Research in Agronomy* 7(6), 545–549. DOI: 10.33545/2618060x.2024.v7.i6h.922.

Kumar, M., Prusty, M.R., Pandey, M.K., Singh, P.K., Bohra, A., Guo, B. and Varshney, R.K. (2023) Application of CRISPR/Cas9-mediated gene editing for abiotic stress management in crop plants. *Frontiers in Plant Science,* 14, 1157678. DOI: 10.3389/fpls.2023.1157678.

Kumar, S. and Mohapatra, T. (2024) Epigenetic modifications in genome help remembering the stress tolerance strategy adopted by the plant. *Frontiers in Bioscience-Landmark* 29(3), 126. DOI: 10.31083/j.fbl2903126.

Kumar, S., Chinnusamy, V. and Mohapatra, T. (2018) Epigenetics of modified DNA bases: 5-methylcytosine and beyond. *Frontiers in Genetics*, (9), 640.

Laha, A., Chakraborty, P., Banerjee, C., Panja, A.S. and Bandopadhyay, R. (2020) Application of bioinformatics for crop stress response and mitigation. In: *Springer eBooks*, pp. 589–614. Springer International Publishing: Cham. DOI: 10.1007/978-3-030-45669-6_25.

Lämke, J. and Bäurle, I. (2017) Epigenetic and chromatin-based mechanisms in environmental stress adaptation and stress memory in plants. *Genome Biology* 18(1), 124. DOI: 10.1186/s13059-017-1263-6.

Langridge, P. and Chalmers, K. (2004) The principle: identification and application of molecular markers. In *Molecular marker systems in plant breeding and crop improvement*, pp. 3–22. Springer: Berlin, Heidelberg.

Lassalle, G. (2021) Monitoring natural and anthropogenic plant stressors by hyperspectral remote sensing: Recommendations and guidelines based on a meta-review. *The Science of the Total Environment* 788, 147758. DOI: 10.1016/j.scitotenv.2021.147758.

Lin, M., Lynch, V., Ma, D., Maki, H., Jin, J. and Tuinstra, M. (2022) Multi-species prediction of physiological traits with hyperspectral modeling. *Plants* 11(5), 676. DOI: 10.3390/plants11050676.

Ly, D., Huet, S., Gauffreteau, A., Rincent, R., Touzy, G. *et al.* (2017) Whole-genome prediction of reaction norms to environmental stress in bread wheat (Triticum aestivum L.) by genomic random regression. *Field Crops Research* 216, 32–41. DOI: 10.1016/j.fcr.2017.08.020.

Mahood, E.H., Kruse, L.H. and Moghe, G.D. (2020) Machine learning: A powerful tool for gene function prediction in plants. *Applications in Plant Sciences* 8(7), e11376. DOI: 10.1002/aps3.11376.

Meher, P.K., Sahu, T.K., Gupta, A., Kumar, A. and Rustgi, S. (2022) ASRpro: A machine-learning computational model for identifying proteins associated with multiple abiotic stress in plants. *The Plant Genome* 17(1), e20259. DOI: 10.1002/tpg2.20259.

Merrick, T., Bennartz, R., Jorge, M.L.S.P., Pau, S. and Rausch, J. (2022) Evaluation of plant stress monitoring capabilities using a portable spectrometer and Blue-Red grow light. *Sensors* 22(9), 3411. DOI: 10.3390/s22093411.

Miryeganeh, M. (2021) Plants' epigenetic mechanisms and abiotic stress. *Genes* 12(8), 1106. DOI: 10.3390/genes12081106.

Montesinos-López, O.A., Montesinos-López, A., Crossa, J., Gianola, D., Hernández-Suárez, C.M. and Martín-Vallejo, J. (2018) Multi-trait, multi-environment deep learning modeling for Genomic-Enabled prediction of plant traits. *G3 Genes Genomes Genetics* 8(12), 3829–3840. DOI: 10.1534/g3.118.200728.

Mosa, K.A., Ismail, A. and Helmy, M. (2017) Omics and system biology approaches in plant stress research. In: *SpringerBriefs in Systems Biology*, pp. 21–34. Springer: Cham. DOI: 10.1007/978-3-319-59379-1_2.

Mozgová, I., Mikulski, P., Pecinka, A. and Farrona, S. (2019) Epigenetic mechanisms of abiotic stress response and memory in plants. In: *Epigenetics in Plants of Agronomic Importance: Fundamentals and Applications*, pp. 1–64. Springer: Cham. DOI: 10.1007/978-3-030-14760-0_1.

Mushtaq, M., Bhat, J.A., Mir, Z.A., Sakina, A., Ali, S. *et al.* (2018) CRISPR/Cas approach: A new way of looking at plant-abiotic interactions. *Journal of Plant Physiology* 224–225, 156–162. DOI: 10.1016/j.jplph.2018.04.001.

Nazar, R., Iqbal, N. and Umar, S. (2017) Heat stress tolerance in plants: Action of salicylic acid. In *Salicylic Acid: A Multifaceted Hormone*, pp. 145–161. Springer: Singapore.

Nazir, R., Mandal, S., Mitra, S., Ghorai, M., Das, N. *et al.* (2022) Clustered regularly interspaced short palindromic repeats (CRISPR)/CRISPR-associated genome-editing toolkit to enhance salt stress tolerance in rice and wheat. *Physiologia Plantarum* 174(2), e13642. DOI: 10.1111/ppl.13642.

Nejat, N. and Mantri, N. (2017) Plant immune system: Crosstalk between responses to biotic and abiotic stresses the missing link in understanding plant defence. *Current Issues in Molecular Biology*, 23(1), 1–16. DOI: 10.21775/cimb.023.001.

Nishio, H., Kawakatsu, T. and Yamaguchi, N. (2023) Beyond heat waves: Unlocking epigenetic heat stress memory in Arabidopsis. *Plant Physiology* 194(4), 1934–1951. DOI: 10.1093/plphys/kiad558.

Pan, L., Ma, J., Li, J., Yin, B. and Fu, C. (2022). Advances of salt stress-responsive transcription factors in plants. *PubMed* 38(1), 50–65. DOI: 10.13345/j.cjb.210135.

Pandey, P. and Senthil-Kumar, M. (2019) Plant-pathogen interaction in the presence of abiotic stress: What do we know about plant responses? *Plant Physiology Reports* 24(4), 541–549. DOI: 10.1007/s40502-019-00483-7.

Pandita, D. (2021) CRISPR/Cas-mediated genome editing for improved stress tolerance in plants. In: *Frontiers in Plant-Soil Interaction*, pp. 259–291. Academic Press, USA. DOI: 10.1016/b978-0-323-90943-3.00001-8.

Parreira, J.R., Branco, D., Almeida, A.M., Czubacka, A., Agacka-Mołdoch, M. *et al.* (2016) Systems Biology Approaches to Improve Drought Stress Tolerance in Plants: State of the Art and Future Challenges. In: Hossain, M., Wani, S., Bhattacharjee, S., Burritt, D., Tran, LS. (eds) *Drought Stress Tolerance in Plants*, Vol 2, pp. 433–471. Springer, Cham. DOI: 10.1007/978-3-319-32423-4_16.

Phelix, C.F. and Feltus, F.A. (2015) Plant stress biomarkers from biosimulations: the Transcriptome-To-MetabolomeTM (TTMTM) technology – effects of drought stress on rice. *Plant Biology* 17(1), 63–73. DOI: 10.1111/plb.12221.

Piasecka, A., Kachlicki, P. and Stobiecki, M. (2019) Analytical methods for detection of plant metabolomes changes in response to biotic and abiotic stresses. *International Journal of Molecular Sciences* 20(2), 379. DOI: 10.3390/ijms20020379.

Poór, P., Nawaz, K., Gupta, R., Ashfaque, F. and Khan, M.I.R. (2021) Ethylene involvement in the regulation of heat stress tolerance in plants. *Plant Cell Reports* 41(3), 675–698. DOI: 10.1007/s00299-021-02675-8.

Rai, K.K., Pandey, N. and Rai, S.P. (2019) Salicylic acid and nitric oxide signaling in plant heat stress. *Physiologia Plantarum* 168(2), 241–255. DOI: 10.1111/ppl.12958.

Ramakrishnan, M., Zhang, Z., Mullasseri, S., Kalendar, R., Ahmad, Z. *et al.* (2022) Epigenetic stress memory: A new approach to study cold and heat stress responses in plants. *Frontiers in Plant Science* 13, 1075279. DOI: 10.3389/fpls.2022.1075279.

Ramesh, S., Kumar, R.R. and Praveen, S. (2020) Plant transcriptional regulation in modulating cross-tolerance to stress. In *Elsevier eBooks,* pp. 231–245. DOI: 10.1016/b978-0-12-817892-8.00015-5.

Rana, M.M., Takamatsu, T., Baslam, M., Kaneko, K., Itoh, K. *et al.* (2019) Salt tolerance improvement in rice through efficient SNP marker-assisted selection coupled with speed-breeding. *International Journal of Molecular Sciences* 20(10), 2585. DOI: 10.3390/ijms20102585.

Rurek, M. and Smolibowski, M. (2024) Variability of plant transcriptomic responses under stress acclimation: a review from high throughput studies. *Acta Biochimica Polonica* 71, 13585. DOI: 10.3389/abp.2024.13585.

Sadiku, M.N., Ashaolu, T.J. and Musa, S.M. (2020) Smart food: an introduction. *International Journal of Trend in Scientific Research and Development*, 4(2), pp.121–124.

Saeed, F., Chaudhry, U.K., Bakhsh, A., Raza, A., Saeed, Y., Bohra, A. and Varshney, R.K. (2022) Moving beyond DNA sequence to improve plant stress responses. *Frontiers in Genetics* 13, 874648. DOI: 10.3389/fgene.2022.874648.

Şahin, E.C., Aydin, Y. and Uncuoglu, A.A. (2024) Molecular marker applications in the selection of elite genotypes for plant stress tolerance and genetic fidelity. *OBM Genetics* 08(03), 1–25. DOI: 10.21926/obm.genet.2403260.

Sahoo, J.P., Behera, L., Sharma, S.S., Praveena, J., Nayak, S.K. and Samal, K.C. (2020) Omics studies and systems biology perspective towards abiotic stress response in plants. *American Journal of Plant Sciences* 11(12), 2172–2194. DOI: 10.4236/ajps.2020.1112152.

Saijo, Y. and Loo, E.P. (2019) Plant immunity in signal integration between biotic and abiotic stress responses. *New Phytologist* 225(1), 87–104. DOI: 10.1111/nph.15989.

Saikanth, D.R.K., Ragini, M., Tripathi, G., Rajesh, K., Giri, A., Pandey, S.K. and Verma, L.K. (2024) The impact of emerging technologies on sustainable agriculture and rural development. *International Journal of Environment and Climate Change* 14(1), 253–263. DOI: 10.9734/ijecc/2024/v14i13830.

Saikia, B.S.R., Debbarma, J., Maharana, J., Sastry, G.N. and Chikkaputtaiah, C. (2024) CRISPR/Cas9-based genome editing and functional analysis of SlHyPRP1 and SlDEA1 genes of Solanum lycopersicum L. in imparting genetic tolerance to multiple stress factors. *Frontiers in Plant Science* 15, 1304381. DOI: 10.3389/fpls.2024.1304381.

Sanches, P.H.G., De Melo, N.C., Porcari, A.M. and De Carvalho, L.M. (2024) Integrating molecular perspectives: Strategies for comprehensive Multi-Omics integrative data analysis and machine learning applications in transcriptomics, proteomics, and metabolomics. *Biology* 13(11), 848. DOI: 10.3390/biology13110848.

Sanchez-Muñoz, R., Depaepe, T., Samalova, M., Hejatko, J., Zaplana, I. and Van Der Straeten, D. (2024) The molecular core of transcriptome responses to abiotic stress in plants: A machine learning-driven meta-analysis. *bioRxiv (Cold Spring Harbor Laboratory), 2024-01*. DOI: 10.1101/2024.01.24.576978.

Sarkar, B., Ghosh, A., Saha, I., De, A. K. and Adak, M. K. (2020) Transcriptomics in deciphering stress tolerance in plants. In: *Plant Ecophysiology and Adaptation under Climate Change: Mechanisms and Perspectives II*, pp. 531–542. *Springer*: Cham. DOI: 10.1007/978-981-15-2172-0_18.

Satapathy, R.R. (2020) Remote sensing in plant disease management. *Journal of Pharmacognosy and Phytochemistry* 9, 1813–1820.

Shafi, A. and Zahoor, I. (2019) Bioinformatics and plant stress management. In: *Essentials of Bioinformatics, Volume III* pp. 47

Shameer, S., Ratcliffe, R.G. and Sweetlove, L.J. (2019) Leaf energy balance requires mitochondrial respiration and export of chloroplast NADPH in the light. *Plant Physiology*, 180(4), 1947–1961. Springer: Cham. DOI: 10.1007/978-3-030-19318-8_3.

Shanker, A3.K., Bhanu, D. and Maheswari, M. (2020) Epigenetics and transgenerational memory in plants under heat stress. *Plant Physiology Reports* 25(4), 583–593. DOI: 10.1007/s40502-020-00557-x.

Shikha, D., Sindhura, K.A.V., Rastogi, M., Saritha, B., Satapathy, S.N., Srivastava, S. and Kurdekar, A.K. (2024) A review on propelling agricultural practices with biotechnology into a New Era. *Journal of Advances in Biology and Biotechnology* 27(3), 99–111. DOI: 10.9734/jabb/2024/v27i3725.

Sinha, D., Maurya, A.K., Abdi, G., Majeed, M., Agarwal, R. *et al.* (2023) Integrated genomic selection for accelerating breeding programs of climate-smart cereals. *Genes* 14(7), 1484. DOI: 10.3390/genes14071484.

Skendžić, S., Zovko, M., Lešić, V., Živković, I. P. and Lemić, D. (2023). Detection and evaluation of environmental stress in winter wheat using remote and proximal sensing methods and vegetation indices—A review. *Diversity* 15(4), 481. DOI: 10.3390/d15040481.

Soda, N., Wallace, S. and Karan, R. (2015) Omics study for abiotic stress responses in plants. *Advances in Plants and Agriculture Research*, 2(1), 00037.

Sun, A. and Guo, F. (2016) Chloroplast Retrograde regulation of heat stress responses in plants. *Frontiers in Plant Science* 7, 398. DOI: 10.3389/fpls.2016.00398.

Tajima, Y., Loo, E.P. and Saijo, Y. (2020). Plant physiological and molecular mechanisms in cross-regulation of biotic-abiotic stress responses. In: Hossain, M.A., Liu, F., Burritt, D.J., Fujita, M. and Huang, B. (eds) *Priming-Mediated Stress and Cross-Stress Tolerance in Crop Plants*, pp. 21–34. Academic Press, USA. DOI: 10.1016/b978-0-12-817892-8.00002-7.

Tricker, P.J. (2015) Transgenerational inheritance or resetting of stress-induced epigenetic modifications: two sides of the same coin. *Frontiers in Plant Science* 6, 699. DOI: 10.3389/fpls.2015.00699.

Turgut-Kara, N., Arikan, B. and Celik, H. (2020) Epigenetic memory and priming in plants. *Genetica* 148(2), 47–54. DOI: 10.1007/s10709-020-00093-4.

Umar, O.B., Ranti, L.A., Abdulbaki, A.S., Bola, A.L., Abdulhamid, A.K., Biola, M.R. and Victor, K.O. (2021) Stresses in plants: biotic and abiotic. In: Ansari, M. (ed.) *Current Trends in Wheat Research*. IntechOpen *eBooks*. DOI: 10.5772/intechopen.100501.

Upadhyay, J., Joshi, R., Singh, B., Bohra, A., Vijayan, R. *et al.* (2017) Application of bioinformatics in understanding of plant stress tolerance. In: Hakeem, K., Malik, A., Vardan-Sukan, F. and Ozturk, M. (eds) *Plant Bioinformatics*, pp. 347–374. Springer: Cham. DOI: 10.1007/978-3-319-67156-7_14.

Van Dijk, A.D.J., Kootstra, G., Kruijer, W. and De Ridder, D. (2020) Machine learning in plant science and plant breeding. *iScience* 24(1), 101890. DOI: 10.1016/j.isci.2020.101890.

Voelckel, C., Gruenheit, N. and Lockhart, P. (2017) Evolutionary transcriptomics and proteomics: Insight into plant adaptation. *Trends in Plant Science* 22(6), 462–471. DOI: 10.1016/j.tplants.2017.03.001.

Wang, X. (2019) Protein and proteome atlas for plants under stresses: New highlights and ways for integrated omics in post-genomics era. *International Journal of Molecular Sciences* 20(20), 5222. DOI: 10.3390/ijms20205222.

Wang, X., Xu, C., Cai, X., Wang, Q. and Dai, S. (2017) Heat-responsive photosynthetic and signaling pathways in plants: Insight from proteomics. *International Journal of Molecular Sciences* 18(10), 2191. DOI: 10.3390/ijms18102191.

Wang, X., Li, N., Li, W., Gao, X., Cha, M., Qin, L. and Liu, L. (2020) Advances in transcriptomics in the response to stress in plants. *Global Medical Genetics* 07(02), 030–034. DOI: 10.1055/s-0040-1714414

Wang, T., Xun, H., Wang, W., Ding, X., Tian, H. *et al.* (2021) Mutation of GMAITR genes by CRISPR/CAS9 genome editing results in enhanced salinity stress tolerance in soybean. *Frontiers in Plant Science* 12, 779598. DOI: 10.3389/fpls.2021.779598.

Wang, X., Li, J., Sun, J., Gu, S., Wang, J. *et al.* (2022) Mining beneficial genes for salt tolerance from a core collection of rice landraces at the seedling stage through Genome-Wide association mapping. *Frontiers in Plant Science* 13, 847863. DOI: 10.3389/fpls.2022.847863.

Wani, S.H., Choudhary, J.R., Choudhary, M., Rana, M. and Gosal, S.S. (2020) Recent advances in genomics assisted breeding for drought stress tolerance in major cereals. *Journal of Cereal Research* 12(1), 1–12. DOI: 10.25174/2582-2675/2020/100821.

Wu, M., Liu, R., Gao, Y., Xiong, R., Shi, Y. and Xiang, Y. (2020) PheASR2, a novel stress-responsive transcription factor from moso bamboo (Phyllostachys edulis), enhances drought tolerance in transgenic rice via increased sensitivity to abscisic acid. *Plant Physiology and Biochemistry* 154, 184–194. DOI: 10.1016/j.plaphy.2020.06.014.

Yan, Q., Zhang, G., Zhang, X. and Huang, L. (2024) A review of transcriptomics and metabolomics in plant quality and environmental response: From bibliometric analysis to science mapping and future trends. *Metabolites* 14(5), 272. DOI: 10.3390/metabo14050272.

Ying, Z., Fu, S. and Yang, Y. (2025) Signaling and Scavenging: Unraveling the Complex Network of Antioxidant Enzyme Regulation in Plant Cold Adaptation. *Plant Stress*, 16, 100833.

Yoon, Y., Seo, D.H., Shin, H., Kim, H.J., Kim, C.M. and Jang, G. (2020) The role of Stress-responsive transcription factors in modulating abiotic stress tolerance in plants. *Agronomy* 10(6), 788. doi: 10.3390/agronomy10060788.

Zhang, J., Huang, Y., Pu, R., Gonzalez-Moreno, P., Yuan, L., Wu, K. and Huang, W. (2019) Monitoring plant diseases and pests through remote sensing technology: A review. *Computers and Electronics in Agriculture* 165, 104943. DOI: 10.1016/j.compag.2019.104943.

Zheng, Y., Wang, Q., Chen, D., Chen, D., Zeng, R. and Song, Y. (2022) Research advances in histone modification-regulated plant stress memory and defense priming.. *PubMed* 33(3), 844–854. DOI: 10.13287/j.1001-9332.202202.039.

Zheng, S., Zhao, W., Liu, Z., Geng, Z., Li, Q. *et al.* (2024) Establishment and maintenance of Heat-Stress memory in plants. *International Journal of Molecular Sciences* 25(16), 8976. DOI: 10.3390/ijms25168976.

Zilberman, D., Wesseler, J., Schmitz, A. and Gordon, B. (2018) Economics of agricultural biotechnology. In: Cramer, G., Paudel, K. and Schmitz, A. (eds) *The Routledge Handbook of Agricultural Economics*, pp. 670–686. Routledge: London. DOI: 10.4324/9781315623351-36.

Zong, J., Li, X., Zhou, Y., Wang, F., Wang, N. *et al.* (2016) The AaDREB1 transcription factor from the cold-tolerant plant adonis amurensis enhances abiotic stress tolerance in transgenic plant. *International Journal of Molecular Sciences* 17(4), 611. DOI: 10.3390/ijms17040611.

Index

Note: Page numbers in **bold** type refer to figures, page numbers in *italic* type refer to tables

Printed and bound by CPI Group (UK) Ltd, Croydon, CR0 4YY
13/09/2026
14961088-0001